*SAT II Math
FOR
DUMMIES®

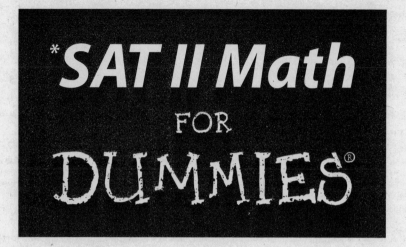

*SAT II Math
FOR
DUMMIES®

by Lisa Hatch, M.A.
and Scott Hatch, J.D.

WILEY

Wiley Publishing, Inc.

*SAT II Math For Dummies®

Published by
Wiley Publishing, Inc.
111 River St.
Hoboken, NJ 07030-5774
www.wiley.com

WILEY

About the Authors

Scott and Lisa Hatch have prepared students for college entrance exams for over 25 years. While in law school in the late '70s, Scott Hatch taught LSAT preparation courses throughout Southern California to pay for his education. He was so good at it that after graduation, he went out on his own. Using materials he developed himself, he prepared thousands of anxious potential test-takers for the SAT, ACT, PSAT, LSAT, GRE, GMAT, and even the unassuming Miller's Analogy Test (MAT).

Scott and Lisa have a special place in their hearts for standardized tests; they kindled their romance in the classroom. Lisa took one of Scott's LSAT preparation courses at the University of Colorado and improved her love life as well as her LSAT score. Lisa's love for instructing and writing allowed her to fit right in with Scott's lifestyle, teaching courses, and preparing course materials. They married shortly thereafter.

Since then, Scott and Lisa have taught test preparation to students worldwide. Currently over 300 universities and colleges offer their courses through live lectures and online. They have written the curriculum for both formats. The company they have built together, The Center for Legal Studies, not only provides standardized test preparation courses but also courses for those who desire careers in the field of law, including paralegals, legal secretaries, legal investigators, victim advocates, and legal nurse consultants.

Scott has presented standardized test preparation courses since 1979. He is listed in *Who's Who in California, Who's Who Among Students in American Colleges and Universities*, and has been named one of the Outstanding Young Men of America by the United States Jaycees. He was a contributing editor to *The Judicial Profiler* (McGraw-Hill) and the *Colorado Law Annotated* (West/Lawyers Co-op) series, and editor of several award-winning publications.

Lisa has been teaching legal certificate and standardized test preparation courses since 1987. She graduated with honors in English from the University of Puget Sound, and received her master's degree from California State University. She and Scott have co-authored numerous law and standardized test texts, including *Paralegal Procedures and Practices* (Thomson/West Publishing), *A Paralegal Primer* (The Center for Legal Studies), and *Preparing for the SAT, Preparing for the ACT, Preparing for the LSAT, Preparing for the GRE,* and *Preparing for the GMAT,* all of which are available online in an interactive classroom format.

Dedication

We dedicate our *For Dummies* series books to Alison, Andrew, Zachary, and Zoe Hatch. Rather than file missing persons' reports on their parents with local law enforcement agencies, they demonstrated extreme patience, understanding, and assistance while we wrote these books, which made this innovative, comprehensive, informative, and entertaining standardized test preparation series possible.

Authors' Acknowledgments

This book would not be possible without the research and writing contributions of standardized test prep and mathematics experts, Martin D. Rollins and Benjamin A. Saypol. Their efforts greatly enhanced our writing, editing, and organization, and we are deeply grateful to them.

We also need to acknowledge the input of the thousands of high school students and other college applicants who have completed our test preparation courses over the last 25 years. The classroom and online contributions offered by these eager learners have provided us with lots of information about what areas require the greatest amount of preparation. Their input is the reason we are able to produce accurate and up-to-date test preparation. Many of our students, including Zach Hatch and Tyler Polhill, have also taken the sample tests that we set forth in this book and have assisted us in perfecting the question and answer choices.

Our meticulous scholarship and attempts at wit were greatly facilitated by the editing professionals at Wiley Publishing. Our thanks go out to Tere Stouffer for her unflagging support and encouragement and Tracy Boggier for initiating us to the process and being available whenever we had questions. Many thanks to Doug Drenth for serving as our technical editor. And to the Composition Services staff at Wiley: our many thanks, especially to proofreader Christine Pingleton for her amazing eagle eye.

Finally, we wish to acknowledge our literary agents, Bill Gladstone and Margo Maley Hutchinson, at Waterside Productions in Cardiff for introducing us to the innovative *Dummies* series. Wiley Publishing has to be commended for its pioneering efforts to make preparing for compulsory entrance exams fun. We thrive on positive reinforcement and feedback from our students and encourage our readers to provide comments and critiques at feedback@legalstudies.com.

Publisher's Acknowledgments

We're proud of this book; please send us your comments through our Dummies online registration form located at www.dummies.com/register/.

Some of the people who helped bring this book to market include the following:

Acquisitions, Editorial, and Media Development

Project Editor: Tere Stouffer

Acquisitions Editor: Tracy Boggier

Technical Editor: Douglas Drenth

Editorial Supervisor: Carmen Krikorian

Editorial Manager: Michelle Hatcher

Editorial Assistant: Nadine Bell

Cartoons: Rich Tennant, www.the5thwave.com

Composition

Project Coordinator: Adrienne Martinez

Layout and Graphics: Carrie Foster, Denny Hager, Joyce Haughey, Clint Lahnen, Barry Offringa, Rashell Smith

Proofreaders: Joe Niesen, Christine Pingleton

Indexer: Lynnzee Elze

Publishing and Editorial for Consumer Dummies

> **Diane Graves Steele,** Vice President and Publisher, Consumer Dummies

> **Joyce Pepple,** Acquisitions Director, Consumer Dummies

> **Kristin A. Cocks,** Product Development Director, Consumer Dummies

> **Michael Spring,** Vice President and Publisher, Travel

> **Kelly Regan,** Editorial Director, Travel

Publishing for Technology Dummies

> **Andy Cummings,** Vice President and Publisher, Dummies Technology/General User

Composition Services

> **Gerry Fahey,** Vice President of Production Services

> **Debbie Stailey,** Director of Composition Services

Contents at a Glance

Table of Contents

Introduction

· ·

There you are, merrily skimming through the admissions requirements for your favorite college, when all of a sudden, you're dealt a shocking blow. Your absolute top college choice — that is, you'll die if you don't get in — requires not only the SAT I and the ACT but also the SAT II. That means yet another set of tests for which you need to sharpen more number-2 pencils and awaken at the crack of dawn on an otherwise sleepy Saturday.

The SAT II Math is just one of many possible SAT II subject tests that you can take, and most colleges and universities that require you to take particular subject tests require you to take the SAT II Math. So you'll be in good company on test day. Many of your fellow SAT II test takers will also be sitting down to graph functions and find the volume of spheres right along with you.

You know you have to get ready for the challenge, but what are you going to do? If you can find them, you can get out your spiral notebooks from all the high school math courses you've taken over the past two or three years (or possibly several years ago) and sift through those years' worth of doodles.

We can just see it now. You know you recorded the formula for the Law of Cosines somewhere in your notes. (There it is! No . . . wait. That's Chris's phone number.) And even if you could read your handwriting, the scribbled formulas are concealed by your little brother's dirty handprint left while he was looking through your stuff for paper airplane material.

Maybe reading through old notes isn't such a good idea.

Clearly, you need a readable, more concisely structured resource. You've come to the right place. *SAT II Math For Dummies* puts at your fingertips everything you need to know to conquer the SAT II Math test. We give you a complete review of the math concepts it tests and provide insights into how to avoid the pitfalls that the people who create the SAT II Math test want you to fall into (and we try to make it as enjoyable as a book that devotes itself to numbers and geometric shapes can be).

About This Book

We suspect that you aren't eagerly anticipating sitting through the SAT II, and you're probably not looking forward to studying for it, either. Chances are, your parents or some other well meaning authority figure bought you this book for your own good, and we know just how much human beings enjoy doing things that are good for them!

Therefore, we attempt to make the study process as painless as possible by giving you clearly written advice in an easy-to-swallow, casual tone. We realize that you have a bunch of things you'd rather be doing, so we break down the information in easily digested bites. If you have an extra hour before basketball practice or clarinet lessons, you can devour a chapter or a particular section within a chapter.

We pepper the math reviews with plenty of sample questions, so you can see just how the SAT II tests a particular concept. Our sample questions read like the actual test questions, so that you can get comfortable with the way the SAT II phrases questions and expresses answer choices. And to further enhance your comfort with the test questions, this book contains four practice tests. Ultimately, the best way to prepare for any standardized test is to practice on lots of test questions, and this book has over 200 of them.

This book also gives you time-tested techniques for improving your score. We show you how to quickly eliminate incorrect answer choices and make educated guesses. You also discover how to manage your time and calculator wisely. And we give you suggestions for creating a relaxation routine to employ if you start to panic during the test. In a nutshell, we've included all kinds of information to help you do your best on the SAT II.

Conventions Used in This Book

You should find this book to be easily accessible, but there may be a few things that require explanation. A few of the chapters contain sidebars (a paragraph or two in a shaded box). Sidebars contain quirky bits of information that we think may interest you, but that aren't essential to your performance on the SAT II. If you're trying to save time, you can skip the sidebars.

The book highlights information you should remember in several ways. Lists are bulleted and marked with a solid bar to the left of the lists. Words to remember are italicized, so that you can develop vocabulary lists from them if you wish. And icons appear in the margins to emphasize particularly significant information in the text (see the "Icons Used in This Book" section for details). You can use these highlighting tools to focus on the most important elements of each chapter.

Foolish Assumptions

Although we guess it's possible that you've picked up this book just because you have an insatiable love for math, we're betting it's more likely that you are reading this book because you've been told you have to take the SAT II Math test. And because we're pretty astute, we figure this means that you intend to apply to undergraduate programs that either require you to take the test for admissions purposes or suggest that you take it for a variety of other considerations.

Generally, the schools that require the SAT II are highly selective, so we're thinking that you're a pretty motivated student with your sights on competitive institutions of higher learning. You probably know a lot about math already, and you've probably taken the equivalent of high school algebra, geometry, algebra II, trigonometry, and maybe some pre-Calculus.

Your math courses may be fresh in your mind, and you need this book only to know what to expect when you arrive at the test site. This book has that information for you. It's also possible that you've been out of high school for awhile, and your math knowledge may have left for parts unknown. If this is you, this book provides you with a basic math refresher to give you a foundation for the more challenging concepts you need to know to excel on the SAT II.

How This Book Is Organized

The first part of this book introduces you to the nature of the SAT II beast and advises you on how to tame it. A comprehensive math review follows in three additional parts. When you feel ready, you can take the four practice tests that come after the review and find out your score.

Part 1: Putting the SAT 11 Math Test into Perspective

Read this part if you want to know more about what information the SAT II Math test covers and how you can handle it in the best way.

Part 11: Addition, Subtraction, and Beyond: Arithmetic and Algebra

You probably have a pretty good grip of math fundamentals, but just in case you're a little fuzzy on exponents and percentages, for example, this part reviews basic concepts as well as some of the more challenging issues in algebra and functions.

Part 111: It Has Curves, and You Have Angles: Geometry and Trigonometry

The SAT II Math test covers the math you learn in your second and third years of high school, so this part concentrates on not only plane geometry but also on three-dimensional and coordinate geometry and on trigonometry.

Part 1V: Highly Unlikely: Statistics, Probability, and Sets

You have to know a little about statistics for the SAT II Math test. This part gives you just what you need to know about the tendencies of numbers, essential formulas for probability, and a review of sets.

Part V: Practice Makes Perfect

When you feel comfortable with your math knowledge, you can practice on the four tests found in this part. There are four tests for your enjoyment, and each comes complete with a scoring guide and explanatory answers.

Part VI: The Part of Tens

This part finishes up the fun with a summary of the most important formulas and definitions you should know for the test.

Icons Used in This Book

One exciting feature of this book is the icons that highlight especially significant portions of the text. These little pictures in the margins alert you to areas where you should pay particularly close attention.

Throughout the book, we give you insights into how you can enhance your performance on the SAT II. The tips give you juicy timesavers and point out especially relevant concepts to remember for the test.

This icon points out those little tricks contained in the SAT II that the testmakers use to tempt you into choosing the wrong answer.

Your world won't fall apart if you ignore our warnings, but your score may suffer. Heed these cautionary anecdotes to avoid making careless mistakes that can cost you points.

Whenever you see this icon in the text, you know you're going to get to practice the particular area of instruction covered in that section with a question like one you may see on the test. Our examples include detailed explanations of how to most efficiently answer SAT II questions and avoid common pitfalls.

Some of the information in this book is relevant only if you're taking the Level IIC exam (see Chapter 1). Whenever we discuss information that is only on the Level IIC exam, we mark it with this icon. In fact, when you see the icon at the beginning of a section, you know that the entire section is relevant to only the Level IIC. If you plan to take only the Level IC exam, you can ignore the sections marked Level IIC, unless, of course, you want to see how the other half lives.

Where to Go from Here

We know that everyone who uses this book has different strengths and weaknesses, so this book is designed for you to read in the way that best suits you. If you're a math whiz and need only to brush up on a few areas, you can go right to the chapters and sections that cover those topics.

We suggest that you take a more thorough approach, however. Familiarize yourself with the general test-taking process in the first two chapters, and then work through the complete math review. You can skim through sections that you know a lot about by just reading the Tips, Warnings, and Traps & Tricks icons and working through the examples.

Some students like to take a diagnostic test before they study. This is a fancy way of saying that they take one of the practice tests in Part V before reading the rest of the book. Taking a preview test shows you which questions you seem to cruise through and which areas need more work. If you take this approach, you can take another practice test after you read through the math review, and then compare your score to the one you got on the first test.

Part I

Putting the SAT II Math Test into Perspective

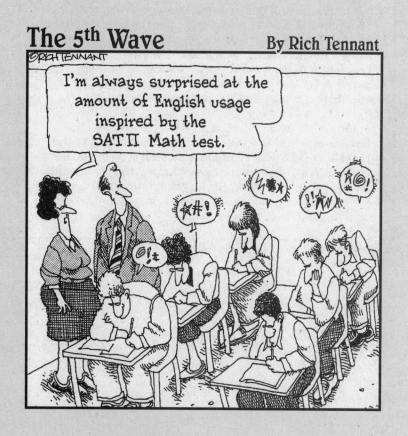

In this part . . .

The first part of this book initiates you to the wonders of the SAT II. Its chapters introduce the format of the test and explain how to take it seriously (but not too seriously). You may be tempted to skip this part and jump headlong into the math review. If you do, we strongly suggest that you come back to this part later. We include information in here that you may not get otherwise.

Among other things, you'll learn what to expect on the test, how the test is scored, whether you should take the Level IC or Level IIC exam, and what stuff is tested on each of them. You'll also discover some helpful tips for organizing your time and relaxing if you get nervous. So, do yourself a favor and give it the 20 minutes or so that it takes to read through these two chapters.

Chapter 1

Getting the Lowdown on the SAT II Math Test

....

....

*E*ven if you're familiar with the way the general SAT I test is set up, read this chapter. The SAT II is significantly different from the SAT I in the way that it's formatted and the way you approach it.

Justification: How Colleges Use the SAT II

How colleges and universities evaluate your SAT II Math score really depends on the particular college. Some colleges use your SAT II score as an additional admissions tool, equal in value to your other standardized test scores (like the SAT and the ACT), your GPA, and other factors, like the difficulty of your high school course load. Your SAT II test scores may also allow you to test out of taking specific first-year college courses at some schools.

The SAT II as an admissions requirement

Before you fill out the application form, research the admissions requirements of your top schools. The most up-to-date information is usually located on the college's Web site. Read the requirements carefully, because there are all sorts of ways that schools apply the information they get from your SAT II score.

Most of the schools that require you to take the SAT II want you to take two or three of them. Some schools tell you exactly which subjects they expect you to take, while others leave the choice up to you. Those colleges that mandate which particular tests you should take usually include the SAT II Math in the requirements.

What part the SAT II plays in the admissions decision also varies from school to school. Some schools don't even consider your SAT II score when they decide whether to sign you on or not. Other colleges give your SAT II score the same weight as they do your SAT or ACT score and your GPA. And other colleges may not put the same amount of importance on your SAT II score as on the other standardized tests and GPA, but they use it for additional consideration like they would your extracurricular activities.

Find out how your college choices evaluate your score, and if you're confused by their Web sites, call the admissions office directly. It's nice to know just what's at stake when you walk into the test.

Each SAT II subject test falls into one of five general categories. It is important to know these categories because some colleges won't accept two tests from the same category. The five categories are as follows.

- **English:** As of the spring of 2005, there is only one English test, and that's the SAT II Literature exam.

- **History and Social Studies:** The two history tests make up this category: SAT II U.S. History and SAT II World History.

- **Science:** The SAT II Biology E/M, SAT II Chemistry, and SAT II Physics are all considered to be science tests. Go figure!

- **Mathematics:** The SAT II Math has two separate math tests, the Level IC and the more advanced Level IIC.

- **Languages:** The SAT II has thirteen language tests. Some of the languages also have separate one-hour listening tests in addition to the hour-long written test.

The SAT II as a placement tool

Colleges that use your SAT II score as an admissions requirement, and even those that don't, *may* consider it for placement purposes. Many colleges and universities allow first-year students to meet a core course requirement. So, if you did really well on the SAT II Literature test, for example, you may not have to take English 101 in college. Specific information about using the SAT II for placement purposes is not usually spelled out for you on the college Web sites, so call the admissions office directly if you need more information about whether your SAT II score can get you out of taking one of those huge freshman lectures with 500 other students.

A case study: How the University of California system uses the SAT II for admissions

The University of California requires students to take the SAT II. According to Ravi Poorsina of the UC Office of the President, the primary concern of the university is that an incoming student comes to the campus ready to handle the rigors of the college curriculum. The university needs to know what students know about particular subjects. The SAT II is so important to the University of California that it will probably weigh it equally with the regular SAT or ACT score and your GPA.

Students in the freshman class of 2006 and beyond have to take two SAT II subject tests. Each test has to be from a different category, like one science and one history, or a language test and the literature test. If you submit two scores from the same subject area, you won't meet the requirement. The university system does not require any one particular test, as long as you take two from different categories.

In addition, because the new SAT I contains math questions that are similar in difficulty to those in the SAT II Math Level I, the University of California will no longer accept the Level I math test. Therefore, if you want to use a math test for acceptance to a University of California undergraduate program, you must take the Level II test.

Even with the advent of a new SAT I, the University of California system still values the information it gets from a student's SAT II score. We guess you'll just have to keep on taking those standardized tests!

Decisions, Decisions: Determining Which Subject Test You Should Take

Until the regular SAT changed in 2005, it used to be a little easier to decide which SAT II subject area tests you needed to take. Most of the colleges that required the SAT II made you take the SAT II Math, the SAT II Writing, and one other. The only decisions your overworked brain had to make were whether to take Level I or Level II for math and which of the other subjects would provide the least amount of torture for your third choice.

Things have changed, however. Beginning in Fall 2006, students entering college will take a new SAT I that incorporates the old SAT II Writing test and some of the elements of the Level I Math test. The result is that colleges will probably no longer mandate particular subject tests, and you will have to decide which tests are best for you.

Figuring out the right subject area tests for you

To determine which of the 21 different subject tests are best for you, consider your strengths. Ignore what your friends are doing and choose the subjects that you do best in and that you know the most about. If you can recite the Declaration of Independence in your sleep, the SAT II U.S. History test is probably a good bet. If all of your friends seek you out for help with their trigonometry homework, you'd do well taking the SAT II Math test. Maybe you spend all of your free time expanding your bug collection; the SAT II Biology test is the one for you. Knowing what you'd like to major in can also help direct you toward a particular subject test. If you're still feeling really conflicted, it may help you to know that you can decide which tests you're taking on test day. We highly suggest, however, that you are certain about which tests you want to take before you step into the testing room.

Harvard University, for example, requires its applicants to take three SAT II subject tests to be considered for admission. Although the requirements don't mandate that each test come from a different general subject area (like those of the University of California), the admissions committee suggests that you show your range of understanding by taking as broad a mix of subjects as you feel comfortable with. So, it's probably best to take your SAT II tests in a variety of subjects.

Knowing which level of the SAT II Math you should take

The SAT II Math test comes in two varieties, Level I and Level II, and you're supposed to decide which is best for you. Some universities make the decision easy for you. Because the math sections on the new SAT I test covers much of the same information that Level I tests, some colleges won't accept the Level I test. If you're going to take the math subject test for them, you have to take Level II. Other schools aren't that picky. They accept either the Level I or the Level II. Most schools won't accept both tests from one person, so if you're a real math fan, it doesn't pay to take both levels and expect to use them as two of your required tests.

If you've taken algebra, algebra II, and geometry and no other math courses, focus on the Level I test. The Level II test is for those torture seekers who have gone on to take trigonometry or pre-Calculus (or both!), in addition to two years of algebra and geometry. A Level II score impresses admissions committees more than a Level I score does, unless, of course, the Level II score is below average. If you like math enough to choose it on your own as one of your subject test options, you're probably taking advanced courses and can handle Level II. But you'll know exactly where you stand after you've worked through the math reviews and practice tests in Part V of this book.

It's All in the Timing: When to Take Your SAT II Math Exam (and What to Bring)

You're not done making decisions yet. After you've figured out what subject tests are best for you (see the preceding section), you need to determine when the best time to take them is and what you should bring along with you to the test.

The best time to take the SAT II

Most of the SAT II subject tests are offered in January, May, June, October, November, and December. When in your high school career you choose to take a particular subject test depends on what the subject test covers. SAT II subject tests that cover the material you learned in a two-semester course (like U.S. history, world history, biology, chemistry, and physics) are best taken as soon after you've taken the course as possible. This means that if you take U.S. history and biology in your sophomore year, you may be wise to take the SAT II U.S. History and SAT II Biology after you complete your sophomore year on either the May or June test date. This is much earlier than you would normally take other admissions tests; most students don't take the regular SAT or ACT until the end of their junior years. The math, literature, and foreign language subject tests, however, cover information that you learn over the span of several years of coursework. You'll probably get a better score on these subjects if you wait until you've had at least two or three years of high school courses. You could take these subject tests at the end of your junior year or even in the fall of your senior year.

You can take up to three SAT II subject tests in one day, or you can take only one or two on any given day if that's what works for you. Some people like to get them over with all at once. Others like to be able to concentrate on just one subject. You're less likely to suffer burnout if you take only one or two tests on any one testing day.

When you've settled upon the date for your SAT II, you need to register; the deadline is usually a little over a month before the test date. Everything you need to know about registering for the SAT II is located in the *Registration Bulletin* published by the College Board. It has the test dates, instructions, registration deadlines, fees, test center codes, and other related information. You can get the bulletin from your school counselor, or you can get the same information and register online at the College Board's Web site at www.collegeboard.com.

Items to take with you to the SAT II

Regardless of when you take the SAT II and how many subjects you take in one day, you need to bring certain items along with you. The absolute essentials include the following:

✔ **Your admission ticket:** When you register for the SAT II, the College Board sends you a form that you must bring to the test with you. It proves you're registered.

✔ **A photo ID:** You have to prove that you are you and not your really smart neighbor coming in to take the test for you. Any form of identification that doesn't have your picture on it is unacceptable, so bring along one of the following:

- A driver's license

- A government or state-issued identification card

- A school identification card

- A valid passport

- A school identification form prepared by your school counselor if you don't have any of the other forms of ID

✔ **Several number-2 pencils, a big eraser, and a little pencil sharpener:** Avoid anxiety by carrying a bunch of pencils with you, plus a pencil sharpener, just in case they all break. A large eraser comes in handy, especially if you need to make clean erasures on your answer sheet.

✔ **A scientific or graphing calculator:** For both levels of the SAT II Math test, you need to have a good calculator, preferably a graphing calculator (see Chapter 2). You can't use the calculator on any of the other subject tests, though.

✔ **A quiet watch:** Chapter 2 relates the importance of having your own watch with you during the test. Just make sure that your watch doesn't make any sounds at all.

If you plan to take more than one test, also bring in a quick, light snack (like a Power Bar) to eat during the short break between tests. You'll need energy for the other tests.

Don't expect to be able to eat or drink during the test. Keep your snack in your pocket. And don't eat anything too heavy before a test. You want the blood pumping to your brain, not your stomach.

Don't bring any scratch paper, highlighters, protractors, or books with you. Pretty much anything that isn't in the preceding list isn't allowed. As nice as it would be to tune out your neighbor's hacking cough with your CD player, listening and recording devices are taboo (unless you're taking a language listening test).

First Impressions: The Format and Content of the SAT II Math

The SAT II Math test actually has two separate tests: the Level IC and the Level IIC. You choose one. You can take each of these tests during separate administrations, but most colleges and universities accept only one. Each test contains 50 questions that you have to answer within 1 hour. While the questions tend to get harder as you move through the test, there may be questions toward the end of the test that are easier for you because they test a math subject that you're better at. For example, if you're better at trigonometry than you are at three-dimensional geometry, you wouldn't want to spend a bunch of time trying to answer a tough geometry question early on in the test, because it may keep you from getting to a trigonometry question later on. (For more on managing your time wisely during the test, check out Chapter 2.)

The answer sheets for the SAT II have places for 100 questions, but you mark answers for only 50, because there are only 50 questions on the test. Put a little pencil mark under question 50 on your answer sheet. If, for some reason, you mark an answer after your pencil mark, you know you've done something wrong. Be sure to erase the pencil mark before you turn in your answer sheet.

The two levels of the math test cover most of the same subjects, but the concepts on the Level IIC test are more advanced. Table 1-1 provides a breakdown of the general subject areas of the SAT II Math tests and how they are specifically tested on each of the levels.

Table 1-1			Topics Covered on the SAT II Math Test				
		Geometry					
	Numbers and Operations	**Algebra and Functions**	**Plane**	**Coordinate**	**Three Dimensional**	**Trigonometry**	**Statistics and Probability**
Level I	Basic operations, ratios and proportion, series, number theory, complex numbers, matrices, and sequences	Equations, expressions, inequalities, and properties of functions	Lines, triangles, circles, quadrilaterals, and other two-dimensional shapes	Lines, parabolas, circles, symmetry, transformations	Solids, surface area, volume	Right triangles and identities	Mean, median, mode, range, graphs, and plots
Level II	Same as Level I	All of Level I plus more advanced properties of functions	Not directly tested	All of Level I plus ellipses, hyperbolas, and polar coordinates	All of Level I plus coordinates in three dimensions	All of Level I plus Law of Sines, Law of Cosines, and equations	All of Level I plus standard deviation

Both tests devote about 12 percent of their questions to numbers and operations, and about 8 percent of their questions to statistics and probability. Level II has a greater percentage of algebra and function problems (about 50 percent) than does the Level I test (about 40 percent). The Level I focuses more heavily on geometry (about 40 percent, half of which are plane geometry questions) than does the Level II test (about 30 percent, with more trigonometry and coordinate geometry questions than the Level I test has). So, for more fun with plane geometry, take the Level I test. For more trigonometry and algebra action, sit for the Level II test.

Where You Stand: Scoring Considerations

When it comes down to it, the thing you're probably most concerned about is your final score, the number the colleges see when they get your report.

How the SAT II testers figure out your score

The SAT II is scored similarly to the SAT I. Each multiple-choice question has five possible answer choices. If you pick the correct answer, you get 1 full point for that question. If you pick the wrong answer, the SAT II deducts ¼ point from your raw score. So, one right answer covers four wrong answers. If you skip a question, you don't get any points but you don't lose any points. It's still better to guess if you can eliminate at least one answer choice. For more on guessing strategies, see Chapter 2.

You determine your raw score by taking the total number of questions you answered incorrectly times 4. Then you subtract that number from the total number of questions you answered correctly. The SAT II doesn't stop there, however. To try to make every test measure students equally, the SAT II develops a scale for each test. Where your raw score lands on the scale determines your final score, the one that the colleges get to see.

Every SAT II subject test has a final score value of 200 to 800 points. You get 200 points for knowing your name and recording it correctly on your answer sheet, but colleges aren't going to be satisfied with a 200, so you're going to have to work harder than that. Generally, if you answer at least 60 percent of the questions on the test correctly, you'll get a score of around 600. The mean score on an SAT II test is a little higher than it is on the SAT I. That's because most of the students who take an SAT II test take it because they think they know something about the subject.

Check with the schools you want to attend regarding acceptable SAT II scores; we think you'll find that most of the more selective schools are looking for numbers in the upper 600s and beyond. This means you can still skip about ten questions and get a good score, as long as you answer the other questions correctly. Don't skip more than ten questions on any SAT II test, though.

Here is a rough idea of how many questions you need to answer correctly on the SAT II Math test to achieve a particular scaled score as long as you don't answer too many of the other questions incorrectly. Our assumption in each case is that of the remaining answer choices you skipped 10 questions and answered the rest incorrectly.

Level I:

✔ To get a 500, you need to answer about 24 of the 50 questions correctly.

✔ To get a 600, you need to answer about 31 of the 50 questions correctly.

✔ To get a 660, you need to answer about 36 of the 50 questions correctly.

✔ To get a 710, you need to answer about 42 of the 50 questions correctly.

✔ To get a 750, you need to answer about 46 of the 50 questions correctly.

Level II:

✔ To get a 500, you need to answer about 19 of the 50 questions correctly.

✔ To get a 600, you need to answer about 26 of the 50 questions correctly.

✔ To get a 660, you need to answer about 32 of the 50 questions correctly.

✔ To get a 710, you need to answer about 36 of the 50 questions correctly.

✔ To get a 750, you need to answer about 39 of the questions correctly.

✔ To get an 800, you can skip 6 questions, as long as you answer all of the other 44 correctly.

Why you should never cancel your SAT II score

The fine folks at the College Board allow you to cancel your SAT II score either on the test day itself or in writing by the Wednesday after you take the test. Here are some reasons why you should never take them up on their offer:

- You'll never know what your score would have been if you hadn't cancelled it. It is really difficult to know how well you've done on an exam when you're in the middle of it, and you can miss a fair amount of questions and still do well on the SAT II. So it may seem like you messed up when, in fact, you really shined.

- If you cancel one SAT II test score, the scores for any and all other SAT II tests you take on that day are cancelled, too.

- It is highly unlikely that you will do significantly better on one administration of the test over another. Only unusual circumstance, like getting the stomach flu in the middle of a test, will cause you to perform much differently on an SAT II test.

- Most colleges and universities consider only your top scores, so if you do pretty poorly on one test, you can retake it. The yucky score will be reported with the others, but most schools will ignore it in favor of the good ones.

There are only a few circumstances where canceling a score may be a good idea. The flu scenario mentioned is one. The others involve mechanical failure. If your calculator malfunctions on the math test or your cassette player fails on a language listening test, you can cancel your score on that particular test without canceling the scores for the other tests you took that day. Be prepared, however, and have a backup calculator or cassette player with you just in case you run into this problem.

How SAT II testers report your score

Unless you choose to cancel your score within three days of taking the SAT II, your score will be reported to the colleges you choose. If you take more than one SAT II subject test, all the scores will be reported together. The colleges get to see them all, and usually they choose the top scores to use in their admissions calculations.

All Over Again: Retaking the SAT II

Because most colleges consider only your top scores, it may be in your best interest to retake a subject test if you aren't happy with your first score. The SAT II administrators let you take a test over and over again if you want (that's pretty big of them considering you have to pay for it every time). The SAT II reports to the colleges scores for up to six administrations of the same test, but we don't suggest you take the same test as many as six times.

If you do retake the test, make sure you take it seriously. You want to show improvement. A college will be much more impressed to see a rising score than a falling one. And most colleges are turned off if they see that you've taken one particular subject test more than two or three times. The key is to prepare to do your best on the first try.

Chapter 2

Maximizing Your Score on the SAT II Math Test

- -

In This Chapter

▶ Getting good at guessing

▶ Taking advantage of what you do know to eliminate wrong answer choices

▶ Using your precious time wisely

▶ Escaping some of the pitfalls of standardized tests

▶ Playing it cool with some relaxation techniques

- -

You could have been sleeping, but instead you have to give up a cozy, quiet down comforter with fluffy pillows for a claustrophobic classroom, a bundle of number-2 pencils, and a neighbor with a chronic cough. If you have to endure this agony, you may as well reward yourself with the best SAT II score you can get. This chapter contains the tools you need to put together a winning strategy. Using what you know to your full advantage and avoiding common test-taking hazards can help you almost as much as brushing up on your coordinate geometry can.

Knowing When to Skip: Guessing Strategies

It may surprise you that we mention guessing at all given that the SAT II penalizes you at ¼ point for every wrong answer. The SAT II folks created that penalty to keep you from using your testing hour fabricating fantastic artwork on your answer sheet with random sweeps of your number-2 pencil instead of methodically reading through questions and answer choices. But don't let that punitive point deduction keep you from guessing. You get points only if you fill in your bubbles; and if you mark an answer only when you are absolutely sure you're right, you may not have enough bubbles filled in to get the score you want.

Get the lead out

How do you know when you should mark your best guess and when you should skip? It all depends on how many answer choices you can cross out. The first thing you learn is to cross out wrong answers.

One of the biggest favors you can do for yourself when you take a multiple-choice test is to use your pencil to cross out wrong answers in the test booklet. Crossing out wrong answers serves several purposes.

✔ You'll never waste time rereading a wrong answer choice. You have a limited amount of time to answer questions, so you don't want to spend it reading answers you've already decided are wrong.

✔ Crossing out answers gives you a psychological boost. When you look at the question, you don't see five possible answer choices anymore. You've made your test booklet look more manageable.

✔ It will be easier to determine whether or not you have eliminated enough choices to mark a bubble on your answer sheet.

Now, you may think that all that crossing out takes up too much precious time. Or maybe you're concerned about a shortage of pencil lead. These are lame excuses. It takes less than a second to push your pencil through a line of text. And if you're concerned about your pencil supply, take in a bundle with you. Arrive at the test with at least ten finely sharpened number-2 pencils at your service. Heck, you may even take in a sharpener just in case. Just make sure it is a manual model and not some noisy battery-operated machinery to make the proctor scowl.

If crossing out answer choices doesn't come naturally to you, practice at home. Use your practice tests to not only hone your knowledge but also train yourself to mark through your question booklet.

X marks the spot

When you've mastered the fine art of lead spreading, you'll use it to determine whether you should go one step further with your pencil and fill in a bubble for a question on your answer sheet. Here's a little system that helps you decide when you should guess and provides a code to let you know which questions are best to return to if you have time at the end of the test. The approach outlined in this section helps make you a better test taker because you'll think and write while most people are just thinking.

1. **Read the question carefully.**

2. **Examine each answer choice.**

3. **Use your pencil to cross out answer choices that are obviously incorrect. (We know you've heard this before, but repetition is a powerful learning tool.)**

4. **If you are able to eliminate four answer choices and, therefore, know the correct answer, fill in the appropriate bubble on your answer sheet, mark a big X next to the question in your test booklet, and go on to the next question.**

5. **If you are able to eliminate at least three of the five answer choices, choose one of the remaining answer choices and fill in its bubble on the answer sheet.**

   Put a big X next to this question in your test booklet and go on to the next question.

6. **If you are able to eliminate only two of the five answer choices, go ahead and choose one of the remaining answer choices and fill in its bubble on the answer sheet.**

   Write a large 3 next to the question and go on.

7. **If you can cross out only one of the five answer choices, put a big 4 next to the question in your test booklet.**

   Leave the answer bubbles blank for this question. Go on to the next question.

8. **If you can't cross out any of the five answer choices, put a big 5 next to the question in your test booklet.**

   Leave the answer bubbles blank. Don't worry about the question and go on.

9. **If you finish before the proctor calls time, you know which questions in the section are the best to take another look at.**

   First go to the questions with 3s by them, then those with 4s, and then (only as a last resort if you have lots of time) the 5s.

10. **When you read through a question again, you may find that you are able to eliminate more answer choices.**

    Mark a new answer if you left the question blank, or change the one you previously marked if you come up with a different answer.

Remember that this is a technique. And just like when learning any other technique, it may be awkward for you at first. With practice, though, it can become your greatest asset.

Eliminating Choices: How to Recognize a Wrong Answer

Although using the process of elimination is more effective for other SAT II subjects, you can still eliminate answer choices to your advantage on the SAT II Math. There may be questions that stump you. You may not find the correct answer, but you may spot incorrect answers that you can cross out with your pencil and get yourself in good guessing range. The strategy also helps with questions you know you could find the right answer to if it just weren't so time consuming. Eliminate as many answer choices as you can without going through the lengthy problem-solving process, and use the time you save for questions you can answer more quickly.

Use common sense

Eliminate answer choices that don't make sense. Senseless answer choices are easy to spot if you read carefully. For example, say the test asks you to determine how many marbles Jacob has if he began with 40 and lost 10 of them to Susan. (It won't. That question is too easy for the SAT II, but humor us while we try to make our point.) Without doing any calculations at all, you could eliminate an answer choice of 40. You know that if Jacob loses some marbles (he probably had been studying for the SAT II all night!), he will not end up with the same amount of marbles he began with.

Likewise, if the test asks you to find Tracy's weight in pounds, you can bet the right answer isn't 25. Tracy couldn't live at that weight! It's just common sense.

Rely on what you do know

If you come across a problem that stumps you, don't panic. Use simple stuff you do know to help you at least eliminate a few answer choices. For example, if the test wants you to choose an answer that is an absolute value, you know you can eliminate any negative answer choices because absolute values are positive. So, even if you can't remember everything you need to know to answer a difficult problem, don't give up. You may be able to eliminate enough answer choices using things you do know and make a good guess.

Winning the Race Against the Clock: Wise Time Management

Questions on the SAT II Math tend to get harder as you go through the section. Most people think this means you should hurry through the easy beginning questions so you have time to work on the hard questions at the end. Actually, the opposite is true. You get the same amount of points for correctly answering an easy question as you do for a hard one, and some of the final questions may be too difficult to answer even if you do have time for them. It's silly to hurry on the easy questions and risk making careless mistakes. You should spend enough time on the early questions to make sure you get right answers. If you have to skip the last questions because of time constraints, at least you can console yourself knowing that they may have been too hard to answer correctly anyway.

Pace yourself

Chances are you're taking two or three different subject tests in one morning, so you need a plan to help you stay focused throughout each test. Each subject test lasts for exactly one hour, and each math test has 50 questions. If you punch the numbers, you realize that you have a little over a minute to spend on each question. Now, before you panic, go get your watch with the sweep second hand. Take a deep breath and hold it for 70 seconds. Unless you're an underwater distance swimmer or a tuba player, you probably had a pretty hard time holding your breath that long. Suddenly, 70 seconds seems like a pretty long time. Plus, some questions will be easy to answer in less time than that, so you'll have more time to spend on the others.

You'll need to keep track of your timing, but you don't want to waste precious seconds continually glancing at your watch. So go in with a plan. Check your progress at quarter intervals. With 50 questions, that means every 12 questions or so you should check your time. At question 12, you should be about 15 minutes into the section, at question 25 about a half hour, at question 36 you should be 45 minutes into the test. If, at any time, you find yourself significantly off this pace — say by three or more questions in either direction — you need to make an adjustment. If you are behind, you're probably spending too much time on hard questions. You need to encourage yourself to move on. If you're ahead, you may be moving through questions too quickly at the risk of reading carelessly. You'll be much less likely to find yourself in either of these positions if you experiment with practice tests (see Part V) using the same strict timing as the test.

Use your own timepiece

One of our students took the test in a room with two clocks. The two clocks looked exactly alike except that they displayed different times. The proctor timed from the clock propped up on the table, and the other clock hung on the wall. Our student noted the time when the proctor began the test, but when she checked her pace at question 12, she couldn't remember which clock she began timing from, and she wasn't sure how much time had gone by.

You can avoid this same frustration by ignoring the clocks in the testing room. Get your hands on a watch with a sweep second hand that you can take with you to the test, and use your watch rather than the clock provided to time yourself during the test. At the beginning of the test, set your watch to the top of the hour, any hour. You don't care about the actual time, just the passing of one hour. It is much easier to remember that your test started when the big hand was at the twelve than it is to remember that it started at exactly 9:17 and 32 seconds. You have enough to remember for the SAT II without having to keep in mind what time it was when the proctor said, "Go." Set your watch for noon, and you know you need to be finished by 1:00 (even if it is really only 8:43 and 11 seconds).

Employing Technology: Tips for Using the Calculator

You need a calculator for the SAT II Math tests. Don't leave home without one. You'll actually be able to answer about half of the questions without using a calculator, but there are some questions on the tests that you simply can't answer in the time allotted if you don't use a calculator.

The right calculator

Most types of scientific or graphing calculators are fine, but your calculator needs to be at least a scientific model, and we strongly suggest you use a graphing calculator, especially if you take the Level IIC test. Graphing calculators are expensive, so if you don't have one of your own, consider borrowing one from a friend who isn't taking the math test that day. Or you may be able to plead with your math teacher to let you borrow one from the school.

The SAT II does *not* allow you to bring in any of the following, so don't even try.

- ✔ Pocket organizers
- ✔ Minicomputers or laptops
- ✔ Devices with keypads, writing pads, or pens
- ✔ Calculators that have paper tapes or make noise

You won't be able to plug your calculator in anywhere, so make sure you have fresh batteries and a full set of backups. Some folks even bring in a backup calculator in case something goes wrong.

The right way to use it

You should be familiar with the calculator before you use it on the test. If at all possible, use the same calculator you plan to use on test day when you take your practice tests at home, so you're comfortable with it.

For about half the test, you won't even use your calculator. Many questions are best answered in your head or with pencil and paper. You can probably leave your calculator on the desk for some questions that ask you to analyze graphs. And for a question like this one, you'll spend less time if you figure out that the exponents are the same than if you use a calculator to work out the problem.

If $42^c = 6^3 \cdot 7^3$, what is the value of c?

The rest of the test consists of questions like the following, which are better answered or only answered with a scientific or graphing calculator.

What is the degree measure of the largest angle of a triangle that has sides of length 2, 5, and 5?

Most of the time, you will use your calculator in degree mode, but for some questions, you will need to switch to radian mode. These problems will usually have the word *radian* in the question. For example, if you see the following question, switch your calculator to radian mode.

If $\sin x = \cotan x$, which of the following is a possible radian value of x?

Remember, in almost all cases on the SAT II Math test, you should use your calculator in degree mode and not radian mode. If you use your calculator in radian mode for one question, make sure you switch it back to degree mode after you're done.

Playing it Smart: A Few Things You Shouldn't Do

The majority of this chapter focuses on ways to perform your best on test day. But there a few equally helpful *don'ts* you should know about to make sure you get your best score.

Don't lose track of the numbers on your answer grid

Skipping questions is normal on the SAT II. Just be really careful that if you do skip question 5, you mark the answer to question 6 in the bubble for question 6 and not the bubble for question 5. If you do, you could mess up your answer sheet from that point forward. And if you don't notice until the end of the test, yikes! All that hard work for nothing! You could cancel your score (what a waste!) or try to change your answers before the proctor calls time (make sure you have a big eraser!), but the best plan is to avoid the problem altogether.

Get in the habit of checking your answer grid every fifth question to make sure that you're marking the right answer in the proper spot. And always circle the right answer in your question booklet before you fill in the answer sheet. That way, if you do discover that you've messed up your answer sheet, you won't have to reread the questions to figure out the right answers.

Don't lose your focus

It may surprise you to learn that one hour of graphs and equations may get a little boring. Don't use this as an excuse for your mind to wander. The test is too important to let yourself get distracted daydreaming about what it would be like to share a piece of chocolate mousse cake with the hottie sitting in front of you. Keep your perspective, focus on the task at hand, and promise yourself you'll get the cutie's number after the test is over.

Don't judge your performance by looking around at others

Sometimes in the frenzy of the exam, you forget that not everyone in the room is taking the test that you are. So, it can be disconcerting to see others put down their pencils while you're still slaving away. You can really psych yourself into thinking you're a loser if you compare yourself to those around you, and feeling like a loser doesn't do much for your test score. Keep your eyes on the test and your watch until the proctor signals that time's up.

Don't waste your time on hard questions

You've heard it before, but we're going to say it again. An easy question is worth the same amount of points as a hard question is. While in the classroom your teacher may reward you for knowing the answers to the hard questions, you don't get more Brownie buttons for answering a hard question on the SAT II. So, discipline yourself to know when to give it up. As tempting as it may be to ponder a question until you see the light, you can't afford that luxury. There may be a simpler question that you don't even get to because you've been stubbornly trying to work out a hard problem. If you need to spend more than 2 minutes on any one question, you probably won't get it right anyway. Guess if you can eliminate an answer choice, or skip it and go on.

Don't stew about how you performed on another test

Each SAT II subject test that you take on test day is a separate entity. Keep it that way. Don't let what you think may have been a poor performance on an earlier subject test affect how you do on the Math test. Every subject test has its own score, and the next test is a chance to start fresh. Forget the last one, and move on.

Don't fail to use extra time at the end of the section

You may finish before the testing time is up. Use those last few minutes wisely. You've already marked your test booklet so that you know which questions are the best to return to in order to double-check your answers. Go back to the questions with 3s next to them and double-check your answers. You may have learned something from the questions you answered later in the test that will help you with the questions you had trouble with earlier. Or you may have an "Aha" moment where the answer suddenly becomes clear to you. Or you could catch a careless error. Use your extra time to pick up additional points.

Don't cheat

Cheating isn't the right thing to do and it's also just plain illogical. Your hand simply isn't big enough to hold all the information you need to master the math tests, and the person next to you is probably working on the Chinese language test, so peeking won't help. This book gives you everything you need to do your best on the exam, so don't take the risk.

Curbing a Case of Nerves: Relaxation Techniques

If all this talk about what you should and shouldn't do is making you nervous, relax! After you have read this book, you'll be plenty prepared for whatever the SAT II Math exams dish out. You may feel nervous on test day, however. That's normal and even a little beneficial. The extra shot of adrenaline keeps you alert, but too much anxiety isn't good for you or your test performance. Sometimes, a frustrating question can paralyze you, so arrive at the testing site with a practiced relaxation plan at the ready in case you get caught with a major case of nerves.

It can happen. You'll be joyfully filling in answer bubbles when all of a sudden a seemingly monstrous question comes along. You know this stuff, so you're probably just missing something. But the question makes you so tense, you can't think. At the first sign of freaking, take a quick timeout. The trick is to forget about that nasty question for just long enough that when you go back to it, you'll have that "Aha!" experience and suddenly see the right answer. Or you'll get enough perspective to realize that it's just one little test question and not worth your anguish, so you can merrily skip it and leave it for later if you have the time. Just don't let one or two yucky questions ruin you for the rest of the test.

Inhale

When you stress out, you tighten up and take quick breaths, which doesn't do much for your oxygen intake. Your brain needs oxygen to think straight. Stressing out restricts the oxygen flow to the brain, and you need to do something about it. Fortunately, the solution is easy. All you need to do is breathe. Deeply. Feel the air all the way down to your toes. Hold it. Then let it all out slowly and do it again, several times. (Just don't blow out your air too loudly; you don't want to attract the proctor's attention.)

Stretch

Anxiety causes tension. Your muscles get all tied up in knots, and it helps to untie them. While you're breathing deeply, focus on reducing your muscle tension. Most people feel stress in their necks and shoulders, so do a few stretches in these areas to get the blood flowing.

- Shrug your shoulders up towards your ears; hold it for a few seconds, and then release.
- Roll your head from front to back and from side to side.
- With your hands together, stretch your arms straight up over your head as high as you can. Relax. Then do it again.
- Stretch your legs out in front of you. Move your ankles up and down, but don't kick the person in front of you!
- Shake your hands vigorously as if you just washed your hands in a public restroom with no paper towels.
- Open your mouth wide as if to say "Ahhh." (Don't actually say it out loud, however.) Close your mouth quickly to avoid catching flies.

These quick stretches shouldn't take you more than ten seconds, so don't invest in a full workout. You need to get back to work!

Give yourself a mini massage

If you still feel tense after all that stretching, play masseuse and give yourself a little rub down. Rub your right shoulder with your left hand and your left shoulder with your right hand. Use both hands to massage your neck. Then move up to your scalp.

Don't get carried away, though. This should only take about ten seconds out of your testing time. (Let's see: At today's going rate for a masseuse, you'll owe yourself about 30 cents.)

Take a little vacation

Have ready a place in your imagination that makes you feel calm and comforted. Maybe it's the beach. Or perhaps a ski slope. Wherever it is, sit back in your chair, close your eyes, and visit your happy place for a few seconds to get away from the question that's bugging you. Just make sure you come back!

You can't afford to tune out forever, so don't use this technique unless you are really tense. You can stay longer in La La Land after the test is over.

Think positive thoughts

Give yourself a break. Realize that the SAT II test is tough, and you'll probably not feel comfortable about every question. But don't beat yourself up about it. That is a sure path to disaster. If, after you've tried these relaxation efforts, you still feel frustrated about a particular question, fill in your best guess if you are able to eliminate an answer choice or just skip it if you can't eliminate any answers. Mark the question in case you have time to review it at the end, but don't think about it until then. Put your full efforts into answering the remaining questions.

Focus on the positive. Congratulate yourself for the answers you do feel confident about and force yourself to leave the others behind.

Part II
Addition, Subtraction, and Beyond: Arithmetic and Algebra

The 5th Wave   By Rich Tennant

"WHAT EXACTLY ARE WE SAYING HERE?"

In this part . . .

A little over ten percent of the SAT II Math questions test your knowledge of the simple things, so make sure you study the basics. You don't want to miss the easy points. The chapters in Part II cover everything from basic operations, like addition and subtraction and the order of operations, to the properties of functions. About 40 percent of the Level IC and 50 percent of the Level IIC test concentrates on algebra and functions. There's a lot to remember in these two areas, and the tips and examples in this part will refresh your memory about quadratic equations, logarithms, simple and complex functions, and much more. It's a virtual cornucopia of numbers!

Chapter 3

Mastering the Basics: Numbers and Operations

• •

In This Chapter

▶ Becoming (re)acquainted with numbers — their types and concepts

▶ Reviewing basic operations and how numbers interact

▶ Reviewing the order of operations

▶ Understanding basic graphs

• •

*B*efore you get too immersed in complicated problem-solving, take a brief refresher course and play it by the numbers. Like anything else in life, math builds on information you already know, so some of this should come fairly easily. While many of the things we need to know we really *did* learn in kindergarten, it's a safe bet that most of us were not taught quadratic equations and trigonometric functions in between show-and-tell and nap time in Ms. Marm's preschool classes. Just as reading and writing build on the A-B-Cs, you may need to review your 1-2-3s before tackling some of the more complex workings of math.

In fact, about 10 to 14 percent of both SAT II Math (Level IC and IIC) tests cover topics related to numbers and operations. So you want to know, for example, the difference between natural numbers and whole numbers before launching into some of the more basic problems. Otherwise, you may do all the calculations exactly right for *some* problem, but you could still end up with a completely wrong result if, for example, you used whole numbers when the question referred to integers. This will set you back in trying to get the best score you possibly can. Some students may end up kicking themselves for missing clues as to just what is being asked in problems that should be relatively simple.

It's also a fairly safe bet that you won't be asked something as simple as the following question:

What is the square root of 25?

(A) 5

(B) 7

(C) 10

(D) 125

(E) 625

You're far more likely to encounter a question such as this:

Twenty-five percent of 48 is what percent of 6?

(A) 8

(B) 12

(C) 42

(D) 100

(E) 200

With that perspective tucked away in the back of your mind and fully armed to tackle the numbers racket, look at the different types of number systems and what they mean.

Type Casting: Types of Numbers

Since the stone age, humans have found it necessary to rely on numbers in order to get through daily living. In hunter/gatherer cultures, the people made notches in bones to count, for example, the number of days in a lunar cycle or perhaps to indicate how long the nomadic tribe spent in a particular location until they got food. But humankind soon realized over the millennia that these numbers would become large and unwieldy, especially if a cave-woman were to ask the Neanderthal of the house to bring home two dozen carcasses of elk while she went out and drew eight gallons of water from the pond. Things would have gotten downright silly.

So, in a sense, modern numbers and arithmetic have simplified matters for us, right? While our mathematical operations may have burst prehistoric man's cerebral cortex, our number systems may have made more sense to them in the long run. Here are the more common types of numbers that mathematicians and real people deal with every day.

Figures you can count on: Natural numbers

What could be more natural in our discussion of numbers than to begin with natural num-bers? Where the cave man made notches on bones to note the passing of the days in the month, the modern day kindergartner counts on her fingers. The *natural numbers* are those numbers starting with 1, 2, 3, 4, 5, and so on. Natural numbers are also known as counting numbers because in counting, we begin with the number 1 and continue in a series. (0 is *not* a counting number, naturally!) At the risk of causing confusion at this early stage, natural numbers can also be referred to as positive integers. Wouldn't it be great if everything else were as easy as 1, 2, 3?

Add the zero: Whole numbers

Whole numbers are a whole lot like natural numbers, but they also include the number 0. In other words, whole numbers are all numbers in the following series: 0, 1, 2, 3, 4, 5, and so on. Whole numbers can also be referred to as non-negative integers. Remember that 0 is neither positive nor negative, but it is one of the whole numbers.

Numbers with a little integrity: Integers

Integers belong to the set of all positive and negative whole numbers, and they also include the number 0. Integers are not fractions or decimals or portions of a number. They really have it all together, and that's what gives them their integrity. Integers can be counted as . . . –5, –4, –3, –2, –1, 0, 1, 2, 3, 4, 5. . . . Integers greater than 0 are called natural numbers or *positive integers*. Integers less than 0 are called *negative integers*. Remember that 0 is neither positive nor negative.

Splitting hairs: Rational numbers

A *rational number* can be expressed as the *ratio* of one integer to another; that is, a number that can be expressed as a fraction. Rational numbers behave rationally. Rational numbers include all positive and negative integers, plus fractions and decimal numbers that either end or repeat. For example, the fraction ⅓ can be expressed as 0.33333. . . . Rational numbers do not include numbers such as π or a radical such as $\sqrt{2}$, because such numbers cannot be expressed as fractions consisting of only two integers.

Having it all: Real numbers

Real numbers cast the widest net of all. They include all numbers that we normally think of and deal with in everyday life. For real! *Real numbers* belong to the set that includes all whole numbers, fractions, and rational as well as irrational numbers. Think of real numbers as those numbers represented by all the points on a number line, either positive or negative. Also think of real numbers as those numbers that you can use to measure length, volume, or weight.

In fact, it's hard to imagine a number that is not a real number, because if a number weren't real, it would be imaginary. We simply assume, when we refer to any number, that it is a real number without stating it in so many words. If you are ever asked on the SAT II Math test to give an answer expressed in terms of real numbers, you can probably guess that your answer should be the number you were planning on choosing anyway. Don't be taken in by the gratuitous red herring when they throw in million-dollar words like real number. It's more information than you probably need to solve the problem.

A league of their own: Prime numbers

Prime numbers are all positive integers that can be divided only by themselves and 1. The number 1 is not a prime number. The smallest prime number is 2, and it is also the only even prime number. This does not mean, however, that all odd numbers are prime numbers. 0 can never be a prime number either, because you could divide 0 by every natural number there is you would still come out with 0. To get the prime numbers, just think of a series of numbers beginning with 2, 3, 5, 7, 11, 13, 17, 19, 23, 29 and so on. What makes them so special is that the only two factors for these numbers will always be the number 1 and the prime number.

Not to confuse things any further, but a *composite* number is any natural number that is *not* a prime number, and it does not include the number 1. In other words, a composite number is *composed* of more than two factors. It is the product of more than simply itself and the number 1.

Which of the following expresses 90 as a product of prime numbers?

(A) $2 \times 2 \times 3 \times 5$

(B) $2 \times 2 \times 2 \times 15$

(C) $2 \times 3 \times 3 \times 5$

(D) $2 \times 3 \times 5$

(E) $1 \times 2 \times 5 \times 9$

This question tests your knowledge of prime numbers. Remember that prime numbers are those numbers you can divide by 1 and the value of the number. (The first prime number is 2.) And remember that 1 and 0 are not prime numbers. You can easily eliminate a couple of answers, B and E, because 15 and 9 are not prime numbers. Also, E has the number 1, which is also not a prime number. So cross out B and E.

The product of A is 60, so that's not right. The product of the numbers in D is even less, 30, so that can't be right either. C is the correct answer; it contains the only numbers that are prime, and they equal 90 when you multiply them together.

It never ends: Irrational numbers

Just like it sounds, an *irrational number* is any real number that is not rational. Some help, eh? Just think of the definition of rational number, and realize that an irrational number is one that can't be expressed as a fraction or *ratio* of one integer to another. Irrational numbers are numbers such as π or any radical such as $\sqrt{2}$ that cannot be simplified any further. An irrational number, if expressed as a decimal, will go on forever without repeating itself.

Not all there: Imaginary numbers

An *imaginary number,* just like it sounds, is any number that is not a real number. I can see that you are getting a kick out of this circular reasoning.

Suffice it to say that an imaginary number is a number such as $\sqrt{-2}$. As you know, any real number, whether positive or negative, when multiplied by itself (squared) results in a positive number. So you can't find the square root of a negative number unless it's simply *not* a real number. So, an imaginary number is the square root of any negative number, or any number containing the number i, which represents the square root of -1. (P.S. You also won't be asked to graph the location of i or $\sqrt{-1}$ on a number line, because it isn't there. Really!)

A large slice of pi

Did you know that a team of computer engineers in Japan recently calculated π out to over 1.24 *trillion* decimal digits? That may be irrational just to attempt such a feat. (Rest easy. You probably won't be asked to perform that operation on the SAT II Math test, even with the help of a calculator.)

Scalpel Please: Basic Operations

Now that you're a bit more comfortable with knowing exactly what kind of stuff you're working with, take a stab at manipulating these numbers. Learning operations can be quite simple, almost as simple as 1-2-3. But, in a way, playing with numbers can be even more interesting than learning about the numbers themselves. And it doesn't take a brain surgeon to open your mind to endless possibilities.

Addition, subtraction, multiplication, and division and properties of real numbers

The following sections discuss the basic rules of addition, subtraction, multiplication, and division of real numbers.

Addition

Put 2 and 2 together to see how it all adds up. Very simply, addition is the operation of combining two or more numbers and the end result is called the sum. You know that

$$2 + 3 + 4 = 9$$

But addition also has two important properties that you may remember from elementary school, the *associative property* and the *commutative property*.

The *associative* property simply shows how numbers can associate differently with one another and still come out with the same answer.

$$(2 + 3) + 4 = 9$$

or

$$2 + (3 + 4) = 9$$

The *commutative* property shows how it doesn't matter what order we add the numbers. They always add up to the same sum. So

$$2 + 3 = 5 \text{ is the same as } 3 + 2 = 5$$

Subtraction (or "what's the difference?")

Subtraction, as you may have guessed, is just the opposite of addition. The end result from this operation is called the *difference*. Thus,

$$\text{If } 2 + 3 = 5, \text{ then } 5 - 3 = 2$$

But subtraction, unlike addition, does not have the associative property.

$$(2 - 3) - 4 = -5$$

but

$$2 - (3 - 4) = 3$$

And true to form, subtraction does not have the commutative property either.

$$2 - 3 \neq 3 - 2$$

Multiplication

You can think of multiplication as an operation of repeated addition, and the end result is the product:

$$3 \times 5 = 15 \text{ or } 5 + 5 + 5 = 15$$

This operation is indicated by the multiplication sign, ×, or simply a dot, •. In many instances, for example, with two or more variables or a number and variable(s), the factors are placed next to each other:

$$a \times b = ab \text{ or } 2 \times a = 2a$$

Multiplication, like addition, obeys the commutative property:

$$(a \times b) = (b \times a)$$

And the associative property:

$$(a \times b) \times c = a \times (b \times c)$$

Multiplication also has the distributive property in relation to addition:

$$a \times (b + c) = ab + ac$$

We say that the a is *distributed* among b and c.

Division (or "where's my dividend?")

Finally, division can be thought of as the opposite of multiplication, where the end result is the quotient. So if

$$3 \times 5 = 15$$

then

$$15 \div 5 = 3 \text{ or } 15 \div 3 = 5$$

Like subtraction, division does not follow the commutative or associative properties, so that's one less thing to worry about remembering. Also, the *dividend* in the preceding expression is 15; that is, the number at the beginning of any equation using division, or the number of things that are being divided by another number.

Absolute value

Absolute value of any real number is that same number without a negative sign. It's the distance the number is from 0 on a number line. For example,

$$|3| = 3 \text{ or } |-3| = 3$$

Be careful about mixing negative numbers inside and outside the absolute value symbols! The SAT II Math people love to trip students up when dealing with multiple numbers and absolute values. While two wrongs may not make a right philosophically, you are normally pretty comfortable with the notion that arithmetically two negatives make a positive. And most of the time, you'll be right. But the absolute value sign can make even the most positive thinking into a negative outcome.

Try this one on for size:

$$-|-3| = -3$$

Starting inward, we take the absolute value of the negative number –3, which thus becomes positive 3, and then we reverse this positive value because the first negative sign before the | | symbol cancels out all that positive thinking. How negative!

Even and odd numbers

You know that even numbers are whole numbers divisible by 2. So here they are:

2, 4, 6, 8, 10, and so on

And odd numbers are those whole numbers that are not. How odd!

1, 3, 5, 7, 9, 11, and so on

What's important to remember for the SAT II Math test is what happens to even or odd numbers when they are added, subtracted, or multiplied by one another.

- ✔ Anytime you *add or subtract* two even numbers, your result will be an even number.
- ✔ Anytime you *add or subtract* two odd numbers, your result will also be an even number.
- ✔ Anytime you *add or subtract* an even number and an odd number, your result will be an odd number.
- ✔ Anytime you *multiply* an even number times an even number, you will get an even number.
- ✔ Anytime you *multiply* an odd number times an even number, you will also get an even number.
- ✔ Anytime you *multiply* an odd number times an odd number, you will get an odd number.

Unfortunately, when you *divide* even and/or odd numbers by one another, there is no set rule like you have with the other three operations. And we were on such a roll, too! This is because you may not end up with a whole number when you divide. You could get a fraction, and fractions are neither even nor odd. Decimals are also not even or odd.

Positive and negative numbers

Here are some more handy rules to remember when you perform certain operations on positive and negative numbers.

- ✔ When you *multiply or divide* two positive numbers, your result will be a positive number.
- ✔ When you *multiply or divide* two negative numbers, your result will again be a positive number.
- ✔ When you *multiply or divide* a negative number by a positive number, your result will be a negative number.

As you may expect, there also some rules about *adding and subtracting* positive and negative numbers.

- ✔ When you *add* two positive numbers, your result will be a positive number.
- ✔ Whenever you *subtract* a negative number from another number, you end up adding it to whatever you began with. So,

$$x - (-3) = x + 3$$

Here's a sample question to test your knowledge of even and odd numbers, as well as prime numbers:

If *a* and *b* are different prime numbers, which of the following numbers must be odd?

(A) *ab*

(B) $2a + b$

(C) $a + b + 1$

(D) $ab - 1$

(E) $2a + 2b + 1$

This question involves number theory. Think of all the possible values for *a* and *b,* and then look at the answer choices to see if they *must* be odd. Because 2 is a prime number and a possible value choice for *a* or *b,* choice (A) could be even. Similarly, if *a* is odd and $b = 2$, choices (B) and (C) would be even. If both *a* and *b* are odd, choice (D) is even. For choice (E), $2a + 2b + 1$ must be odd. So the correct answer is (E).

Bases and exponents: Power to the people, but cover your bases!

Exponents are mighty powerful. Just as multiplication can be thought of as repeated addition, exponents can be thought of as repeated multiplication. So,

$$4^3 \text{ is the same as } 4 \times 4 \times 4 \text{ or } 64$$

In the preceding example, the 4 is referred to as the *base,* and the superscript 3 is called the *exponent.* If you add a variable into this mix, such as $4b^3$, then the *base* becomes *b,* while the 4 becomes known as the *coefficient.* In our example, the coefficient 4 is simply multiplied by b^3.

As my high school algebra teacher used to scream (usually when we were caught napping): "The power governs only the number immediately below it!" (that is, the base). So the coefficient is *not* affected by the exponent. Only the *base* is squared or cubed or whatever the exponent says.

This brings up some fascinating properties regarding positive and negative *bases* and even and odd *exponents.*

✔ When a positive number is taken to an even power, it remains a positive number.

✔ When a negative number is taken to an odd power, it remains a negative number.

✔ When a negative number is taken to an even power, it turns into a positive number.

So any number taken to an even power either remains or becomes positive. And any number taken to an odd power keeps the sign it began with.

Another interesting tidbit to digest is that any term with an odd power that results in a negative number will have a negative root and only one root, or answer, to the expression or equation. So,

$$\text{If } a^3 = -125, \text{ then } a = -5$$

That is, the cubed root of -125 is -5.

Conversely, anytime you have an even number for a power, there are two potential roots, one positive and one negative, to the expression or equation. For example:

If $a^2 = 64$, then $a = 8$ or -8.

Adding and subtracting exponents

Did you know that you can *add or subtract* terms with exponents? The only catch is that the the base and exponent of each term must be the same. So,

$$4a^2 + a^2 = 5a^2 \qquad 4a^2 - a^2 = 3a^2$$

Notice that the base and exponent remain the same, and the coefficient is the only number that changes in the equation.

Multiplying and dividing exponents

You can also *multiply* and *divide* terms with exponents. You need to make sure that the *base is the same*.

When *multiplying* terms with exponents, simply add the exponents. Thus,

$$a^2 \times a^3 = a^5 \qquad a \times a^2 = a^3$$

If you throw in a *coefficient,* simply multiply the coefficient as you would have done anyway and add the exponents:

$$4a^2 \times 2a^3 = 8a^5$$

When you *divide* terms with exponents, just subtract the exponents.

$$a^5 \div a^3 = a^2 \qquad a^5 \div a = a^4$$

Same deal with coefficients:

$$9a^5 \div 3a^3 = 3a^2$$

You can actually multiply and divide exponential terms with *different bases,* but only if the *exponents are the same*. So:

$$4^3 \times 5^3 = 20^3 \qquad a^3 \times b^3 = (ab)^3$$

Here is the same deal if you divide exponents when the bases are different, but the exponents are the same:

$$20^3 \div 5^3 = 4^3 \qquad (ab)^3 \div b^3 = a^3$$

One more thing about exponents is that you can raise a power to another power, and in this case, you guessed it, you would multiply the exponents. So,

$$(a^3)^5 = a^{15} \qquad (5^4)^5 = 5^{20}$$

If you happen to have a coefficient in the first operation just stated, you have to take it to the same power:

$$(3a^3)^5 = 243a^{15}$$

Powers of zero and one

Some exponents deserve special treatment. Two of those exponents are 0 and 1. You just have to make sure you commit these to memory.

> ↙ Anytime you raise an exponent to the zero power, for example, 7^0, you get the number 1.
>
> ↙ Anytime you raise an exponent to the first power, for example, 3^1, you simply get the number you started with, that is, 3.

An easy way to prove this little rule so that it makes sense is to plug some numbers into an equation following the rules you just learned:

$$4^4 \div 4^3 = 4^1 = 4 \qquad 4^4 \div 4^4 = 4^0 = 1$$

If you do the math in the preceding equations, it will become apparent that the equations are true. To prove this, just start with the first equation:

$$4^4 \div 4^3 = 4^1 = 4$$
$$256 \div 64 = 4$$

Same thing with the second equation:

$$4^4 \div 4^4 = 4^0 = 1$$
$$256 \div 256 = 1$$

Exponents as fractions — not half as hard as they seem!

When you see a number raised to the power of a fraction, don't panic. Just remember that the top number on the fraction (the *numerator*) is your actual exponent, and that the bottom number (the *denominator*) is the root.

To solve $256^{\frac{1}{4}}$, simply take 256 to the first power, which is 256. Then take the fourth root of 256, which is 4, and voilà! (You can also take the square root of 256 and take the square root of the result, which is 16, for the same answer of 4). Mathematically, this expression looks like:

$$256^{\frac{1}{4}} = \sqrt[4]{256^1} = \sqrt[4]{256} = 4$$

Here's another example:

$$a^{\frac{2}{3}} = \sqrt[3]{a^2}$$

You can use a calculator to solve for the first situation just given. Simply enter $256^{\wedge}(\frac{1}{4})$

Negative exponents

A negative exponent works very much like a positive exponent, but the end result will, for the most part, be a number that's a lot smaller. A negative exponent will simply take the positive exponent, and then flip it around so that the exponent becomes its reciprocal.

$$3^{-3} = \frac{1}{3^3} = \frac{1}{27}$$

So you can see that 3^3 is much greater than $\frac{1}{27}$.

The root of all evil

Trivia. Here is a popular expression in the form of a math rhebus you'll not see on the SAT II Math test. Can you guess what it is?

$$\$ = \sqrt{allevil}$$

Another way to see how this works is to divide your exponents. Remember that when you divide powers having the same base, you subtract the exponents. So,

$$3^3 \div 3^6 = 3^{-3}$$

or

$$27 \div 729 = 0.037 = \tfrac{1}{27}$$

We're sure you won't fall for the common trick that many math students fall for, and that is to assume that the negative exponent somehow turns the original number into some kind of negative number. It ain't gonna happen! For example:

$$3^{-5} \neq -243 \text{ or } -\tfrac{1}{243} \text{ or } -\tfrac{1}{15} \text{ or anything like that.}$$

Digging in: Roots

If you like exponents, you'll *love* roots. Roots are simply the opposite of exponents. This means is that if $3 \times 3 = 9$ or if $3^2 = 9$ or if 3 squared is equal to 9, the square root of 9 is 3, or $\sqrt{9} = 3$. In this case, the square root, 3, is the number that you would square to get the number 9. What could be simpler?

There are as many roots as there are powers. Most of the time in the SAT II Math, you'll be working with square roots. You may also see some other roots, but you're not intimidated, right? If you come upon a cube root or fourth root, you'll know what it is by the radical sign.

For example, a cube root might be expressed as $\sqrt[3]{27}$ or written out as the cube root of 27. This expression is asking what number, when raised to the third power, equals 27? Of course, the answer is 3, because 3 cubed is equal to 27, or $3^3 = 27$.

When you see an ugly radical with a big number under the radical sign, do you: a) run for the hills, or b) try to simplify what's under the sign? If you chose b, you're well on your way to taking the mystery out of these little characters.

Remember that radicals, even the ugly ones, can many times be simplified. For example, if you see a number such as $\sqrt{98}$, don't panic! Just look at it calmly, and see whether you can factor out the 98 to see whether there are any factors that you *can* determine the square root of. Know your perfect squares (4, 9, 16, 25, 36, 49, and so on), and ask what perfect square goes into 98.

Simply factor the ugly $\sqrt{98}$, and it becomes a much prettier and more manageable $\sqrt{49 * 2}$. Because 49 is the perfect square of 7, you can take the 49 out from under the radical sign, i.e, 49 is 7^2 so out comes the 7! Finally, you can simplify the expression to say: $\sqrt{98} = 7\sqrt{2}$.

If $\sqrt[n]{128} = 4\sqrt[n]{2}$, then $n = ?$

(A) 1

(B) 2

(C) 3

(D) 4

(E) 5

You can solve this equation most easily by simplifying the radical. The *n* root of 128 is equal to 4 time the *n* root of 2. $128 = 2 \times 64$, and $64 = 4 \times 4 \times 4$ or 4 cubed. Therefore, the cubed root of 128 equals 4 times the cubed root of 2. $n = 3$, and C is the correct answer.

Because roots are the opposite of exponents, they obey the same rules when it comes to using operations on them. You can add and subtract roots as long as the roots are of the same order (for example, square root, cube root, and so on) and the roots are of the same number. So,

$$5\sqrt[3]{7} + 6\sqrt[3]{7} = 11\sqrt[3]{7} \qquad 11\sqrt{a} - 5\sqrt{a} = 6\sqrt{a}$$

$\sqrt{16 + 9} = ?$

(A) 5

(B) 7

(C) $12\frac{1}{2}$

(D) 25

(E) 625

It is important to note in this question that because the line of the square root symbol extends over the entire expression, you are being asked to find the square root of 16 + 9, not the square root of 16 + the square root of 9.

16 + 9 = 25. The square root of 25 is 5. Therefore, A is the correct answer. If you chose 7, you determined the square root of 16 (4) and added the square root of 9 (3). For 7 to be the correct answer, your problem should have been written with 2 square root signs, one over the 16 and one over the 9.

And with multiplication and division, make sure that the roots are of the same order and you're in like flint. With multiplication, simply multiply what's under the radical signs:

$$\sqrt[3]{9} \times \sqrt[3]{3} = \sqrt[3]{9 * 3} = \sqrt[3]{27} = 3$$

With division, you simply divide what's under the radical signs:

$$\sqrt[4]{7} \div \sqrt[4]{3} = \sqrt[4]{\frac{7}{3}}$$

Turning the tables: Reciprocals

Anytime you want a reciprocal of a number, you simply divide 1 by your number. So the reciprocal of 5 is $1 \div 5$ or simply $\frac{1}{5}$. The reciprocal of $\frac{3}{5}$ is $\frac{5}{3}$. The reciprocal of any number *a* is $\frac{1}{a}$, so long as $a \neq 0$. While the SAT II bosses won't ask this, the reciprocal of a number is also known as the *multiplicative inverse*. Now aren't you glad you didn't ask that question?

Cutting portions: Fractions, decimals, and percentages

Fractions, decimals, and percentages are interrelated concepts, and they generally work very well with one another. They all represent a part of a whole. You will likely need to convert from one to the other in order to solve several problems on the SAT II Math test.

Fractions represent the answer to a division problem. If you divide the number *a* by the number *b*, you get the fraction *a/b*. So,

$$1 \div 4 = ¼$$

To convert the fraction to a decimal, you simply do the division according to the fraction bar.

$$¼ = 1 \div 4 = 0.25$$

To convert a decimal back to a fraction, you first count the digits to the right of the decimal point, and then you divide the number over a 1 followed by the same number of zeroes as there were digits to the right of the decimal. Then you simplify. So,

$$0.25 = {}^{25}\!/_{100} = ¼$$

Changing a decimal to a percent is really pretty easy. *Percent* simply means per 100, or $\div 100$. So to perform the conversion, just move the decimal two places to the right, and the number you are left with should be written as a percent. So:

$$0.25 = 25\% \text{ and } 0.925 = 92.5\%$$

Turning a percent back into a decimal is just the reverse. You move the decimal two to the left and lose the percent sign.

$$0.1\% = 0.001$$

For practice, try and convert the missing figures in Table 3-1. You can find the solutions if you need them in the "Solution to the practice exercise for fractions, decimals, and percentages" sidebar near the end of this chapter.

Table 3-1	Practice Exercise for Fractions, Decimals, and Percentages	
Fraction	**Decimal**	**Percent**
½	0.5	50%
		7.8%
	5.2	
⁷⁄₁₆		
	0.37	

Sometimes, working with percentages involves word problems.

The history of zero

The number zero has very interesting properties and for centuries, it did not exist. Mathematicians simply did without and numbers got somewhat confusing without this place holder. Zero was invented by the Hindu mathematicians Aryabhata and Varamihara in India around or shortly after the year 520 A.D. For a number that will never amount to anything, zero sure is something!

215 million tons of garbage are generated in the United States every year, and about 86 million tons of the total are paper products. Express the portion of U.S. garbage that is paper as a fraction, a decimal, and a percent.

✔ As a fraction: $^{86}\!/_{215}$ or $\frac{2}{5}$

✔ As a decimal: $\frac{2}{5} = \frac{4}{10}$ or 0.4

✔ As a percent: 0.4 = 40%

Twenty percent of fifty percent of 100 servings is how many servings?

$$20\% \times 50\% \times 100 = x$$
$$0.2 \times 0.5 \times 100 = x$$
$$0.1 \times 100 = x = 10$$

Numerators, denominators, and other fancy stuff

When you are asked about fractions on the SAT II Math test, the question may refer to the *numerator* or the *denominator* or some other such verbiage that you may have studied at one time but may have forgotten when you began to learn how to ride a bike.

Suppose you slice a pie into eight equal pieces, and you have a family of 7, each of whom has a slice after dinner (or *before* dinner if they are really hungry). You can visualize it as shown in Figure 3-1:

Fraction of a pie.

Figure 3-1:
Fraction
of a pie.

The shaded pieces of pie in Figure 3-1 show how much of the dessert was gobbled up by the family; the unshaded piece shows what's left of it when the family takes no more than each other's share (or what lil' Abner will steal when mom isn't looking).

To put this pie into terms of a fraction, the total number of pieces in the pie to begin with represents the *denominator*, while the number of pieces that got eaten can be the *numerator*. In this case, the number of pieces that were eaten made up $\frac{7}{8}$ of the total pie, with 7 being the numerator and 8 being the denominator. Taking this one step further, you can think of the piece that was left as $\frac{1}{8}$ of what you started with. The numerator and denominator are also known as the *terms* of the fraction.

Proper fractions are those fractions where the numerator is less than the denominator. Examples of proper fractions are: $\frac{3}{4}$, $\frac{7}{8}$, $\frac{13}{15}$, and so on.

You can also show an improper fraction as a *mixed fraction*, where it is made up of a whole number and a fraction. Examples of mixed fractions are $1\frac{1}{2}$, $7\frac{3}{4}$, $2\frac{2}{3}$, and so on.

Improper fractions are those fractions where the numerator is either greater than or equal to the denominator. Examples of these fractions that have no sense of propriety are: $\frac{15}{2}$, $\frac{5}{3}$, $\frac{8}{5}$, and so on.

When you work with fractions on the SAT II Math test, you may have to substitute mixed fractions for improper fractions, and vice versa. You'll find that it's often easier to change a mixed fraction into an improper fraction before doing operations with it.

Another handy tip for working with fractions involves *simplifying* fractions. You're probably saying, "These fractions are simple enough, so how can it get any easier?" All it means to simplify a fraction is to *reduce* it to its simplest terms. Anytime you can take larger terms in a fraction and make them smaller, you are simplifying the fraction. Examples of reducing or simplifying fractions are:

$$\frac{12}{12} = \frac{6}{6} = \frac{3}{3} = \frac{1}{1} \quad \text{or} \quad \frac{10}{20} = \frac{5}{10} = \frac{1}{2}$$

All you've done is take the larger terms and made them smaller if both the numerator and denominator can be divided by the same number. In the first fraction, $\frac{12}{12}$, you can divide each of the terms by 12. In the second fraction, $\frac{10}{20}$, you can divide both the numerator and denominator by 10.

Just as easy as it is to reduce fractions, you may want to make them bigger in some problems on the SAT II Math test. This happens when adding or subtracting fractions with different denominators. The key is to find the least common denominator when adding and subtracting fractions with different denominators.

Adding and subtracting fractions

Because fractions are parts of whole numbers, they're not as easy to put together as adding 2 + 2. But with some ingenuity, you can at least try to make them whole again. In order to add or subtract fractions, you need to have the same denominator. That is,

$$\frac{2}{5} + \frac{4}{5} = \frac{6}{5} \text{ or } 1\frac{1}{5}$$

$$\frac{6}{5} - \frac{4}{5} = \frac{2}{5}$$

When fractions have the same denominator, don't touch it! Just add or subtract the numerators (top numbers).

Suppose you have something like

$$\frac{2}{3} + \frac{3}{4} = x$$

Whatcha gonna do? Who ya gonna call? Simple, really. Just find that proverbial lowest common denominator. Because your denominators here are 3 and 4, you need to find a number that will divide evenly by both of them. Looks like if you multiply 3 by 4, you get the number 12, and that so happens to be the lowest common denominator. So $\frac{2}{3}$ becomes $\frac{8}{12}$, and $\frac{3}{4}$ becomes $\frac{9}{12}$. Now you can add them up, and voilà!

$$\frac{8}{12} + \frac{9}{12} = \frac{17}{12}$$

Multiplying and dividing fractions

In order to multiply fractions, simply multiply the numerators by each other, and then do the same with the denominators. Reduce if necessary:

$$\tfrac{4}{5} \times \tfrac{5}{7} = (4 \times 5)/(5 \times 7) = {}^{20}\!/_{35} = \tfrac{4}{7}$$

An easier way to do this would have been to simply cancel out those fives that appear in the denominator of the first fraction and the numerator of the second one. After you've done that, you're looking at:

$$\tfrac{4}{5} \times \tfrac{5}{7} = \tfrac{4}{1} \times \tfrac{1}{7} = \tfrac{4}{7}$$

Piece of cake! Now to divide fractions, turn the second fraction upside down; that is, get its reciprocal, and then multiply the numerators and denominators that are left. So

$$\tfrac{2}{7} \div \tfrac{3}{5} = \tfrac{2}{7} \times \tfrac{5}{3} = (2 \times 5) \div (7 \times 3) = {}^{10}\!/_{21}$$

$\tfrac{1}{2} + (\tfrac{3}{8} \div \tfrac{2}{5}) - (\tfrac{5}{6} \times \tfrac{7}{8}) = ?$

(A) $\tfrac{1}{8}$

(B) $^{15}\!/_{16}$

(C) $^{17}\!/_{24}$

(D) $1\tfrac{3}{8}$

(E) $\tfrac{5}{6}$

To solve this problem, you need to do all four operations with fractions. First, figure out the operations inside the parentheses.

$\tfrac{3}{8}$ divided by $\tfrac{2}{5}$ is the same as $\tfrac{3}{8} \times \tfrac{5}{2}$. It equals $^{15}\!/_{16}$.

$$\tfrac{5}{6} \times \tfrac{7}{8} = {}^{35}\!/_{48}$$

Now the equation reads:

$$\tfrac{1}{2} + {}^{15}\!/_{16} - {}^{35}\!/_{48}$$

The common denominator is 48.

To convert the denominator 2 in the first fraction to 48, you multiply by 24. $1 \times 24 = 24$, so $\tfrac{1}{2}$ is the same as saying $^{24}\!/_{48}$.

To convert the denominator 16 in the second fraction to 48, you need to multiply by 3. $3 \times 15 = 45$, so $^{15}\!/_{16} = {}^{45}\!/_{48}$.

Now you have the following expression:

$$^{24}\!/_{48} + {}^{45}\!/_{48} - {}^{35}\!/_{48} \text{ or } {}^{34}\!/_{48}$$

$^{34}\!/_{48}$ is not one of your options in this problem, so you need to simplify the fraction. Divide the numerator and denominator by 2 to get $^{17}\!/_{24}$, which is your answer choice C.

What is 25% of $5\tfrac{3}{4}$?

(A) $^{37}\!/_{130}$

(B) $5\tfrac{3}{4}$

(C) $1\tfrac{7}{16}$

(D) $2\tfrac{3}{16}$

(E) 12

This question asks you to determine the percentage of a fraction. Note that the answers are in fraction form rather than decimal form. Therefore, you need to work out the problem so that it leads to a fraction rather than a decimal.

What is 25% of 5¾? Converted to a fraction, 25% is ¼. Therefore, you are trying to find the answer to ¼ × 5¾. Converting 5¾ to a mixed fraction means that you get ²³⁄₄, so the answer is 1/4 × 23/4, or ²³⁄₁₆.

There is not an answer choice for ²³⁄₁₆, so you need to convert it back to a mixed fraction. ²³⁄₁₆ = 1⁷⁄₁₆, which is answer C.

You can easily eliminate B and E. Obviously 25% of 5¾ is not equal to or greater than 5¾.

Percent change

Percent change is the amount a number *increases* or *decreases* expressed as a percentage of the original number. So if a store normally sells tennis shoes for $72 and has them on sale for $60, then what is the percent of the markdown? Fairly simple, right? To get the percent decrease, simply take the difference in price, that is, $12, and divide that number by the original price. So

$$12 \div 72 = 0.1667 \text{ or } 16⅔\%$$

But now, what if the store *increases* the marked down price back up by 16⅔%? You're back up to the original price tag, right? Wrong! Guess again! Instead, if you increase the lower price of $60 by 0.1667, you get just about a $10 increase. The price jumps from $60 to just about $70.

$$60 \times 1.1667 = 70.002$$

How can that be? The reason the numbers don't seem to add up is because when you dropped the price the first time, you took 16⅔% of $72, which is a bigger number to take a percent of than when you multiply that same percent by the lower sale price.

So what percent of the marked down price of $60 must you increase it by in order to get the original price of $72? Again, take the difference in price, $12, and find out what percent that is of the sale price of $60.

$$12 \div 60 = ¹²⁄₆₀ = 0.20 = 20\%$$

So it's a 20% increase from 60 to 72.

If we know what the percent increase or decrease of an original number is and want to find out how that increase or decrease changes the original number, you need to keep in mind a very important detail.

✔ To find the amount of increase, you need to multiply the original number by 1 plus the rate.

✔ To find the amount of decrease, you need to multiply the original number by 1 minus the rate.

If you increase 100 by 5%, you multiply 100 by (1 + 0.05).

$$100 \times (1 + 0.05) = 100 + 5 = 105$$

If you decrease 100 by 5%, you divide 100 by (1 − 0.05).

$$100 \times (1 - 0.05) = 100 - 5 = 95$$

Notice that when you start from the same number, the *amount* of increase and decrease is the same when you use the same percent of the original number.

A dress that originally cost $52 is on sale at 15% off. If the sales tax on clothing is 5% of the purchase price, how much would it cost to buy the dress at its sale price?

(A) $7.80

(B) $40.00

(C) $44.20

(D) $46.41

(E) $48.23

This word problem asks you how to deal with two percentages, the subtraction of the percentage discount and the addition of the percentage sales tax. First, calculate the discount.

You can use your calculator, or you can figure 15% in your head by knowing that 10% of 52 is 5.20 and half of that (5%) is 2.60, so the discount is $7.80. Now subtract the discount from the original price. $52.00 − $7.80 = $44.20. The discount price for the dress is $44.20.

You still need to calculate the sales tax, so don't choose C! 5% of 44.20 is half of 4.42 (10%) or 2.21. You add $2.21 to $44.20. The only answer that ends in 1 is choice D. You can do the math to verify your guess, but D is the correct answer. $44.20 + $2.21 = $46.41. Not bad for prom night, eh?

Thirty of the 70 boys in a school participate in sports. Thirty of the 50 girls in the school also participate in sports. What percentage of these boys and girls participate in sports?

(A) 40%

(B) 50%

(C) 42⅗%

(D) 60%

(E) 66⅔%

The trick to solving this word problem is to determine the total number of people from which the percentage is derived. You know that there are 70 boys and 50 girls, so there is a total of 120 people in the school. 30 boys and 30 girls (60 total) participate in sports. What percentage of 120 is 60? If it isn't immediately obvious, you can determine the answer as follows:

Evaluate the question: What percentage of 120 is 60?

✔ "What" means ? (the unknown) or what you are trying to find out.

✔ "Percentage" means %

✔ "Of" means you multiply

✔ "Is" means equals

So, the following equation results from the question:

$$?\% \times 120 = 60$$
$$?\% = 60 \div 120$$
$$?\% = 0.50$$
$$? = 50\%$$

The correct answer is B.

Repeated percent change: Who wants to be a millionaire?

Suppose you want to show a percent change repeated over a period of time. One example where this is used is when you want to figure out how much interest accrues on a bank account after several years.

Suppose you have $100 in a bank account at the end of 1992 and you want to know how much money will be in that same account at the end of 2002 and the interest rate is 5%. No fair pulling it out when the stock market is making a bull run! One way to figure this is by simply using the formula just given, and the first step would look something like this:

$$100 \times (1 + 0.05) = 105$$

So you have $105 at the end of the first year. Then, all you have to do is multiply by 10 and you get $1,050 right? Oops, sorry, wrong number. That's almost as bad as getting hit with the millennium bug. This type of question will try and sucker anyone who isn't paying attention every time.

To get the correct answer, you take the same formula and tweak it a bit by adding an exponent. The exponent will be the number of times the original number is changed. This can be expressed as

final amount = original number $\times (1 + \text{rate})^n$, where n is the number of changes

Therefore,

$$100 \times (1 + 0.05)^{10}$$

This quickly multiplies out to 100×1.05^{10}. So,

$$100 \times 1.05^{10} = 100 \times 1.6289 = 162.89$$

So the final answer is $162.89.

If you wanted to use the same formula to show a *repeated percent decrease* over time, you tweak the preceding formula like so:

final amount = original number $\times (1 - \text{rate})^n$, where n is the number of changes

Ratios: Goin' to surf city 'cause it's 2:1

A ratio is the relation between two like numbers or two like values. A ratio may be written as a fraction, such as ¾, as a division expression, $3 \div 4$, with a colon, $3 : 4$, or can be stated as "3 to 4."

Because a ratio can be regarded as a fraction, multiplying or dividing both terms of a ratio by the same number does not change the value of the ratio. Thus, ¼ = ⅛ = ¹⁄₁₆. To reduce a ratio to its lowest terms, treat the ratio as a fraction and reduce the fraction to its lowest terms.

Suppose an auto manufacturer ships 160 cars to two dealerships, and the ratio of shipment to the dealers is $3 : 5$. How many cars does each dealership receive? First thing you do is to add the terms of the ratio, or $3 + 5$, to get the total number of fractional parts each dealership will get. So $3 + 5 = 8$. Therefore, the first dealership will receive ⅜ of 160 cars, or ⅜ × 160 = 60. The second dealership receives ⅝ of 160 cars, or 100. As long as the total number of *things* in this type of problem can be evenly divided by the total number of fractional parts, the answer is workable.

If the ratio of 4*a* to 9*b* is 1 to 9, what is the ratio of 8*a* to 9*b*?

(A) 1 to 18

(B) 1 to 39

(C) 2 to 9

(D) 2 to 36

(E) 3 to 9

This problem may appear to be more difficult at first than it actually is. If 4*a* is to 9*b* is a 1 to 9 ratio, then 8*a* to 9*b* must be a 2 to 9 ratio, because 8*a* is 2 times 4*a*. If 4*a* = 1, then 8*a* must = 2. The answer is therefore C.

It is very important to remember to keep the elements of your ratios and proportions consistent. For example, if your proportion is 3 is to 4 as 5 is to ?, you must set the problem up in the following manner:

$$\tfrac{3}{4} \text{ as } \tfrac{5}{?}$$

and not:

$$\tfrac{3}{4} \text{ as } \tfrac{?}{5}$$

Proportions: Trying to relate

A *proportion* is a relationship between two ratios where the ratios are equal. It may be written as the proportion sign, ::, or with an equals sign, =. Thus, 1 : 4 :: 2 : 8, which can read 1 is to 4 as 2 is to 8.

In a proportion, the first and last terms are called the *extremes,* and the second and third terms are called the *means.* If you multiply the means together and multiply the extremes together, and then compare the products, you find that the products are the same.

$$1 \times 8 = 2 \times 4$$

Thus, any time you know three terms of a proportion, you can find the missing term first by multiplication of either the two means or the two extremes — whichever is available — and then divide your product by the remaining term.

Scientific notation: Playing the big numbers

Scientific notation is simply the way you write out humongous (technical term) numbers so they are more manageable, both for yourself as well as your calculator. You can express a number in scientific notation by dividing the entire number so that the decimal point moves to the left and all digits except one are to the right of the decimal point. Then, you put in an exponent to show how many places you moved the decimal point over. So

$$1{,}234{,}567 = 1.234567 \times 10^6$$
$$20 \text{ million } (20{,}000{,}000) = 2.0 \times 10^7$$

The number of organisms in a liter of water is approximately 6.0×10^{23}. Assuming this number is correct and exact, how many organisms are in a covered Petri dish that contains ½₀₀ liter of water?

(A) 6.9

(B) 3.0×10^{21}

(C) 6.0×10^{22}

(D) 3.0×10^{23}

(E) 1.2×10^{26}

This question uses many words to simply ask you to the find the answer to 6.0×10 to the 23rd power divided by 200. If a liter of water contains a certain number of organisms, ½₀₀th liter of water would contain the same number of organisms divided by 200. Try not to let the wording of the question confuse you.

6.0 divided by 200 equals 0.03. The answer is 0.03×10 to the 23rd power. None of the answer choices provides this possibility, but if you move the decimal point two places to the right, you have to change the power by decreasing it by two. The answer is B, 3.0×10 to the 21st power.

Order of Operations: Please Excuse My Dear Aunt Sally

Basic arithmetic requires that you perform the operations in a certain order from left to right. Okay, so maybe you don't have an aunt named Sally, but this acronym is a helpful mnemonic to remember the order in which numbers are manipulated when you have several operations to deal with. What that means is that if you have an expression that contains addition, subtraction, multiplication, division, exponents (and roots), and has parentheses to boot, it helps to know which operation is performed first.

The acronym Please Excuse My Dear Aunt Sally can help you to remember to perform operations in the following order:

- ✓ **P**arentheses
- ✓ **E**xponents (and roots)
- ✓ **M**ultiplication and **D**ivision
- ✓ **A**ddition and **S**ubtraction

Try it out and see what happens:

$$20(4 - 7)^3 + 15(\%)^1 = x$$

First, evaluate what's inside the parentheses:

$$20 \times (-3)^3 + 15(3)^1 = x$$

Then evaluate the exponents (any number to the power of one is equal to that number):

$$20 \times (-27) + 15(3) = x$$

Then multiply where you can:

$$-540 + 45 = x$$

Finally, do the addition and subtraction from left to right:

$$-540 + 45 = -495$$

$|8 - (7 - 1)| - |8 - 10| = ?$

  (A) −2

  (B) 0

  (C) 2

  (D) 4

  (E) 14

This problem tests your knowledge of absolute value as well as the order of operations. First determine the value inside the parentheses, or $(7 - 1) = 6$. Then subtract this value from 8. $8 - 6 = 2$. Then determine the difference of $8 - 10$. $8 - 10 = -2$. The absolute value of 2 is 2, and, because the absolute value is always positive, the absolute value of −2 is also 2. $2 - 2 = 0$. Therefore, the answer is B. All that for nothin'!

Get the Picture: Graphs

Graphs make it possible to visualize equations or the relationship between two or more measurements. The graph of an equation is the set of points that make the equation true.

One type of graph you may encounter on the SAT II Math test commonly shows the relationship between time and distance. This graph, also known as a two-axes line graph, consists of a bottom axis and an axis on the left side. Typically, the left axis may measure intervals of time, while the bottom axis may measure distance traveled. Thus, such a graph helps you visualize how much distance is traveled in a certain time frame. A line on this graph may vary from another line on the same graph according to the rate or speed of travel.

Solution to the practice exercise for fractions, decimals, and percentages

If you actually completed the practice exercise on fractions, decimals, and percentages and want the answers, here is the solution table:

Fraction	Decimal	Percent
½	0.5	50%
³⁹⁄₅₀₀	0.078	7.8%
5⅕	5.2	520%
⁷⁄₁₆	0.4375	43.75%
³⁷⁄₁₀₀	0.37	37%

Another type of graph that appears frequently on SAT II Math testing is the graph of the coordinates on a plane. This graph also has two axes, the x-axis running horizontally, and the y-axis going vertically. These graphs are useful in showing coordinates for a particular location, or to show how an algebraic equation appears in coordinate space. A typical graph showing x- and y-coordinates is shown in Figure 3-2.

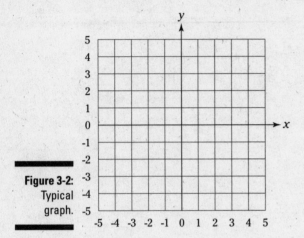

Figure 3-2:
Typical
graph.

This two-axes coordinate graph is also useful for graphing functions such as lines, parabolas, and so on. Functions are discussed in subsequent chapters, but suffice it to say that a graph gives useful information about a function. The *range* of a function is the set of all the y-values of that function. The *domain* is the set of all x-values of the function.

Chapter 4

Contemplating the Variables: Algebra

Algebra is the study of properties of operations carried out on sets of numbers. That may sound like mumbo-jumbo, but understand that algebra is really a generalization of arithmetic in which symbols, usually letters, stand for numbers. You study algebra to solve equations and to find the value of a variable. For example, how often have you heard the command, "Solve the equation for x"? The SAT II Math test has a liberal dose of algebra problems in both Level IC and Level IIC.

Facing the Unknowns: Variables

Variables are symbols that stand for numbers. Generally, when letters are used as variables, they represent a certain value for a number. Just like their name implies, the valuable of variables can change, depending on the equation they are used in. And now, to introduce the most famous variables of all time, let us present: x, y, and z.

Constants, on the other hand, and as their name implies, are numbers that don't change their values in a specific problem. In algebra, while some letters may also refer to constants, they don't change their values in an equation as a variable may. To distinguish constants from variables, and because we have lots of letters in the alphabet to play with, algebraic problems generally designate constants with the letters a, b, or c.

Terms are any set of numbers and/or constants you can multiply or divide together to form a single unit in an equation. You can combine these single parts of an equation by addition or subtraction to become a sum or difference. As an example, the following algebraic expression has three terms.

$$ax^2 + bx + c$$

The first term is ax^2, the second term is bx, and the third term is c. In plain English, think of *terms* as the nouns in a sentence, and the sentence itself is known as the *equation*.

What's it all about, Algie?

Algebra is nothing new. In fact, the earliest record of the subject comes from a treatise written by Diophantus of Alexandria in the 3rd century AD. But it didn't stop there. The classic Greek algebra may have been lost had it not been for the efforts of an Arab mathematician in the 9th century AD. The term algebra comes from the Arabic *al-jabr*, which literally means ``the reunion of broken parts'' (which is a fancy way of saying, "Solve for *x*!") Aside from putting together the broken parts of an equation, the word also refers to the treatment of fractures. Algebra had its origins in ancient Babylon, Egypt, and India. Arabs brought it to Europe via Italy. It really caught on through a book with a rather long-winded title: *ilm al-jabr wa'l-mukabala*, or "The Science of Restoring What is Missing and Equating Like with Like." The mathematician Abu Ja'far Muhammad (c. 800–847) wrote this book. This man eventually became known as al-Khwarazmi, the man of Kwarazm. He is called the father of algebra because he was the first to teach the subject in its elementary form for its own sake. Many of his early calculations still survive to this day. The term *algorithm*, in fact, comes from this name.

An *expression* in algebra is a collection of terms that are combined by addition or subtraction, and the terms are often grouped by parentheses, such as $(x + 2)$, $(x - c^3)$, $(2x - 3y)$, and the like. While an expression can contain as few as one term, it's more common to think of an expression combining two or more terms.

Think of variables as abbreviations for discrete things. For example, if you buy two apples and four oranges, the clerk would not simply ring them up together by adding 2 + 4 to get 6. After all, that would be like mixing apples and oranges, wouldn't it!

So in algebra, you could actually make an expression for combining 2 apples and 4 oranges, and it may look something like

$$2a + 4o$$

In this expression, you know that the variables are a and o. The numbers 2 and 4 are called *coefficients*. So the coefficient of the variable a is 2, and the coefficient of the variable o is 4. Likewise, in the expression $ax^2 + bx + c$, you can say that the coefficient of the variable x^2 in the term, ax^2, is a, while b is the coefficient of the variable x in the term bx. Finally, c is not a coefficient because it doesn't multiply another number. Instead, c is just a constant with a fixed numerical value.

In an algebraic expression, terms involving the same variable, even if they have different coefficients, are called *like terms*. For example, in the expression

$$3x + 4y - 2x + y$$

$3x$ and $2x$ are like terms for obvious reasons. Likewise, $4y$ and y are also like terms. Like terms can be combined with each other, while other terms may not.

In the preceding expression, you can subtract the like terms, or $3x - 2x = x$; and you can also add the like terms $4y + y = 5y$. The end result of the expression, by performing these operations of adding and subtracting like terms, will become $x + 5y$, a much simpler expression to work with.

$$3x - 2x + 4y + y = x + 5y$$

Coming Together: Polynomials

A *monomial* is an expression containing only one term, such as $4x$ or ax^2. A monomial is also, therefore, called a term in an algebraic expression.

A *polynomial* is an expression that has more than one term that can be added together or subtracted from one another. A couple of examples of polynomials are

$$a^2 + 2ab + b^2$$

and

$$ax^2 + bx + c$$

A polynomial is not a name for a parrot; in this sense, *poly* means many, and *nomial* means name, so there are many names for the terms in a polynomial.

A *binomial* is a polynomial that contains two terms, such as $a + b$ or $2a + 3$.

Finally, and just as you may imagine, a *trinomial* is a polynomial that contains three terms. An example of a trinomial would be

$$4x^2 + 3y - 8$$

This expression is also known as a *quadratic* polynomial, too, because one of its terms is squared; that is, it has the power of 2, and there is no higher power than 2 in the polynomial. The classic form of a quadratic polynomial is the trinomial

$$ax^2 + bx + c$$

This quadratic polynomial is so famous that you will be at the top of your game and will surely have it memorized by the time you take the Sat II Math test. But you probably already remember it from math courses gone by.

Keeping Order: Algebra Operations

Algebra is full of symbols, and it's best to get a handle on how these symbols interact. For example, the expression $a + b$ uses symbols a and b to represent numbers, while the $+$ indicates that you combine the two symbols a and b by addition. Simple as that, eh?

Symbols such as $+, -, \times, \div, \sqrt{}$ are common to arithmetic and algebra. They are used for operations on numbers. Arithmetic is more down to earth in that it uses numbers with known values, such as $5 + 7 = 12$, while algebra may use the following expression: $x + y = z$.

While the former gets an exact number, the algebraic equation cannot give you an exact numerical value, because you don't know what x and y represent, let alone z. But that doesn't matter, because algebra is more general than simple arithmetic.

Here are some of the more common symbols used in algebra to signify equality and inequality.

=	equal to (you knew that)
≠	not equal to (probably knew that)
≈	approximately equal to (you may have remembered that)
>	greater than
<	less than
≥	is greater than or equal to (can't make up its mind, but definitely not less than)
≤	is less than or equal to (definitely has an inferiority complex regarding the other side)

Adding and subtracting algebraic expressions

From arithmetic, you know that 3 dozen plus 6 dozen is 9 dozen or

$$(3 \times 12) + (6 \times 12) = (9 \times 12)$$

In algebra, you could write a somewhat similar, but simpler, equation using variables.

$$3x + 6x = 9x$$

You can also use the same like terms and subtract to get the opposite result.

$$9x - 6x = 3x$$

Notice that you can add and subtract like terms, or terms with the same variable. You simply add or subtract the coefficients in the expression; that is, those numbers that multiply the variables. You cannot combine terms with different variables in this manner. That is, if you have an expression containing $3x + 5y$, you cannot combine or simplify any further unless you know the value of either x or y. That's where algebra helps out.

Remember to combine signed numbers (positive or negative) according to the rules of arithmetic. If you add two or more numbers in an expression, they keep their positive sign. If you add a positive to a negative number, it's the same as subtracting.

As an example, add $7x$, $-10x$, and $22x$ and what do you get? Remember you can add the two positive signed numbers and leave the negative number aside for a moment.

$$7x + 22x = 29x$$

Then *add* $-10x$. Adding a negative number is the same as subtracting a positive number. So,

$$29x - 10x = 19x$$

That's fine for adding and subtracting like terms, you may say, but what about working with *unlike* terms? You can't combine terms with different symbols or variables the same way you can where the variables are the same. Check this out:

$$7x + 10y + 15x - 3y$$

If you simply combine the whole expression by adding and subtracting without accounting for the different variables, you could get something very sloppy, and very wrong. In other words, if you just add 7, 10, 15, and –3, you may get something like:

$$7x + 10y + 15x - 3y = 29xy$$

which is incorrect. In order to avoid that kind of error on the SAT II Math test, you first separate the x's from the y's to get something more manageable like this:

$$7x + 15x = 22x$$
$$10y - 3y = 7y$$

The final expression you end up with is

$$22x + 7y$$

Suppose you have two or more expressions and you want to add them together. You can do that by setting them up just like you may set up an addition problem in arithmetic. Remember, only like terms can be combined together this way.

$$3x + 4y - 7z$$
$$2x - 2y + 8z$$
$$\underline{-x + 3y + 6z}$$
$$4x + 5y + 7z$$

For all x and y, $(2x^2 - 3xy - 6y^2) - (4x^2 - 6xy + 2y^2) = ?$

(A) $-2x^2 - 9xy - 8y^2$

(B) $-2x^2 + 3xy - 8y^2$

(C) $-2x^2 + 3xy - 4y^2$

(D) $2x^2 - 3xy + 8y^2$

(E) $6x^2 - 9xy - 4y^2$

The easiest way to begin solving the equation is by applying the negative to the second expression and combining the two expressions, keeping like terms together.

$$(2x^2 - 3xy - 6y^2) - (4x^2 - 6xy + 2y^2) = ?$$
$$2x^2 - 4x^2 - 3xy + 6xy - 6y^2 - 2y^2 = ?$$
$$-4x^2 + 3xy - 8y^2$$

The correct answer is therefore B.

The important thing to keep in mind after you have combined like terms is to double-check that you have used the correct signs, such as changing all signs in the second term, canceling out negatives when two of them occur, and so on. As you can see from the given answer choices, many answers are the same, but the signs have been switched. These answers exist to trap you in case you have made an addition or subtraction error dealing with signs. Add and subtract carefully, and you won't fall for it.

Multiplying and dividing algebraic expressions

Multiplying and dividing two or more variables works just as if you are using these same operations on numbers with a known value.

So if $2^3 = 2 \times 2 \times 2$, then $x^3 = x \cdot x \cdot x$.

And likewise, if $2^2 \times 2^2 = 2^4$, then $x^2 \times x^2 = x^4$.

Similarly, if $2^6 \div 2^4 = 2^2$, then $y^6 \div y^4 = y^2$.

When you multiply a number by a binomial, you need to multiply the number by each term in the binomial. In the next equation, you multiply $4x$ by each term inside the parentheses:

$$4x(x - 3) = 4x^2 - 12x$$

Similarly, with division you do the same operation in reverse:

$$(16x^2 + 4x) \div 4x = 4x + 1$$

Stacking

One easy way to multiply polynomials is by stacking the two numbers to be multiplied on top of one another. Suppose you have an expression like the following:

$$(x^2 + 2xy + y^2) \times (x - y)$$

You can calculate this expression just like an old-fashioned arithmetic problem. Just remember to multiply each of your terms in the second line by each term in the first line.

Like signs (that is, positive and positive), when multiplied (or divided) by each other give a positive result. Unlike signs, when multiplied (or divided) by each other give a negative result.

$$\begin{array}{r}
x^2 + 2xy + y^2 \\
\underline{x - y} \\
x^3 + 2x^2 y + xy^2 \\
\underline{-x^2y - 2xy^2 - y^3} \\
x^3 + x^2y - xy^2 - y^3
\end{array}$$

It also helps to line up your numbers during the first round of multiplication as in the preceding example so that like terms match up before you add your first two products together.

For all x, $26x - (-10x) - 3x(-x + 3) = ?$

 (A) $10x$

 (B) $-3x^2 + 25x$

 (C) $3x^2 + 7x$

 (D) $3x^2 + 25x$

 (E) $3x^2 + 27x$

This question tests your ability to add, subtract, and multiply terms in an algebraic expression. First, multiply $(-x + 3)$ and $3x$.

$$-3x \times -x = 3x^2$$

$$-3x \times 3 = -9x$$

Therefore:

$$26x - (-10x) + 3x^2 - 9x = ?$$

You can add those terms that contain the x variable:

$$26x + 10x - 9x = 27x$$

So the answer to the equation is E:

$$3x^2 + 27x$$

What is the sum of all the solutions of the equation: $\dfrac{2x}{(4+2x)} = \dfrac{6x}{(8x+6)}$

(A) –3

(B) 0

(C) 2

(D) 3

(E) 6

This question is a bit complex. The easiest way to solve this problem is to make the numerators equal to each other. To do this, multiply the first fraction by ⅗ (or 1), which does not change its value. The result is the following equation:

$$\frac{6x}{3(4+2x)} = \frac{6x}{(8x+6)}$$

Because the two numerators are equal, the denominators must also be equal to each other. Now you just place the two denominators equal to each other and solve for x:

$$3(4 + 2x) = 8x + 6$$
$$12 + 6x = 8x + 6$$
$$6x = 8x - 6$$
$$-2x = -6$$
$$x = 3$$

The correct answer is D.

You could also solve this by cross-multiplying opposite numerators and denominators, but that is more complicated and time-consuming. You're in a race against the clock on the SAT II Math test, and using shortcuts like this gives you the edge.

FOIL method: Shiny solutions

When you multiply binomials — that is, algebraic expressions with two terms — it's a cinch you won't get foiled when you use the FOIL method.

FOIL is an acronym for first-outside-inside-last, and that's exactly the order you use to multiply the terms from one binomial by the terms of the second binomial before adding their products.

For example:

$$4x - 5$$
$$\underline{3x + 8}$$

First, multiply the *first* terms in each binomial — $4x$ and $3x$.

$$
\begin{array}{r}
4x - 5 \\
3x + 8 \\
\hline
12x^2
\end{array}
$$

Then you multiply the *outside* terms ($4x$ and 8) together, and the *inside* terms ($3x$ and -5) together, and you can add the products at this point because these are like terms.

Make sure you keep like terms in their own proper columns to insure accuracy when you add this total with the other products later on.

$$
\begin{array}{r}
4x - 5 \\
3x + 8 \\
\hline
+32x \\
-15x \\
\hline
+17x
\end{array}
$$

Last, but not least, multiply the *last* terms: $-5 \times 8 = -40$.

$$
\begin{array}{r}
4x - 5 \\
3x + 8 \\
\hline
-40
\end{array}
$$

Now that you've multiplied the terms, you can add the products together

$$12x^2 + 17x - 40$$

The foil method comes in handy for several problems on the SAT II Math test, although you'll use it more often on the Level IC test. The following is a typical Level 1C example.

When the polynomials $6x + 8$ and $2x - 10$ are multiplied together and written in the form $12x^2 + kx - 80$, what is the value of k?

(A) 6

(B) 8

(C) –10

(D) –44

(E) –60

For this question, all you really have to do is multiply the outer and inner numbers of the two expressions and then add them. Remember with FOIL, you multiply the first, outer, inner, and last. The "first" is given to you: $12x^2$. The "last" is also given: -80.

Multiply the outer numbers:

$$6x \times (-10) = -60x$$

Multiply the inner numbers:

$$8 \times 2x = 16x$$

Therefore:

$$-60x + 16x = -44x$$

k must equal -44, which is the answer found in D.

Suppose you want to (or more importantly, *have to*) divide a polynomial by a monomial. Simply divide each term of the polynomial by the monomial. This one's kind of cute and works out to be a pretty even number in the end

$$\frac{60x^4 - 20x^3}{5x} = \frac{60x^4}{5x} - \frac{20x^3}{5x}$$

$$= \frac{60}{5} * \frac{x^4}{x} - \frac{20}{5} * \frac{x^3}{x}$$

$$= 12x^{4-1} - 4x^{3-1}$$

$$= 12x^3 - 4x^2$$

By dividing the numbers out, you get rid of that ugly fraction bar, plus it's so much simpler to work with when the numbers more manageable.

What is the average of these four expressions: $2n + 8$, $3n - 2$, $n + 4$, and $6n - 2$?

(A) $3n + 2$

(B) $3n + 8$

(C) $12n + 8$

(D) $48n + 32$

(E) $2n$

You figure this average in the same way that you would with arithmetic. Divide the sum of the expressions by the number of expressions.

$$Average = \frac{(2n + 8) + (3n - 2) + (n + 4) + (6n - 2)}{4}$$

$$Average = \frac{(2n + 3n + n + 6n) + (8 - 2 + 4 - 2)}{4}$$

$$Average = \frac{12n + 8}{4}$$

Divide both terms in the numerator by 4:

$^{12}\!/\!_4 = 3n$, and $^8\!/\!_4 = 2$. Therefore:

$$Average = 3n + 2$$

Choice A is the answer.

Extracting Information: Factoring

Factors are two or more numbers multiplied together that result in a product. So if factors add up (that is, multiply together) to form a product, factoring means you take those factors out of the big number to get the two factors that make up that big number.

Try an easy one and factor a trinomial, which also happens to be a quadratic equation because one of its terms is taken to the second power.

$$x^2 + 5x + 6 = 0$$

To factor this equation, you need to get rid of those ugly exponents, combined terms, and so on. You'll want to take them out of the picture and make the whole thing easier to work with. In fact, you're really just doing reverse multiplication. Factoring a quadratic equation means you're going to get rid of the exponent and come up with a couple of factors that happen to be binomials, for example $(x + a) \times (x + b) = 0$.

So look at the preceding equation and ask yourself, what two numbers are factors of the number 6 (the third term), yet add up to the number 5 (the coefficient of the second term)? Process of elimination will give you the numbers +2 and +3. No other numbers will work here. Try it.

$$(x + 2) \times (x + 3) = 0$$

Notice that the equation still has the exact same value as it did before, because you haven't really changed anything at all. You have merely rearranged things to suit your purposes.

If one of the two factors of 6 were a negative number (for example, –2 or –3), your product would have ended up a negative number; you would not have gotten the +6 you so dearly wanted, but the downside is that you'd have to start over again, somewhat frustrated but also a bit wiser from the process of elimination. So a negative and a positive number would not work as factors of +6.

And if the two factors of +6 were both negative numbers (for example, –2 and –3), their product may very well have been +6, but their difference, instead of resulting in the coefficient +5 in the second term 5x, would have given you a negative number.

You may have noticed that by factoring the terms of this equation, you did just the opposite of what you do when you multiplied binomials using the FOIL method.

See what happens if you use FOIL and multiply these terms you just factored.

- The first two terms are $x \times x$, or x^2.
- The outer terms are $3 \times x$, or $3x$.
- The inner terms are $2 \times x$, or $2x$.
- The last terms are 2×3, or 6.

Putting them together gives you, $x^2 + 3x + 5x + 6$, or $x^2 + 5x + 6$

When you break down the trinomial into two polynomials by factoring, you see how much quicker and easier it is to find solutions to algebraic problems.

Another way to think of factoring with any kind of polynomial is to take out the common factor from an expression or equation to simplify things.

See how many common factors you can find in the next expression and just pull them out.

$$-14x^3 - 35x^6$$

First, you can pull out the –7. That not only gets rid of the minus sign but it also makes the numbers smaller and easier to think about and work with.

$$-7(2x^3 + 5x^6)$$

Next, you can take out the common factor of x^3, because x^3 or a multiple of it is common to both terms. This gets you an even simpler expression. Remember that you can divide an exponent with the same base, and when you do that, you subtract the exponents.

$$-7x^3(2 + 5x^3)$$

Distribution

While factoring is nice when you can use it, keep in mind that distribution is another one of your tools when factoring doesn't cut it. Distribution is one of those properties of division in which the factors are already there and you want to combine them. In fact, you may say that distribution is simply going about factoring in a backward manner.

Use distribution when you see two or more terms inside a pair of parentheses acting as one factor, and the whole expression inside the parentheses is being multiplied by another factor. Try this on for size:

$$5(4x + 2x^2) = 10x^2 + 100$$

Wouldn't it be so much nicer if that 5 dangling outside the parentheses weren't just fouling up what may be an otherwise perfectly good expression? Well, now you can do something about it by *distributing* that 5 among the two terms inside the parentheses. Go ahead and multiply those terms by 5 and see what happens.

$$20x + 10x^2 = 10x^2 + 100$$

Now, you're able to start getting rid of some of those terms that are making the equation more complex than it needs to be. Subtract $10x^2$ from both sides, and that gets rid of the exponent.

$$20x + 10x^2 = 10x^2 + 100$$
$$20x + 10x^2 - 10x^2 = 10x^2 + 100 - 10x^2$$
$$20x = 100$$
$$x = 5$$

Not all the factoring and distribution problems on the SAT II Math exam will be so straightforward or come with such nice round numbers. However, you may be pleasantly surprised that mastering these simple little tricks can give you the confidence and know-how you need to solve most of the problems that come your way.

Problem Solving

This is what you have been waiting for, and this is what algebra's all about. You may need to solve for x or perhaps solve an equation or inequality. You can manipulate expressions quite easily with just a few simple steps.

You may also be asked to solve word problems on the SAT II Math test. This means you have to translate the language of the words in those problems into numbers that are arranged in a way that makes algebraic sense, whether you are setting up an equation, an inequality, or whatever.

Table 4-1 has some of the more common words you will see in an algebra word problem that must be translated into math symbols.

Table 4-1	Common Words as Math Terms
Plain English	*Math Equivalent*
More than, increased by, added to, combined with, total of, sum of	Plus (+)
Less than, fewer than, decreased by, diminished by, reduced by, difference between, taken away from	Minus (−)
Of, times, product of	Times (×, ·)
Ratio of, per, out of, quotient	Divide (÷, /)
What percent of	÷ 100
Is, are, was, were, becomes, results in	Equals (=)
How much, how many	Variable (x, y)

Be sure to isolate the variable in the equation or inequality you're trying to solve. Also remember that whatever you do to one side of the equation, you must do to the other side.

Linear equations

A *linear equation* has an unknown variable such as x for you to solve and there is no exponent greater than 1. In other words, you don't have to work with variables taken to the power of 2, 3, or more.

In its simplest form, a linear equation can be expressed $ax + b = 0$, where x is the variable and a and b are constants. An easy way to look at this is to plug in some numbers for the constants and solve the equation for x. The solution for the equation is also called the *root*. An easy example may look like this:

$$4x + 10 = -38$$

To isolate the variable x means to get that variable on one side of the equation all by itself, and then see what shows up on the other side. To begin to isolate this variable, you need to subtract 10. And remember that if you do something to one side of the equation, you need to do the same thing to the other side. Otherwise, your math teacher is liable to rap you on the knuckles with a slide rule. Here's what happens when you subtract 10 from both sides:

$$(4x + 10) - 10 = (-38) - 10$$
$$4x = -48$$

Next, divide by 4, and you're down to the answer.

$$(4x) \div 4 = (-48) \div 4$$
$$x = -12$$

How about a slight curve on the next pitch? Solve for x where it's part of a fraction:

$$\tfrac{x}{4} = -5$$

That's no trouble because you know that you still have to isolate x to the left-hand side of the equation. To do that, you have to pull that 4 out of there somehow. You simply multiply both sides of the equation by 4 and you're there!

$$\tfrac{x}{4} \times 4 = (-5) \times 4$$
$$x = -20$$

Here's another equation with fractions:

$$\tfrac{3x}{5} + \tfrac{8}{15} = \tfrac{x}{10}$$

TIP

To solve an equation with a fraction, get rid of the fraction by multiplying each fraction by the lowest common denominator of the fractional terms. The lowest common denominator is the lowest positive whole number that each denominator (number under the fraction bar) of the fractions divides into evenly.

In the preceding equation, the lowest common denominator is 30. Thus,

$$(\tfrac{3x}{5}) \cdot 30 + (\tfrac{8}{15}) \cdot 30 = (\tfrac{x}{10}) \cdot 30$$
$$(30 \cdot 3x)/5 + (30 \cdot 8)/15 = (30 \cdot x)/10$$
$$\tfrac{90x}{5} + \tfrac{240}{15} = \tfrac{30x}{10}$$
$$18x + 16 = 3x$$

Now that you got rid of the fractions, you need to isolate x on the left-hand side. So subtract $3x$ from both sides, and subtract 16 from both sides as well.

$$18x + 16 = 3x$$
$$(18x + 16) - 3x = 3x - 3x$$
$$(15x + 16) - 16 = 0 - 16$$
$$15x = -16$$

Finally, you can solve for x by dividing by 15.

$$15x = -16$$
$$\tfrac{15x}{15} = -\tfrac{16}{15}$$
$$x = -\tfrac{16}{15} = -1.0667$$

Try one more pitch, this time a change-up:

$$11 + 6x - 14 = 7x - 3 - 2x$$

First things, first. You need to combine like terms to see whether that simplifies matters.

$$11 + 6x - 14 = 7x - 3 - 2x$$
$$11 - 14 + 6x = 7x - 2x - 3$$
$$-3 + 6x = 5x - 3$$

Now begin to isolate the variable that appears on both sides of the equation.

$$-3 + 6x = 5x - 3$$
$$-3 + 6x - 5x = 5x - 5x - 3$$
$$-3 + x = -3$$

Finally, make sure the x is all by itself on the left-hand side of the equation by adding 3 to both sides.

$$-3 + x = -3$$
$$(-3 + x) + 3 = (-3) + 3$$
$$x = 0$$

So x is equal to 0. Seems kind of funny, going through all those steps for nothing, eh? But don't laugh, because 0 is every bit as much of a solution as any other number. The variable x has some value in the equation, and in this case the value of x just happens to be 0.

Substitution: Even better than the real thing!

Suppose you have two algebraic equations with two different variables. An example may be to solve for x, when

$$4x + 5y = 30 \text{ and } y = 2$$

All you really need to do here is to substitute 2 for the value of y in the first equation and you're off to the races.

$$4x + 5y = 30$$
$$4x + 5(2) = 30$$
$$4x + 10 = 30$$
$$4x + 10 - 10 = 30 - 10$$
$$4x = 20$$
$$x = 5$$

So don't get discouraged if you are confronted with two different equations having two different variables. If that happens just ask yourself, wouldn't you rather be an x than a y in a situation like that?

Inequalities: All numbers are not necessarily created equal

An inequality is a statement such as "x is less than y" or "x is greater than or equal to y."

The signs for inequalities are:

>	greater than
<	less than
≥	greater than or equal to
≤	less than or equal to

Also, the arrow on the inequality sign always points to the smaller number. And remember the *gobbling rule,* which says that the large side of the inequality sign gobbles up the greater number.

Treat inequalities a lot like equations for the most part. Isolate the variable. And whatever you do to one side of the inequality, you must also do the other side. The only wrinkle with that last statement is that if you *multiply* or *divide* by a *negative* number, you need to *reverse* the inequality sign.

Take a look at how these inequalities work. When you multiply both sides by +25, your inequality still remains true.

$$5 > 2$$
$$5 \cdot 5 > 2 \cdot 5$$
$$25 > 10$$

Now try multiplying the numbers in the inequality by a negative number (–3) and see what happens. Notice that you need to switch the inequality sign to make it work.

$$5 > 2$$
$$-3 \cdot 5 < -3 \cdot 2$$
$$-15 < -6$$

Try it yourself by dividing both sides by positive or negative numbers and see if you still should reverse the inequality sign. (*Hint:* It still works!)

How about when you add terms together with an inequality?

$$x + 5 < 0$$
$$(x + 5) - 5 < 0 - 5$$
$$x < -5$$

If $x^2 - 1 \leq 15$, what is the smallest real value x can have?

(A) –16

(B) –5

(C) –4

(D) 0

(E) Cannot be determined from the information provided

This problem asks you to determine the smallest real value of x if $x^2 - 1$ is less than or equal to 15. If $x^2 - 1$ can be equal to 15, then x^2 can be equal to 16 ($16 - 1 = 15$).

Now you need to solve for x. Remember that the square root of a number may be positive or negative. The square root of 16 is either 4 or –4. –4 is the smallest real value of x.

Look at your answer choices: –16 would make x^2 equal 256, and –5 would make x^2 equal 25. Therefore, neither could be a solution for x. 0 is a solution for x (0^2 is 0), but it is not the smallest solution; we already know that –4 is a possibility. E cannot be the correct answer because the smallest real value can be determined. C is the correct answer.

Ranges: Not just for the deer and the antelope

You can also use inequalities to show a range of numbers rather than just one single value. For example, you may want to show the range of numbers between –6 and +12 as an algebraic inequality:

$$-6 < x < 12$$

Now suppose you want to show the range of –6 and +12, which also includes those numbers:

$$-6 \leq x \leq 12$$

You can add or subtract with ranges and still have the same equation. Take the first inequality and add 5 to x:

$$-6 < x < 12$$
$$(-6) + 5 < (x) + 5 < (12) + 5$$
$$-1 < x + 5 < 17$$

And the inequality keeps all its values intact.

You may want to add two ranges together. For example,

If $4 < x < 15$ and $-2 < y < 20$, then what is the range of values of $x + y$?

To find the sum of two ranges, follow these steps:

1. **Add the smallest values of each range.**

2. **Add the largest values of each range.**

3. **Add the smallest value of the first range to the largest value of the second range.**

4. **Add the largest value of the first range to the smallest value of the second range.**

Use this formula on the two ranges x and y in the preceding example.

$$4 < x < 15 \text{ and } -2 < y < 20$$
$$4 + (-2) = 2$$
$$15 + 20 = 35$$
$$4 + 20 = 24$$
$$15 + (-2) = 13$$

The sum of these two ranges is, thus, greater than 2 (the smallest sum) and less than 35 (the largest sum), displayed algebraically as:

$$2 < (x + y) < 35$$

Ranges can also be subtracted from one another as well. Much like you did when you added ranges, simply subtract the values of endpoints of the two ranges in the same manner as you did before. The lowest and highest end results give you the new range that is the difference of the two ranges.

You can also do the same thing with multiplication. Just multiply the values of the endpoints of the ranges as you did with adding or subtracting them. The product of the ranges will run between the lowest and highest numbers you end up with. Naturally!

Variation: the spice of life

Many algebra problems deal with relationships between numbers, or how one quantity is related to another.

Direct variation

Direct variation shows how an increase in one number results directly in a corresponding increase in another value. And vice versa. If one number decreases, the other number will decrease in direct proportion to the first.

A good example may be that the amount of electrical current is directly related to the amount of voltage in the circuit. The greater the voltage, the greater the current. You can say that the voltage is an independent variable and the current is a dependent variable, because the amount of current *depends* on the amount of voltage.

Actually, pretend we didn't get into all that technical stuff, because all the rule of direct variation says is that two quantities will maintain their same ratio of one to the other. If one quantity increases, so does the other one, and it does so in the same proportion.

The mathematical relationship between current and voltage may be:

$$current = voltage \div resistance$$

While you won't be asked to memorize this equation, you will be expected to know, if you see an equation like this, that if the voltage increases, the current will increase. If the voltage doubles, the current will double. That's what makes it a direct relationship or direct variation between voltage and current.

As another example, you could say that a company's profits are directly related to its productivity, so if productivity is up, then so are the profits. If productivity doubles, then so do the profits.

The mathematical relationship between two quantities that are directly related is $x \div y = k$, and k is the constant of direct proportionality.

If $x = 5$ and $y = 10$, what will y be equal to if $x = 10$? Just eyeballing this one will tell you that you had to double x to get from 5 to 10. So, if you double y, you come up with 20.

Another common example: If a map is drawn to scale and 1 inch = 1 mile, you know that 12 inches = 12 miles. Simple as that.

Inverse variation

Inverse variation is just the opposite of direct variation. It simply means that if two quantities are inversely related to one another, their product will always be the same. That is, if one number increases, the other number will decrease and always in a proportionate amount.

The formula for inverse variation is $xy = k$, and k is the constant of inverse proportionality.

Look again at the equation showing the direct relationship between current and voltage:

$$\text{current} = \text{voltage} \div \text{resistance}$$

You can see that if current and voltage are *directly* related, current and resistance are *inversely* related. In other words, if the resistance doubles on the right side of the equation, the current will have to decrease by one-half on the left side. And if the resistance decreases, the current will increase in the same proportion. Nothing shocking about that, is there?

16 tons and what do you get: Work problems

Well, we hope it's not another day older and deeper in debt, but see what the work product is for two miners going after coal.

The formula for doing work problems in algebra is

$$\text{production} = \text{rate of work} \times \text{time}$$

The production is the same as saying the amount of work that gets done. Because you get that quantity by multiplying two other numbers, you may say that production is the *product* of the rate times the time.

Suppose you have two coal miners, Alf and Bob. Alf can mine 16 tons of coal per day, while Bob can mine 20 tons per day. If they each work 8 hour days, how much coal can two of them mine in one hour, assuming they maintain a steady rate?

This question asks you to find the amount of production and gives you the rate and the time. To calculate the rate properly, you must state hours in terms of days. So if Alf mines 16 tons in 8 hours, he mines 2 tons in 1 hour. And Bob will mine ⅛ of 20 tons in one hour, or 2½ tons. An hour in this problem will be ⅛ of a day. The equation for this problem will read:

$$\text{production} = (16 \cdot \tfrac{1}{8}) + (20 \cdot \tfrac{1}{8})$$

$$\text{production} = 2 + 2.5$$

$$\text{production} = 4.5$$

Going the distance: Going for speed

Distance problems are a lot like work problems, but somehow your muscles don't get as sore.

The formula for doing distance or speed problems in algebra is

$$\text{distance} = \text{rate} \times \text{time}$$

Any problem involving distance or speed or time spent traveling can be boiled down to this equation. The important thing to remember is to make sure you have your variables and numbers plugged in right.

Abby can run a 7-minute mile. How long does it take her to run $\frac{1}{10}$ of a mile at that speed?

Her rate is $\frac{1}{7}$, because you need to remember to set the problem up as 1 mile per 7 minutes. The distance is $\frac{1}{10}$. The problem is asking what the time is:

$$\text{distance} = \text{rate} \times \text{time}$$

$$\tfrac{1}{10} = \tfrac{1}{7} \cdot t$$

You need to isolate t on one side of the equation, so divide by $\frac{1}{7}$ or multiply by 7. Try division, because that's a bit tougher.

$$(\tfrac{1}{10}) \div (\tfrac{1}{7}) = t$$

$$\tfrac{1}{10} \cdot \tfrac{7}{1} = t$$

$$t = \tfrac{7}{10}$$

Beware of a Trojan horse–type question that asks you to find an average rate of travel. Look for it when you see something like this:

> John drives 50 miles to work each day and returns by the same route in the evening. He is able to drive only 25 miles per hour during rush hour in the morning. He decides to come home early and take advantage of the light traffic in the early afternoon. He makes it back home in half the time. What is his average speed?

The answer sounds a lot like 37.5 miles an hour is the average speed, because that is the midpoint between 25 miles an hour in the morning and 50 miles an hour in the afternoon, right? Wrong!

To get the average rate of travel, use the following formula:

$$\text{average rate} = \text{total distance} \div \text{total time}$$

The total distance is given to you: 100 miles. You have to calculate the total time. If John drives 50 miles in the morning at a speed of 25 mph, then it would take him 2 hours to get to work:

$$50 = (25) \cdot t$$

Divide both sides by 25

$$\tfrac{50}{25} = t = 2$$

Therefore, the time it takes him to drive home in the afternoon (half the time as in the morning) is 1 hour. So, back to the formula for average speed of travel:

$$\text{average rate} = \text{total distance} \div \text{total time}$$

$$\text{average rate} = 100 \div 3$$

$$= 33\tfrac{1}{3} \text{ mph}$$

Joe must travel a total of 225 kilometers to visit his aunt. He rides his bike 5 kilometers to the bus station. He travels by bus to the train station. He then takes the train 10 times the distance he traveled by bus. How many kilometers did Joe travel by bus?

(A) 20

(B) $227\frac{1}{11}$

(C) 22

(D) $447\frac{1}{10}$

(E) $227\frac{1}{10}$

This is an algebraic word problem. The trick here is that this question is not asking you to determine rate or time. So you would simply designate the unknown as "x."

Joe travels a total of 225 kilometers:

$$225 =$$

Part of the trip consists of a 5-kilometer bike ride:

$$225 = 5 +$$

Joe travels by bus, but we don't know what distance. Go ahead and designate the bus distance with the unknown "x":

$$225 = 5 + x$$

He then takes a train for 10 times the distance he traveled by bus (x):

$$225 = 5 + x + 10x$$

Now, solve for x:

$$225 = 5 + x + 10x$$
$$220 = x + 10x$$
$$220 = 11x$$
$$20 = x$$

Joe traveled 20 miles by bus. The answer is A.

Mixing it up: Mixture problems

Mixtures are word problems that the SAT II throws into the mix to try to confuse you. Don't worry. They are much easier than they look and are usually solved using a simple chart.

Say, for example, you encountered the following mixture problem:

Martin goes to the candy store to purchase his two favorite candies, jelly beans and lemon drops, for a big party. He wants to mix 20 pounds of jelly beans priced at 40 cents a pound with an amount of lemon drops priced at 90 cents a pound. He wants to pay 50 cents a pound for the whole mix. How many pounds of lemon drops should Martin buy?

This question seems confusing at first glance, but if you chart it out it will become manageable. Label the rows with the items to be mixed. Head the columns with the known and unknowns.

	Pounds	Price per Pound	Total Price
Jelly beans	20	0.40	$8.00
Lemon drops	x	0.90	$0.90x$
Mix	$20 + x$	0.50	$0.50(20 + x)$

Think about the information you have charted out. The cost of the jelly beans ($8.00) plus the cost of the lemon drops ($0.90x$) equals the cost of the entire mixture [$0.50(20 + x)$].

Now you have your equation. If you want to make it easier to work with you can multiply all the numbers by 100 to get rid of the decimal point.

$$800 + 90x = 50(20 + x)$$
$$800 + 90x = 1{,}000 + 50x$$
$$800 + 40x = 1{,}000$$
$$40x = 1{,}000$$
$$x = 25 \text{ pounds of lemon drops}$$

Wow! That's quite a sweet tooth.

Notice that the chart also helps you keep track of what the x stands for. Drawing a chart may seem time consuming at first, but the time you save in confusion prevention will override the time it takes to create the chart.

The shell game: Solving simultaneous equations

You will most likely encounter problems on the SAT II Math test that call for solving simultaneous equations. This happens when you have a group of two or more equations that must be true at the same time. For this to work, you have to have as many equations as you have variables to solve for. Also, you don't want any numbers with powers of 2 or more.

The way to solve simultaneous linear equations is to combine the equations and knock out one variable at a time. Here is an example:

$$6x + 4y = 66$$
$$-2x + 2y = 8$$

This looks like you can multiply the second equation by 3 and use elimination.

$$6x + 4y = 66$$
$$-6x + 6y = 24$$

Subtract the second simplified equation from the first.

$$10y = 90$$

$$y = 9$$

Now you can plug y into the first equation and solve.

$$6x + 36 = 66$$

$$6x + 36 - 36 = 66 - 36$$

$$6x = 30$$

$$x = 5$$

The solutions, or roots, to the simultaneous equations, then, are

$$x = 5, y = 9$$

Solving quadratic equations

A *quadratic equation* is an equation in which the highest power is 2. That is, one term will be expressed as a square. An example of the classic quadratic form is

$$ax^2 + bx + c = 0$$

where a, b and c are constants and x is a variable that you must solve. Notice that 0 is on one side of the equation and all non-zero terms are on the other side.

Quadratic equations come in various sizes:

$$x^2 = 0$$

$$x^2 - 4 = 0$$

$$3x^2 - 6x + 5 = 0$$

You will very likely be asked problems on the SAT II Math test that require you to solve for x, and generally x will have 2 roots or solutions. A good way to solve a quadratic equation is to factor the equation into two binomials, such as the following:

$$x^2 - 6x + 5 = 0$$

Remember that with factoring the trinomial, you try to see what numbers multiply together to become 5 and, when added together, result in –6.

The two factors of 5 are 5 and 1, or –5 and –1. Here, it looks like you need to go with the negative numbers because when you add them, you'll get –6. So

$$x^2 - 6x + 5 = 0$$

$$(x - 5) \cdot (x - 1) = 0$$

In order to make these equations true statements, you need to know how to make each binomial factor a 0. Because you need to find the value of x in both cases, the solutions or roots to the equation then become clear:

$$x = \{1, 5\}$$

For the quadratic equation $x^2 - 6x - 16 = 0$, there is a postive number solution and a negative number solution. What is the positive value minus the negative value?

(A) –12

(B) –10

(C) –6

(D) 6

(E) 10

First, factor the terms in the quadratic equation to find the values for x

$$x^2 - 6x - 16 = 0$$

The two unknown numbers multiplied together equal –16 and added together equal –6. It is likely that these numbers are –8 and 2. Therefore, this quadratic equation could be factored as follows:

$$(x - 8)(x + 2) = 0$$

Solve for x in both cases (read carefully!):

$$x - 8 = 0$$
$$x = 8$$
$$x + 2 = 0$$
$$x = -2$$
$$x = 8 \text{ or } -2$$

The positive value is 8; the negative value is –2. Now subtract the negative value from the positive value:

$$8 - (-2) = 10.$$

The correct answer is E.

Quadratic identities

There are some quadratic identities that you should commit to memory to answer some of the questions on the SAT II Math test. These formulas are useful to solve questions involving quadratic equations when you can't factor to get two binomials. Here are some common quadratic identities:

$$x^2 - y^2 = (x + y)(x - y)$$

If you're paying attention, you'll notice that the left side of the equation shows the difference of two perfect squares. Here are a couple more:

$$(x + y)^2 = x^2 + 2xy + y^2$$

$$(x - y)^2 = x^2 - 2xy + y^2$$

The quadratic formula

Sure, it's easy solving quadratic equations when the solutions come out to be nice round numbers. But what if the ultimate solutions are harsh looking radicals or perhaps not even real roots? When you can't simply solve a quadratic equation by factoring, you may have to use the quadratic formula, which is a rearrangement of the classic

$$ax^2 + bx + c = 0$$

to become

$$x = \frac{-b \pm \sqrt{b^2 - 4ac}}{2a}$$

While this may look like a mighty unmanageable formula, it may be the only way to find the solution to x in a quadratic equation when nothing else seems to work. Try to solve a quadratic equation using the formula

$$3x^2 + 7x - 6 = 0$$

In this equation, $a = 3$, $b = 7$, and $c = -6$. Plug these numbers into the quadratic formula

$$x = \frac{-b \pm \sqrt{b^2 - 4ac}}{2a}$$

$$x = \frac{-7 \sqrt{7^2 - 4(3)(-6)}}{2(3)}$$

$$x = \frac{-7 \sqrt{49 + 72}}{6}$$

$$x = \frac{-7 \pm \sqrt{121}}{6}$$

$$x = \frac{-7 \pm \sqrt{121}}{6}$$

$$x = \frac{-7 \pm 11}{6}$$

$$x = \frac{-18}{6} \text{ or } -3$$

$$\text{and } x = \frac{4}{6} \text{ or } \frac{2}{3}$$

Thus, the solutions to x are ⅔ and –3. Whew!

Dressing Up Exponents: Logarithms

You know about exponents, those funny little numbers above another number (called the *base*) that raise the base to another power. In other words, an exponent will cause the base to be multiplied by itself as many times as the exponent says. So the number 2^3 represents a base of 2 being raised to the 3rd power. The answer, of course, is 8.

Exponents can also be expressed as a *logarithm*, which is essentially the number of times you multiply the base times itself to get the big number. Another way of thinking of a logarithm is as the inverse of an exponent.

So $\log_4 256$ is essentially the number you of times you multiply 4 by itself to get 256. In this case

$$\log_4 256 = 4$$

The definition of a logarithm can be boiled down to this equation

$$x = a^y$$

which can be rearranged to become

$$y = \log_a x$$

And this is just another way of saying y is the log to the base a of x.

Low-class lumber: Common logs

Most times, you may see a logarithm expressed simply as log 100 (without the base). In that case, the base is assumed to be 10.

$$\log 100 = 2$$

This means that the base 10 must be multiplied by itself twice to turn it into 100. Logarithms with a base of 10 are called *common logarithms.* On your calculator, the "log" function assumes you are looking for a base 10 log. So if you punch in 10,000 and hit "log," the answer will be 4, which is exactly how many times you must multiply 10 by itself to become 10,000.

As you can imagine, you can calculate all kinds of astronomically huge numbers in terms of logarithms and not cause your calculator to start steaming around the edges. Isn't it much simpler to work with a log such as 10, which may represent 10^{10} written out with so many 0s?

Suppose, though, you aren't lucky enough to have a number with a base of 10? Maybe the number you need to figure out is

$$\log_7 23$$

which essentially means log to the base 7 of 23. You can still use your calculator, but it involves an extra step.

For problems like this, use the formula

$$\log_X Y = \frac{\log Y}{\log X}$$

So plug the numbers of our ugly duckling into this formula:

$$\log_7 23 = \frac{\log 23}{\log 7}$$
$$= \frac{1.3617}{0.8451}$$
$$= 1.6113$$

It's also important, at least for the Level IIC test, to understand certain properties of logarithms to any base, which are similar to the properties of exponents. You can bone up on exponents in Chapter 3 at your leisure, but for now, you should know the product rule, the quotient rule, and the power rule for logarithms:

$$\log_a (xy) = \log_a x + \log_a y$$

$$\log_a \frac{x}{y} = \log_a x - \log_a y$$

$$\log_a (x^n) = n \log_a x$$

Natural logarithms

One of the most important numbers is a certain irrational number that has a non-repeating decimal. The number is called *e* and it is very useful in calculating compounded interest on savings, the natural decay of radioactive material, and other scientific measures such as atmospheric pressure at various altitudes.

Besides these fascinating properties in other sundry fields, the number *e* represents an amazing type of logarithm that may very well show up on the Level IIC test. It's called a natural logarithm. The *natural logarithm* of a positive number *x* (written as ln*x*, and pronounced ell enn ex — just like it's spelled!) is a logarithm to the base *e* of *x*, where *e* = 2.71828.

Natural logarithms, just like common logs, can be expressed in several different ways, so memorize the following formulas for the Level IIC exam:

$$\ln a = x$$
$$\log_e a = x$$
$$e^x = a$$

With these formulas committed to memory, you will be able to manipulate logarithms in any number of ways to make your SAT II Math exam more enjoyable, or at least less stressful. In order to take full advantage of your knowledge of logarithms, though, you will need to become thoroughly familiar with the log function of your calculator and know when to use it.

Chapter 5

Minding Your Ps and Qs: Functions

In This Chapter

▶ Understanding math symbols

▶ Synthesizing compound functions

▶ Reverting inverse functions

▶ Comparing domain and range

▶ Analyzing functions with intervals

▶ Recognizing graphs of functions

▶ Discerning degrees of functions

*F*unctions are relationships between two sets of numbers. Functions can describe the relationship between the *x*- and *y*-axes on a graph, but you can also work on them in algebraic formats. If the relationship between two variables, *x* and *y*, is such that there is exactly one value of *y* for each value of *x*, you can say that *y* is a function of *x*.

Additionally, the SAT II Math test may throw in some cryptic-looking symbols to see whether you're awake. Those kinds of functions are just meant to intimidate you, like many of the other questions on the SAT test. They can't scare you, though, if you know the tricks of the trade.

A *function* is simply a rule that turns each member of one set into a member of another set. Usually, you're dealing with numbers, and a function turns one number into another number. The number you want to find the function of is called the *independent variable*. The independent variable is also called the *input,* because you put it into the function to see what happens to it when it comes out the other side of the function.

The resulting *output* for the function is called the *dependent variable.* It's called the dependent variable because, as you may have guessed, the output of the function *depends* upon what the input is as well as what the function is.

The set of all possible values of the independent variable is called the *domain.* The corresponding set of all possible values of the dependent variable is called the *range.*

One thing to keep in mind about functions is that for each value of the input (independent variable), there is one and only one output (dependent variable). When two quantities or numbers are related to one another like this, the *dependent* variable is said to be a *function* of the *independent* variable.

Symbolism: Understanding SAT II Math Symbols

Functions can be displayed in any number of ways, especially on the SAT II Math test.

Sitting in: Symbols as functions

Some functions use any number of different types of symbols to show the relationship between two sets of numbers. Some examples of these kinds of symbols are #, $, &, @, and other even stranger ones that may look more confusing than the Greek alphabet. Don't fret when you see these unusual symbols. The SAT test-makers are throwing these extra red herrings in to throw you off balance. One such expression may appear on the exam like this:

$H(x) = 3x^2 + 5$. What is $H(8)$?

Again, the strange-looking symbol in the initial expression is nothing more than a different way of saying that the function of x is to take x squared, multiply the result by 3 and then add 5. In calculating that function exercise with, say, the number 8, you simply substitute the x with 8, like so.

$$H(8) = 3(8)^2 + 5$$
$$= 3(64) + 5$$
$$= 162 + 5$$
$$= 167$$

See, that wasn't so scary after all! Now you try one.

Find $\&(3)$ if $\&(x) = 2x - 1$.

Once more, all you're doing is finding the value of the function of 3 by substituting 3 for x in the equation that gives you the functional relationship. Here, the functional relationship is to take the old number, x (or 3 where x is given a value), multiply it by 2, and then subtract 1. When you input 3 into the function in place of x, your work will look something like this.

$$\&(x) = 2x - 1$$
$$\&(3) = 2(3) - 1$$
$$\&(3) = 2(3) - 1$$
$$\&(3) = 6 - 1 = 5$$

Keeping the status quo: "Normal" functions

More commonly, you'll see functions displayed with more conventional characters, such as the letters f, F, g, G, and ϕ (another Greek letter, phi). Any type of letter or symbol can represent a function.

So, for example, $f(x)$ is used to indicate the function of x, and is simply stated as "f of x."

Just a word of caution on this convention, the expression $f(x)$ is not the same as $f \times x$. It's merely standard notation for a function.

For any function f, $f(x)$ indicates the value of f at x; that is, the number that f is equal to when x is equal to a given number. This means simply that if you have a value of x as your *input* into the function, the f of x is the value of f (or the y-value on the vertical y-axis) as the *output* of the function. All the possible values for x are known as the *domain*. All the possible values of $f(x)$ are called the *range*. Here are a few examples of various functions.

$$f(x) = 2x^2 + 17$$
$$g(c) = (4c - 1)^3 - 3c^2 + 17c - 10$$
$$f(t) = \tfrac{1}{4}(t - 7)^2$$
$$g(q) = (3q - 5) \div (q + 1)$$

While all these equations may appear to be very, very different in looks, you treat them all pretty much the same. Just do the math! You get a value for the *old* number; that is, the value for the variable in the parenthesis to the left-hand side of each of the preceding equations. Then, you simply find the *new* number, or function, by plugging the old number into the equation that's given and see what the operation kicks out.

So, for the first equation, you are looking for the *f* of (*x*), and you are told that the value of *x* = 12. Therefore, the value of *f*(the function of *x*) is:

$$f(x) = 2x^2 + 17$$
$$f(12) = 2(12)^2 + 17$$
$$f(12) = 288 + 17 = 303.$$

When you do the math on this function, you'll find that the *x*-value is 12, and the *y*-value — the *f*(*x*) — is 303. If you were to graph this on a coordinate plane (see Chapter 7), the ordered pair would be (12, 303). It's that simple.

Occasionally, you may be asked to find a *piecewise function*. A piecewise function is one that is broken into pieces, or different segments. It may also be called a *split function*. A piecewise function is not much different from a "normal" function. You determine the function depending on how you define the possible values of the domain. The function in each "piece" is continuous. Here are some examples of piecewise functions.

$$f(x) = \begin{cases} 2x - 1, x \le 1 \\ x + 7, x > 1 \end{cases}$$

$$y(x) = \begin{cases} x^2, x \le -2 \\ -x, x > -2 \end{cases}$$

Notice that in each of these piecewise functions, the function is split into pieces. The value of the *y* or *f*(the range) is determined by the value of *x* (the domain) just as with normal functions, but the *y* or *f* value of a piecewise function follows a different pattern depending on which of the two rules *x* falls under.

So, in the first piecewise function just given, if *x* = 0, then the first rule goes into play, and *f*(0) = 2(0) – 1, or simply *f* = –1. If *x* = 2, the second rule governs, and *f*(2) = 2 + 7, or, more simply, *f* = 9.

You can also manipulate the value of *y* in the second piecewise function equation just given in much the same way. If *x* = –2, then *y*(*x*) = (–2)², or simply *y* = 4. On the other hand, if *x* = –1, the second rules goes into play and *y*(*x*) = 1.

There's also another kind of strange-looking "normal" function you need to recognize. You should be able to recognize a *paired value* when it's used in a function. Suppose you come across something like the following:

$$t(m, n) = 4m^2 - 3n$$

This looks only slightly stranger than a normal function, so you should have slightly more than no trouble figuring it out, right? In the case of a paired function, you may be asked to find the value of the function of (5, 6), or find $t(5, 6)$. Just work this like you would a normal function. Plug the two paired values right into the spots where they belong in the equation of your function and you won't go wrong.

$$t(m, n) = 4m^2 - 3n$$
$$t(5, 6) = 4(5)^2 - 3(6)$$
$$= 4(25) - 18$$
$$= 100 - 18$$
$$= 82$$

Many questions regarding functions may ask you to just plug in numbers to get the answer, but sometimes you'll be asked to plug in variables as well. Consider the following algebraic functions that will take an extra step or two to solve.

If $f(x) = (x - 4)^2$, find the value of $f(4x - 4)$.

(A) $16x^2 - 16$

(B) $16x^2 + 16$

(C) $16x^2 - 32x + 64$

(D) $16x^2 - 64x + 64$

(E) $16x^2 - 64x - 64$

This is really pretty simple as long as you don't try to do this in your head. Just use a systematic and logical approach, plug in the variables and numbers where they belong, and you're there! The first thing you need to do with this exercise is substitute $4x - 4$ in place of your x in the initial definition of the function. Thus, $f(x)$ becomes $f(4x - 4)$. Then you get the following:

$$f(x) = (x - 4)^2$$
$$f(4x - 4) = (4x - 4 - 4)^2$$
$$f(4x - 4) = (4x - 8)^2$$
$$f(4x - 4) = (4x - 8)(4x - 8)$$
$$f(4x - 4) = 16x^2 - 32x - 32x + 64$$
$$f(4x - 4) = 16x^2 - 64x + 64$$

Your correct answer is D. If you chose A, all you did was square the two terms in the function, and you didn't multiply it out correctly. The answer B is similar, but the operation is just reversed. C is close, but you need to get the correct coefficient in the second term by adding the $32x$ twice to get the $64x$. Finally, E is wrong, because it doesn't switch the sign on the last term as must be done when multiplying two negatives together.

Here is one way to figure this one out if you're in a hurry and you don't want to mess around with all those x's and stuff. Just substitute a number into your equation for x and see what happens if you solve for the equation. You could pick an easy number like 2 and see what happens when you plug it in as the x-value in your $f(4x - 4)$. Thus, if $f(x) = (x - 4)^2$, find the value of $f(4x - 4)$. Plug the 2 in the second f expression.

$$f(4 \times 2 - 4)$$
$$f(8 - 4) = f(4)$$

Now, you should plug your new number (4) into the original function.

$$f(4) = (4 - 4)^2$$
$$f(4) = (0)^2 = 0$$

Your original number 2 has become 0 if it's plugged into the original function. How do you know this is correct? You don't until you see whether this works by plugging that same number 2 into the answer choices to see which one comes up with 36. The only equation where that works, amazingly enough, is choice D. Try it.

$$16x^2 - 64x + 64$$
$$16(2)^2 - 64(2) + 64$$
$$16(4) - 128 + 64$$
$$64 - 128 + 64 = 0$$

That's probably not as much fun when your final answer is 0, is it? Here are some more function questions to see whether you're still having fun. But don't spend too much time on these on the SAT II test because they're not really that difficult. Work on them now mainly to see whether you can do what the SAT II folks want you to do. They mainly want you to plug the numbers in correctly. The answers should flow from that simple operation pretty easily.

Given $h(r) = \begin{cases} 4|r| & \text{if } r \geq 2 \\ -|r| & \text{if } r < 2 \end{cases}$, evaluate $h(-r)$ if $r = -7$.

(A) -28

(B) -14

(C) -7

(D) 7

(E) 28

Don't make the mistake of getting goofed up by the negative signs here. If $r = -7$, then $h(-r)$ is the same as saying $h(7)$.

Because 7 is greater than 2, look to the first rule of the function $h(r)$. You want to find the solution to $h(r) = 4 \times$ the absolute value of 7, or simply 4×7. Your answer for this one, then, is E. You can see that if you miss the sign here and how it translates into the function $h(r)$, you will get a negative of the correct answer, choice A, or any of the others if you follow the incorrect rule.

If $g(x) = x + 1$ and $h(x) = 2x - 1$, what is the absolute value of the difference between $g(2)$ and $h(2)$?

(A) -2

(B) -1

(C) 0

(D) 1

(E) 2

You can find the answer using simple algebra. Simply plug in the number 2 into each function to get the result of each operation. When you plug 2 into the first function, $g(x)$, you get $2 + 1 = 3$. When you plug it into the second function, $h(x)$, you get $2(2) - 1$, or 3. The difference is, quite simply, 0.

You know you can eliminate choices A and B right off the bat, because the question asks what the *absolute* value of the difference is. Absolute values are always positive. Your choices are narrowed already. Your answer choice should be 0 if you correctly subtracted 3 from 3. That was almost too easy, but you still get a gold star. If you chose D, you somehow subtracted 1 from 2, or 2 from 3. Go back and rework your math. If you ended up with E, you may have added the terms in your $h(x)$ function instead of subtracting them. Just be careful with your math.

Adding to the Fun: Compound Functions

Imagine what would happen if you could put double or even triple the "fun" into functions. What would happen if you tried to find out the function of a function? Well, that's exactly what is meant when you calculate a *compound function*. The way it works is to find the first function, and then plug that value into the second function. Take a look at these two functions.

$$f(x) = 3x + 1 \qquad g(x) = x - 4$$

Now, suppose you want to find $g(f(x))$. This function reads "find g of f of x." The way you solve this is to start with the first function you can work with. With typical algebra, you try to find the answer to f of x first, because the f of x is buried most deeply inside the parentheses. After you find the answer to (the output of) the f function, you need to plug it into the function g of x.

Take an example of a number that you may plug in. Say that $x = 2$. Now, you want to find $g(f(2))$. Broken down in steps, you first find $f(2)$:

$$f(x) = 3x + 1$$
$$f(2) = 3(2) + 1$$
$$f(2) = 6 + 1 = 7$$

You now know that $f(x) = 7$. Your next task is to find $g(f(x))$, so you are basically looking for $g(7)$. You can see that the number 7 is now the x that is your input in the g function. Plug 7 in and see what you get:

$$g(x) = x - 4$$
$$g(7) = 7 - 4$$
$$g(7) = 3$$

Easy as pie! Just remember to do the inside part of the compound function first, and then work your way to the outside. It's just a matter of finding these functions in the right order.

The preceding operation in finding $g(f(x))$ is not the same as finding $f(g(x))$. While you want to use the same two functions just given, you need to do them in the reverse order. Again, suppose that $x = 2$, and you need to find $f(g(x))$. Try it out.

$$f(g(x)) = ?$$

First, go to the inside of the parentheses and find out the g of x.

If $x = 2$, then
$$g(2) = 2 - 4 = -2$$

Now, you have the first part done, and see how easy that was! You're on a roll, so go for the second half. Because you've discovered that $g(x)$ is -2, you want to plug -2 as the *new* input or the new x into the second function $f(x)$.

$$f(x) = 3x + 1$$
$$f(-2) = 3(-2) + 1$$
$$f(-2) = (-6) + 1 = -5$$

So, your final answer is $f(g(x)) = -5$. This is quite different from $g(f(x)) = 3$. We can't stress enough how important it is to keep the two functions straight when you're doing these calculations for compound functions. Try one or two yourself.

If $t(x) = 3x - 12$ and $r(x) = (\frac{1}{3})x + 4$, what is $t(r(x))$ where $x = -1$?

(A) -15

(B) -2

(C) -1

(D) 2

(E) $3\frac{2}{3}$

This problem requires you to do the function formula twice. First, you go to the inside parentheses. You need to figure out what the answer to $r(x)$ is and, from there, it goes pretty smoothly with a couple of quick algebraic equations.

$$r(x) = (\frac{1}{3})x + 4$$
$$r(-1) = (\frac{1}{3})(-1) + 4$$
$$r(-1) = -\frac{1}{3} + 4 = 3\frac{2}{3}$$

You now have half the problem solved. To find out $t(r(x))$, simply plug the number $3\frac{2}{3}$ as a new x-value into the $t(x)$ function.

$$t(x) = 3x - 12$$
$$t(3\frac{2}{3}) = 3(3\frac{2}{3}) - 12$$
$$t(3\frac{2}{3}) = 11 - 12 = -1$$

Your correct answer is -1, so you should have chosen C. If you chose A, you found only $t(x)$, and not $t(r(x))$. If you chose B or D, you got the inverse of an incorrect calculation. If you chose E, you merely found $r(x)$, and not $t(r(x))$.

In this last problem, you end up with the same number you started with. If you had to find the function $(r(t(x))$, you may be surprised to find out that you would come out with, of all things, -1. This tells you that the two functions, $t(x)$ and $r(x)$ are *inverse* functions, and the subject of the next section.

Doing the Opposite: Inverse Functions

An *inverse function* is a function that does just the opposite of an ordinary function. There are a couple of different ways you can show an inverse function.

✔ If the function f is the inverse of the function g, and if $y = g(x)$, then $x = f(y)$. In other words, if you put x into the original function and you get y as an outcome, you can put y into the inverse function and get x as your outcome.

✔ Another way of showing inverse functions is to use the following notation. The inverse of the function $g(x)$ is the same as $g^{-1}(x)$. Take a look at how this works in practice.

$$g(x) = 4x - 3 \qquad g^{-1}(x) = (x + 3) \div 4$$

You can see what's happening here. When you take an inverse of a function, you do just the opposite of what you did to get the ordinary function. Try to prove it.

Suppose you are given the preceding function, $g(x) = 4x - 3$ and that $x = 10$. To solve, you use the rule for the function as follows.

$$g(x) = 4x - 3$$
$$g(10) = 4(10) - 3$$
$$g(10) = 40 - 3 = 37$$

Now take the output of the original function (37) and plug that number into the inverse of the original function. Thus, you have $g^{-1}(x) = (x + 3) \div 4$, where $x = 37$. Here's what you get.

$$g^{-1}(x) = (x + 3) \div 4$$
$$g^{-1}(37) = (37 + 3) \div 4$$
$$g^{-1}(x) = (40) \div 4 = 10$$

Get it? You're right back where you began before you started doing any of the functions. Simple as all that!

A function is no more than a rule, and an inverse function is no more than an undoing of that rule by going backward. Remember to undo everything by reversing the operations in exactly the opposite order. In the preceding example, the original function first multiplies by 4, and then subtracts the number 3. When doing the inverse, you need to first *add* the number 3, and *then* divide by the original coefficient 4.

Compare the standard notation for a function as $f(x)$ and the inverse function as $f^{-1}(x)$. Don't be fooled into thinking that the inverse function $f^{-1}(x) = 1/f(x)$. While the latter may be the same as the inverse of the *number* that you get when you find the original function of x, it's *not* the inverse function of $f(x)$. This is a common mistake; make sure you don't fall victim to this trap.

Doing inverse functions can seem complicated, but it isn't as long as you remember that you just take the opposite of the original operations, and you take them in reverse order from the original function. Here's another.

If $f(x) = (x + 17)/5$, then find $f^{-1}(x)$.

The initial function or rule requires you to add 17 and then divide by 5. To get the inverse of the original rule, you must now first multiply by 5, and then subtract 17. So, if you had $f(x) = (x + 17) \div 5$, and $x = 3$, you want to find $f^{-1}(x)$. Here's how.

$$f(x) = (x + 17) \div 5$$
$$f(3) = (3 + 17) \div 5$$
$$f(3) = (20) \div 5 = 4$$

Now find $f^{-1}(x)$, where $x = 4$. Just do the opposite of the original operations in the reverse order. First, multiply by 5, then subtract 17:

$$f^{-1}(x) = 5(x) - 17$$
$$f^{-1}(4) = 5(4) - 17$$
$$f^{-1}(x) = 20 - 17 = 3$$

And that's how you get back to square 1!

These two exercises show you exactly what happens when you take the compound of two inverse functions and put them together. Here is how the combination of two inverse functions, $f(x)$ and $g(x)$, may appear as a compound function.

$$f(g(x)) = x$$

Another way to say the same thing is

$$f(f^{-1}(x)) = x$$

You can see that by placing two inverse functions together as a compound function, your result is that the two functions will *undo* each other, and you end up with the number you started with!

If $f(x) = (2x - 3) \div 2$ and $f(g(x)) = x$, what is $g(x)$?

(A) $2(2x - 3)$

(B) $2(4x - 6)$

(C) $2x + \frac{3}{2}$

(D) $(2x + 3) \div 2$

(E) $4x + 6$

Because you know that the compound function $f(g(x)) = x$, you should realize that $f(x)$ and $g(x)$ are inverse functions. From there, the answer is quite simple. The *rule* for $f(x)$ is to first multiply by 2 and subtract 3, and then divide your answer by 2. To get the inverse of that function — $g(f(x))$ — you need to first multiply by 2 and add 3, and then divide your answer by 2. This gives you an answer choice of D. Answers A, B, and E don't have any division in them, so they cannot be an inverse of an original function that involves multiplication. Choice C is close, but results when x is multiplied by 2, instead of adding 3 and *then* dividing by 2; the product is added to the *quotient* of 3 divided by two. It looks similar on its face, but the two sets of operations yield very different results.

If $g(x) = 2x^2 - 7x - 8$, find $g^{-1}(7)$.

(A) -2

(B) 2

(C) 4

(D) 5

(E) 7

Because the function $g^{-1}(7)$ is the inverse of $g(x)$, you should know that when you calculate the function $g(x)$, your answer is 7.

The best way to find the correct answer is by plugging your various options from the list of answers into $g(x)$, and then seeing what answer is produced. When you come up with the one that results in 7, you've hit the jackpot. It may well take you a couple of tries with the answer choices, but ultimately, you'll come up with 5 as the number to plug into the initial function that gives you 7. Therefore, your correct answer is D. The other answers just won't work.

Setting Limits: Domain and Range

Sometimes, the SAT II Math test may ask you questions that test your knowledge of how to determine what the extent of the domain or the range is for certain functions. These questions are not difficult, but you need to be aware of some basic rules of algebra in order to determine the proper limits of the domain and range.

Domain

The *domain* of a function is the set of all numbers that could possibly be an input (or *argument*) of a function. You normally think of the domain of a function including all real numbers, unless it is limited in some artificial way. The only limitation on whether a number is a real number is if it follows the rules of real numbers.

There are only a few limitations on what a real number cannot be.

- ✔ A real number can't be a fraction where the denominator is 0. Otherwise the number is undefined.

- ✔ A real number can't be an even-numbered root of a negative number. Otherwise, as you know, it's an imaginary number. Any number when squared or else taken to some other even power cannot result in a negative number.

What does this all mean? Simply that the domain of real numbers cannot contain a fraction where the denominator is 0 or a square or other even-numbered root of a negative number.

How does this work with functions? Suppose you have the following function.

$$f(x) = \frac{x+4}{x-2}$$

Normally, the domain of x in a function can contain an unlimited number of values. In the preceding example, though, there is a fraction in the function having the variable x in the denominator. Because your denominator can't add up to 0, the denominator of $x - 2$ cannot be equal to 0. This means that x cannot be equal to 2. In terms of functions, the domain of $f(x)$ is $\{x \neq 2\}$.

Look at another.

$$f(x) = \frac{x-3}{5x+1}$$

You have pretty much the same kind of situation. Because the value of the denominator cannot be 0, the value of $5x + 1$ cannot equal 0. This means that $5x$ cannot be equal to -1, and therefore, x cannot equal $-\frac{1}{5}$. In terms of functions, you would say that the domain of $f(x)$ is $\{x \neq -\frac{1}{5}\}$.

Take a look at functions involving square roots.

$$g(n) = 3\sqrt[4]{n+2}$$

In the preceding function, you have an even-numbered radical sign with the variable x under it. You know that the root of an even-numbered radical — in this case, the 4th root — cannot be a negative number. Otherwise, you won't have a real number as your final answer. That said, the number under the radical sign cannot be less than 0. So n must be greater than or equal to –2. The result is the domain of the function $g(n)$ is $\{n \geq -2\}$.

One more example of how the domain is limited by what's under the radical sign

$$g(y) = 7\sqrt{-2y}$$

What's that, you say? There's *already* a negative number under the radical sign. Well, yes there *seems* to be, but take heart. In order to have a proper domain, you need to make sure the *product* of –2 and y is not negative. If y is a positive number, the number under the radical would be negative, and you don't want that. Because two negative numbers multiplied together become positive, the y in this case needs to be negative to make the domain viable. The resulting domain of the function $g(y)$ is $\{y \leq 0\}$.

Sometimes, your function will be a bit more complicated, and you just need to do another step or two to figure what values to eliminate from the range. Consider the following function.

$$f(x) = \frac{3}{x^2 - 4x - 21}$$

In the preceding function, you need to make sure that nothing in the fraction causes the denominator to equal 0.

This problem is not as easy to figure out as the first two functions just given. The best way to figure out the range in this kind of function is to use good old factoring. This tells you how to find out what values of x can make each factor in the denominator be 0. And those are the terms you want to avoid in the domain.

$$f(x) = \frac{3}{x^2 - 4x - 21}$$

$$f(x) = \frac{3}{(x+3)(x-7)}$$

You can easily see that the two values of x that give you a 0 in the denominator are –3 and 7. If any one of the factors in the denominator is 0, then the entire denominator is 0. This means that the domain of $f(x)$ is $\{x \neq -3, 7\}$.

There's another example where a polynomial in the function makes it a bit tougher to spot what's wrong with the potential domain. When you have a polynomial under a radical sign, be prepared. The following function is one such example.

$$t(a) = \sqrt{a^2 + 5a - 50}$$

You need to find out what values of a under the radical sign turn the entire value into a negative number. Remember, a negative number can't be a square root and still be a real number. So factoring saves the day.

$$t(a) = \sqrt{a^2 + 5a - 50}$$

$$t(a) = (a+10)(a-5)$$

That wasn't too bad, was it? You have to compare the variable a with two numbers, –10 and 5.

This process isn't quite the same as finding out how to make the factors become 0 as you did in the previous two functions involving fractions. It may be tempting to say that the range does not include the numbers –10 and 5. Nothing could be more fatal for the SAT II math test, though.

Keep in mind that when you multiply the two factors together, you cannot come up with a product that is a negative number. That means that the factors *must* either be both positive or both negative. The only way to make that happen is to make sure that a is either less than or equal to –10 or greater than or equal to 5. Thus, the domain of $t(a)$ is $\{a \leq -10\}$ or $\{a \geq 5\}$.

What is the domain of $f(x) = \sqrt[3]{-2x^2 + 5}$?

(A) $x \geq 0$

(B) $x \leq 0$

(C) All real numbers

(D) $-1.71 < x < 1.71$

(E) There is no solution for the domain of $f(x)$

Because your function gives you an odd-numbered root, you can have either a positive or a negative value under the radical sign. Thus, the laws of math do not limit you except that you need a real number to plug in as your domain.

If you decided on C, all real numbers, you would be in good company with those who got this answer correct. If you chose A or B, you are artificially limiting the domain of the function as it may be if you were looking for the square root instead of the cube root. If you answered D, you are getting hung up on finding the cube root of 5 and artificially limiting the domain. Finally, if you chose E, you simply gave up too soon and were looking for an easy way out.

Determine the domain of the function $f(x) = \dfrac{4}{x^2 - x - 2}$.

(A) $\{x \neq -1, 2\}$

(B) $\{x \neq 1, -2\}$

(C) $\{x = -1, 2\}$

(D) $\{x = -4, 2\}$

(E) $\{x \neq -4, 2\}$

This is simple algebra. You know the denominator cannot equal 0, so solve for x in that trinomial. You do that by factoring:

$$(x^2 - x - 2)$$
$$(x + 1)(x - 2)$$
$$x = -1, 2$$

Simple enough, right? If you went on to choose C as your answer, your factoring would have been absolutely right, but your answer would be 100% wrong. C gives you only the factors in the polynomial expression in the denominator. You're not finished.

–1 and 2 are the values of x that make the denominator equal to 0 and, therefore, they cannot be values in the domain. Thus, your correct answer is A. If you chose B, you had the factors switched around with the incorrect sign in front of them. If you chose D, you found the correct factors of the denominator, and then divided the numerator by each fraction. Again, not correct. Finally, if you chose E, you did a hybrid; you divided the numerator by the solution to the factors of the denominator. You also said your domain was not equal to either of those numbers, but this again misses the boat that the solutions to the factors themselves are the numbers you need to exclude from the domain.

Range

The *range* of a function is the set of all numbers that could possibly be an output of a function. In other words, if you think of the domain as the set of all possible independent variables to be put into a function, the range is the set of all possible dependent variables that can come out of any particular function.

Just as the domain of a function is limited by certain laws of mathematics, so, too, is the range. The rules of math that limit the range are few.

✔ An absolute value of a real number cannot be a negative number.

✔ An even exponent or power cannot produce a negative number.

How does that work in real life? Check out some situations where these rules come into play. Look at the following functions.

$$g(x) = x^2 \qquad g(x) = |x|$$

The one thing these two functions have in common is that the range of both of them is limited to the laws of mathematics. Each of these functions can only give you an output that is a positive number. Therefore, in each case, the domain of the function of g is greater than or equal to 0.

The notation format for this expression can be any of the following.

✔ The range of $\{g(x) \geq 0\}$

✔ The range of $g(x)$ is the set $\{g: g \geq 0\}$

✔ The range of $g(x)$ is $\{y: y \geq 0\}$

Suppose the SAT folks decide to throw a 180° curveball at you and do something like this.

$$g(x) = -x^2 \qquad g(x) = -|x|$$

Does this kind of negativity make you go berserk and want to storm out of the test center? We hope not!

All you need to realize is that the negative sign simply flips around the potential values of the range in the opposite direction. Thus, the new range of these functions is g is less than or equal to 0. This can be expressed as follows.

✔ The range of $\{g(x) \leq 0\}$

✔ The range of $g(x)$ is the set $\{g: g \leq 0\}$

✔ The range of $g(x)$ is $\{y: y \leq 0\}$

That last item reads: "The range of g of x is the set of all numbers y, such that y is less than or equal to 0." You see the variable y used to substitute for the function $f(x)$ or $g(x)$, because when you graph the function, you generally have an x- and a y-value.

Now, suppose you see a function that looks a bit scarier and has several operations involved.

$$f(x) = \frac{-x^2 - 7 \cdot |-x|}{-8}$$

Just look at this logically and ask yourself what happens to the x when it gets put into the function. You look at x^2 and know it can't have a negative value, so the lowest possible value of the range of $f(x)$ at this point is 0. But the negative sign in front of x^2 means that it now can't have a positive value. Whatever positive value it had before, the negative sign took care of that and flipped it around. So at this point, the range of $f(x)$ becomes *less than* or equal to 0.

Moving right along, you can see that the absolute value of $-x$ must be positive, so you subtract 7 times what must be a positive number from your initial negative value of $-x^2$. Therefore, the numerator must be a negative number for any value of x. Now you divide the whole thing by the denominator, -8. Of course, when you divide a negative number by a negative number, you end up with a positive number.

Go ahead and plug in a couple of different values for x. Whether you choose a positive or negative value for x, your final answer comes out positive The lowest non-negative number you could possibly get is 0. This means that the lowest possible value of the range of $f(x)$ is either 0 or some positive number.

The way to express this range is

the range of $f(x)$ is $\{f(x) \geq 0\}$

You may well be asked on the SAT II Math test to pick out a possible range from a given function. Try this one on for size.

What is the range of the function $g(x) = 1 - \sqrt{x-2}$?

(A) $g(x) \geq -2$

(B) $g(x) \leq -2$

(C) $g(x) \geq 2$

(D) $g(x) \geq -1$

(E) $g(x) \leq 1$

It's very easy to get tricked and start looking for the *domain* when you should be finding the *range*. If you chose C, you were thinking about the range of x. If you chose A or B, you were hung up on the negative sign when trying to get the number under the radical to be a non-negative number. If you chose D, you somehow got the number 2 out of the radical and subtracted 1. The correct answer is E, where $g(x) \leq 1$.

First, make sure you have figured out how to make the radical a real number. You must have at least 0 under the square root sign to have a real number, so x must be at least 2. If x is 2, the function would be $1 - 0$, or simply 1. Any higher value for x results in a lower value for the output of the function. Thus, $g(x) \leq 1$.

What is the range of $f(x) = -\left|-x^3\right| - x^2$ if the domain of x is all negative numbers?

(A) $\{f(x) < 0\}$

(B) $\{f(x) > 0\}$

(C) $\{f(x) \geq 0\}$

(D) $\{f(x) \leq 0\}$

(E) $\{f(x) = \text{all real numbers}\}$

This can almost be considered a trick question. Remember that 0 is neither negative nor positive, so the *domain* of x must be less than 0. That said, it's easy to figure out the range of the function as long as all those negative signs in the expression don't terrify you. It doesn't matter how great the exponent or how many negative signs there are inside the absolute value symbol. When you see that symbol, everything inside becomes positive. The negative sign outside the absolute value turns the term after it into a negative. Looking at what follows, you know that the x^2 can only be a positive number. So the bottom line is, you are subtracting a positive number from a negative number, and your result will be something less than 0. This gives you the correct answer A.

If you chose B, you got fouled up with all the negative signs along the way. Just remember to take a deep breath when you see a pile of negative signs and be systematic about it. Start with the inside number or variable and work your way out. If you chose C as your answer, you not only got fouled up with all the signs, you ignored the fact that the range cannot equal 0 because your domain is less than 0. Remember, 0 is not a negative number. And if you chose D, you again did not exclude 0 from the domain and range. E is wrong because the limitation on your domain to negative numbers has limited your range accordingly.

If $f(x) = (-x^2 - 7) \div 10$, what is the range of f/x? (Consider using a graphing calculator for this problem.)

(A) $\{y: y = -0.7\}$

(B) $\{y: y = 0.7\}$

(C) $\{y: y \leq -0.7\}$

(D) $\{y: y \geq 0.7\}$

(E) $\{y: y \leq -0.8\}$

You know that x^2 must be a positive number because any number squared yields a positive result (or 0). The negative sign in front of x^2 turns the positive number into a negative number (or 0). After you subtract 7, your numerator is now certainly negative, no greater than –7. And when you divide the negative numerator by the positive denominator 10, you end up with a negative number, –0.7 at the most. This means that the range of the function is either less than or equal to –0.7, and your correct answer is C.

If you chose A as your answer, you may have been thinking that you were trying to solve for the greatest value of *y* without solving for the range. If you chose B, you thought you were solving for the lowest value of *y*, because you did not flip the sign correctly in the numerator. If you chose D, you got a range value, but you did not flip the sign correctly. And if you chose E, you probably did not consider 0 as a possible value of *x* and ended up going to the next integer, thus throwing your answer off by 1.

Functions with intervals

You may occasionally come across a question on the SAT II Math test that gives you a set of values, or an *interval*, for the variables that can be included in the domain of the function. You then have to come up with the set of values for the range of that function. The following question illustrates the type of problem you may encounter that gives the interval of the domain and asks you to find the interval of the range.

A person can earn $5.00 an hour doing janitorial work. What is the set of values that represents how much money that person can earn in any one day?

In this question, you can say that the amount of money that the person earns is a function of the number of hours in the day that he or she works. Thus:

$$f(x) = 5x$$

This means that x is the variable representing the hours that a person works, and for every hour (x) that the person works, he or she earns $5.00. The fact that there are only 24 hours in a day means that the set of values in the domain is limited to the set $\{0 \le x \le 24\}$, which is an interval of x-values.

Another way of asking to find the range within a certain interval of the domain is

If $f(x) = 5x$ for [0, 24], then what is the set that represents the range of $f(x)$?

The question asks you to find the different values of the range when you are given the interval of values from 0 to 24 inclusive as your domain. Finding the set that includes the values of the range is a simple matter of multiplying 5 (that is, $5.00/hour) times the different amounts of hours in a day (the x-value) that the person could possibly work, which is the interval or artificial limitation on the domain. The answer is that the person can earn between $0 and $120 in a day. You get that when you plug the upper and lower limits of the hours in a day into the function that multiplies those values times 5. You express the interval for the range of this function like so:

$$\{y: 0 \le y \le 24\}$$

The preceding example is a simple illustration of how you can find and express the values of the range as limited by the interval of values for the domain. The questions on the SAT II Math test will probably not be nearly so easy, which means more work for you. It's not too bad, though, because these types of exercises build on what you already know about the limitations on domains and ranges.

If $f(x) = \dfrac{1}{x^2 - 5x - 14}$ for [3, 10], then what is the range of $f(x)$?

(A) $\{y: -0.03 \le y \le 0.03\}$

(B) $\{y: -0.03 \le y \le 0.05\}$

(C) $\{y: -0.05 \le y \le 0.03\}$

(D) $\{y: -0.14 < y < 0.50\}$

(E) $\{y: -0.50 < y < 0.14\}$

This question gives you a red herring. Because you have a fraction, your natural inclination may be to solve for x by factoring $x^2 - 5x - 14$ to find out what's going to give you a 0 in the denominator. When you factor the expression in the denominator, you get $(x + 7)$ and $(x - 2)$. You may be thinking to yourself that you should not have $x = -7$ or 2, because either of those numbers would result in one of your factors being 0, and then you would have a 0 in the denominator. This strategy may be okay if you were trying to find the *domain* of x, but in this case you are looking for the *range*. In fact, the information in the problem tells you that the lower and upper bounds of the domain are 3 and 10, respectively. If you took the bait and went barking up that tree, you may have decided to eliminate –⅐ and ½ from the range. If you had done so, you may have chosen either D or E as your answer because those two choices are variations on the theme of eliminating –⅐ and ½ from the range.

Your best bet is to simply plug the upper and lower ends of the given range into the function.

First, plug the lowest value of the domain (3) into the function, and you get

$$f(x) = \frac{1}{x^2 - 5x - 14}$$

$$f(x) = \frac{1}{3^2 - 5(3) - 14}$$

$$f(x) = \frac{1}{9 - 15 - 14}$$

$$f(x) = -\frac{1}{20}$$

This value of the function, $-\frac{1}{20}$, can also be expressed as -0.05. Your next task is to plug in the high value of the domain of $x(10)$ and you end up with

$$f(x) = \frac{1}{x^2 - 5x - 14}$$

$$f(x) = \frac{1}{10^2 - 5(10) - 14}$$

$$f(x) = \frac{1}{100 - 50 - 14}$$

$$f(x) = \frac{1}{36}$$

You have now figured out your "high" value, which in this function turns out to be $\frac{1}{36}$, or 0.03. Not a large number, but at least it's a positive number and it's higher than the low range value you got by plugging in the number 3. Thus, the range y is greater than -0.015, and it is also less than 0.03. A very slim range, but one that is called for by the problem. The correct choice is, therefore, C. If you chose A or B, you found a variation on this theme, but the signs somehow got flipped.

Lining Up: Graphing Functions

You should be able to use algebra to recognize and determine functions along with their domains and ranges, but you should also be familiar with functions as they relate to graphs of functions. By looking at a graph of a function, you should be able to tell something about the function itself. You should also be able to look at the graph of a function and determine something about the domain or range of the same function. You may be asked to look at the graph of a function and determine whether a statement about the function is true or false.

 When you graph a function $f(x)$ on the coordinate plane, you'll notice that the x-value of the function (the input or the domain of the function) goes along the x- or horizontal axis. And you chart the $f(x)$ value of the function along the y- or vertical axis. Anytime you see a coordinate pair that represents a function, for example (x, y), the x-value is the domain, or input, of the function, and the y-value is the output, or range, of the function.

The vertical line test

Remember that a function is a distinct relationship between the x (input) value and the y or $f(x)$ (output) value of the function. For every x-value, there is a distinct y-value that will be different from the y-output value of any other x-value. The vertical line test is one way to look at a graph and tell whether it is a graph of a function. The *vertical line test* says that no vertical line intersects the graph of a function at more than one point.

For example, the graphs in Figure 5-1 show a line that passes the vertical line test and is, therefore, a function.

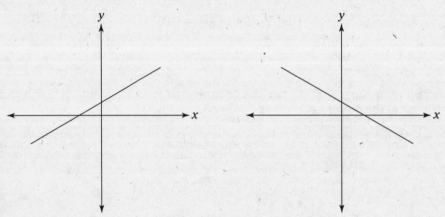

Figure 5-1:
Vertical line
test #1.

The two lines in Figure 5-1 go on in both directions to infinity. For each graph, you could draw a vertical line and the line would only intersect the line in the graph at one point. Thus, the lines in these graphs are functions because they pass the vertical line test. For every *x*-value along the line in each of these graphs, there is a separate and distinct *y*-value that corresponds to it. You know that they are both functions.

You probably already knew that most lines are functions when you think about the equation of a line, $y = mx + b$, but now you can see it for yourself graphically.

The difference between the two lines in Figure 5-2 is obvious. One line is horizontal, while the other is vertical. The graph of the horizontal line is $y = 2$, and the graph of the vertical line is $x = 2$. While these are both technically lines, only one of them is a function. Can you guess which one? Remember the vertical line test. The second graph in Figure 5-2 with the vertical line fails that test (miserably!). There are bazillions of *y*-values along the line having only one *x*-value. The vertical line in Figure 5-2 is not a function.

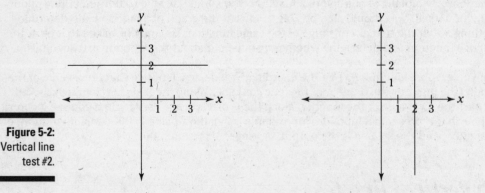

Figure 5-2:
Vertical line
test #2.

Take a look at the graphs in Figure 5-3 and decide which sets of points are functions.

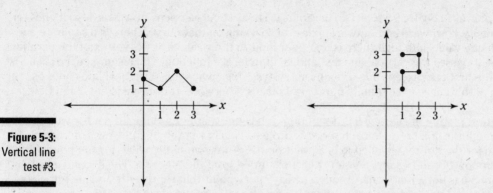

Figure 5-3:
Vertical line
test #3.

As you can see, the first graph in Figure 5-3 shows a function, because each *x*-value has a separate and distinct *y*-value. However, the second graph shows several points that have more than one *y*-value for a discrete *x*-value. Thus, the second graph in Figure 5-3 fails the vertical line test and is therefore not a function.

Look at the curves in Figure 5-4. Which one passes the vertical line test?

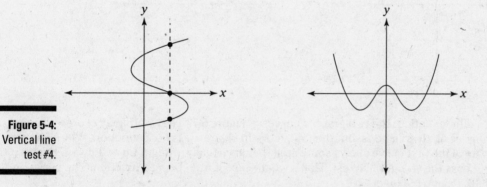

Figure 5-4:
Vertical line
test #4.

The first graph in Figure 5-4 obviously does not pass the vertical line test. It shows where a vertical line crosses the curve in three separate places. The second graph in Figure 5-4, on the other hand, shows a curve wherein, while the *y*-value may be repeated in places, there is still only one distinct *y*-value for every *x*-value on the graph. Thus, the second graph is a function, and it certainly passes the vertical line test.

Take a gander at a couple more graphs of curves in Figure 5-5.

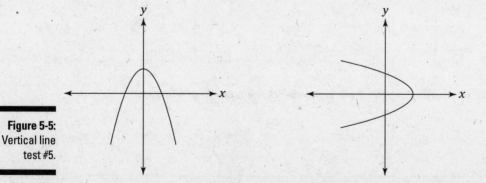

Figure 5-5:
Vertical line
test #5.

The curves in Figure 5-5 look like parabolas. The first curve opens downward, so it goes on to infinity downward and outward. For every x-value on that curve, there is a separate and distinct y-value, although there is an upper limit to the y-values at the vertex of the parabola. Thus, it passes the vertical line test and is, therefore, a function. The second curve is almost like the first one, except that it opens sideways. This means that a vertical line can cross the path of this curve in an infinite number of places. Therefore, this curve is not a function.

The first picture in Figure 5-6 is a function. It has one distinct y-value for each x-value, although the link between each piece is broken. This is a graph of a piecewise function. The value for the dot that is filled in is a point on the graph, while the value for the point with the hollow dot represents a point that is excluded from the function. The second graph in Figure 5-6 is not a function. Although nearly every point that represents a value for x has only one corresponding y-value, there is one place where the x-value has two corresponding y-values. This means the second curve in Figure 5-6 does not pass the vertical line test.

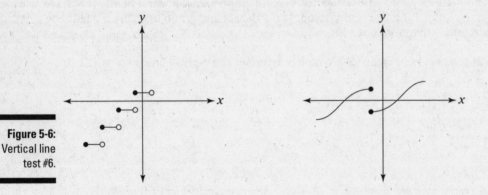

Figure 5-6:
Vertical line
test #6.

What's different about the two graphic curves in Figure 5-7? They both may represent some humongous math equations, but the first curve in Figure 5-7 is *not* a function. Why? The very first part of the first curve shows some area that is virtually straight up and down. Hence, it doesn't pass the straight-line test. That's what keeps it from being a function. The curve on the right, in fact, *is* a function.

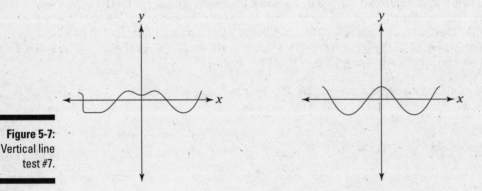

Figure 5-7:
Vertical line
test #7.

Which of the following graphs in Figure 5-8 is not a graph of a function?

A.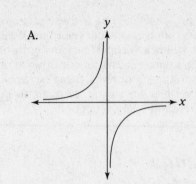

B.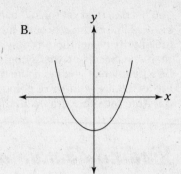

C.  D. 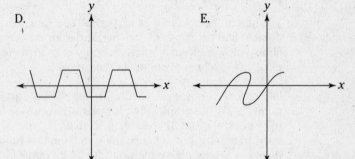 E.

This kind of question is relatively easy, and you just have to use the vertical line test to see which graph fills the bill. Choice E is the correct answer, because it is a curve that sort of doubles back from right to left. It's possible for a vertical line to intersect that curve at more than one point. The other graphs in this question show curves, lines, or some other function where a vertical line never touches the figure at more than one point. Thus, they all pass the test and, of course, are functions.

Which of the following graphs in Figure 5-9 is not a graph of a function?

A.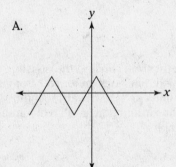

B.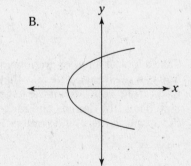

C. 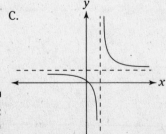 D. 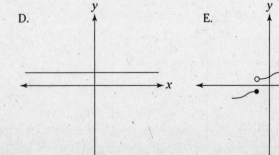 E.

You need to dust off the old eyeglasses and see which one of these graphs does not have a vertical line cutting through more than one point on the curve or line. While some of these look pretty close, there is only one graph where a vertical line can cut through it in more than one place, and that's choice B. While a parabola that opens upward or downward can be a function, if you have one that opens to the left or right, it can never be a function. There are an infinite number of points that the vertical line can cut through the graph. So it stands out from the other graphs in this exercise as not being a function.

Finding domain and range from a graph of a function

You should be able to look at a graph of a function and have a pretty good idea of what the domain and the range of that particular function are. The SAT II Math test may occasionally show you a graph of a function, and you're expected to determine the effective domain and range. Take a look at some of these graphs.

In Figure 5-10, you can see that the function is a parabola, and its vertex is the coordinate point (0, 2). The graph extends outward from side to side so this function contains all possible values of x and, therefore, its domain is "all real numbers." The graph also extends downward to infinity, but because the y-value in this function is limited on the upward side and does not extend above the point (0, 2), its range is $\{y: y \leq 2\}$

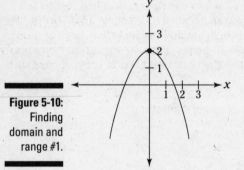

Figure 5-10:
Finding
domain and
range #1.

Take a look at another graph and see whether you can find the possible domain and range. You can see from Figure 5-11 that the straight line goes on forever from left to right. This line also extends upward on the left side to infinity and downward to infinity on the right-hand side. Thus, the domain and range of this linear function is "all real numbers." There is no artificial limit to the x- and y-values in this graph.

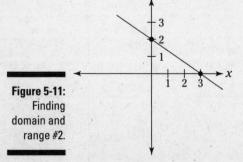

Figure 5-11:
Finding
domain and
range #2.

In Figure 5-12, the horizontal line goes off into infinity from right to left, but it has only one value on the vertical or *y*-axis. Its *y*-value is limited to –3. Of course the equation for this line is *y* = –3. Because the line goes on forever from left to right, it includes every *x*-value there can possibly be. Thus, the domain of this linear function is "all real numbers." The range is limited to simply {*y*: *y* = –3}.

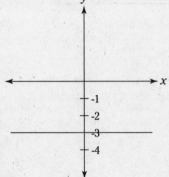

Figure 5-12:
Finding
domain and
range #3.

Evaluating asymptote lines

The equation for the function in Figure 5-13 is $f(x) = \dfrac{1}{x-1}$. As you can see, this hyperbola

approaches two asymptote lines. An *asymptote* is a straight line that the function comes very close to but never quite touches. One asymptote line in this function is the *x*-axis, while the other asymptote is *x* = 1. As a consequence, both the domain and range in this function are limited. The *x* in the function (the domain) never equals 1, and the *f*(*x*) — the *y*-value or the range — never equals 0. Thus, the domain is {*x*: *x* ≠ 1}. The range is {*y*: *y* ≠ 0}.

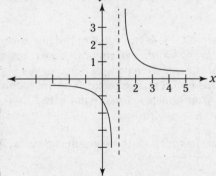

Figure 5-13:
Asymptote
line.

Here are some tips for working with asymptotes:

✔ If you have trouble figuring out an asymptote, try using your graphing calculator. After all, if you're allowed to bring them in for the test, put 'em to use.

✔ You can also plug in the numbers from your answer choices to see which ones fill the bill.

✔ If there is a vertical asymptote, the *x*-value is limited, and your domain is likewise limited. You have to exclude the *x*-value of the asymptote from your domain.

✔ If there is a horizontal asymptote, the *y*-value is limited, and that means your range is likewise limited. You must exclude that *y*-value from your range.

✔ If a function can go "to infinity and beyond" in any direction, its domain and range consist of "all real numbers."

✔ If a function cannot go on forever in any direction, its domain or range is limited.

Which of the following could be the range of the function of the graph shown in Figure 5-14?

(A) $\{y: y \leq 0\}$

(B) $\{y: y \geq 0\}$

(C) $\{y: y = 0\}$

(D) $\{y: y \neq 0\}$

(E) $\{y: y < 0\}$

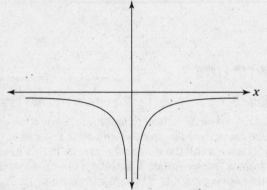

Figure 5-14:
Determine the range of the function for this graph.

Just by looking at the graph of this function you know that the range (the *y*-value) is going to end up being less than 0.

That knowledge easily eliminates answers B and C. (C is the value of the horizontal asymptote line in this function.) If D were true, the graph would extend both above and below, but not touch, the *x*-axis. So that answer is out of the question as well. That leaves either A or E. Because the *x*-axis is an asymptote line for this function, the range does not include the *y*-value of 0. Thus, you've eliminated A from your choices. You are left with E, and, of course, the range in this function is less than 0.

Which of the following could be the domain of the function of the graph shown in Figure 5-15?

(A) $\{x: x \neq 0\}$

(B) $\{x: x \neq 3\}$

(C) $\{x: x = 0\}$

(D) $\{x: x \leq 3\}$

(E) $\{x: x < 0 > x\}$

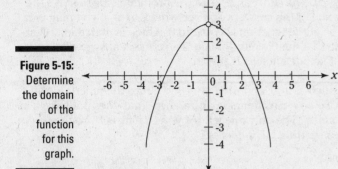

Figure 5-15:
Determine
the domain
of the
function
for this
graph.

Remember that this question is asking for the *domain* and not the *range,* so the fact that the upper limit of the *y*-value is just shy of 3 should not distract you from looking for all the possible *x*-values that make up the domain. You should drop any answer choice that references the value 3 like a hot potato. So get rid of answers B and D right away. They are red herrings. The circle around the point located at (3, 0) on the graph means that you don't count that point in your answer. This means that choice C is exactly the opposite of what you are looking for. C limits your domain to only one value, and that is 0. In this function, the value of 0 is, in fact, *excluded* from the function, so throw out C. Choice E is nonsensical from just about any standpoint. Set your sights on A as the answer of the hour here. The domain, or *x*-value, is not equal to 0.

Which of the following lines is an asymptote of the graph of function $y = 1 \div (x - 2)$?

    I. $x = 2$

   II. $y = 0$

   III. $y = 1$

  (A) I only

  (B) II only

  (C) III only

  (D) I and II only

  (E) I and III only

Your graphing calculator comes in very handy here, or you can draw a graph of the function if you have time. This function looks something like Figure 5-16.

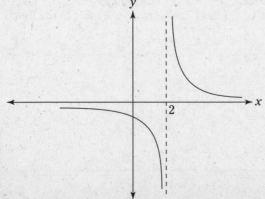

Figure 5-16:
Graph of the
function.

This kind of function, which is basically a quotient involving polynomials in the fraction is called a *rational function*. Seeing this function in graphic form, it's very easy to see that the your asymptote lines are $x = 2$ and $y = 0$. The short answer to the problem is D. To solve the problem algebraically, just plug in some of the answers and see what works. If you plug in 2 for x, you'll find that the denominator of the function is equal to 0, which is undefined. That means you have your asymptote line that limits your domain. The correct answer choice must have I as one of the solutions. Exclude choices B and C.

Now try giving x a very large value, such as 100, and then plugging it into the function. You can see that y begins to approach 0, so you can eliminate any answer that does not have II as a solution, and out go choices A, C, and E. The only one left for you to pick is D. You're on your way to the next one. Aren't these getting terribly easy?

Finding the roots of a function

Every once in awhile, you may be asked to find the roots of a function in a graph. The *root* of a function is the solution to the function that makes the equation equal to 0. Roots are also called *zeroes*. A zero or root of a function is a point where the graph of the equation or function intersects with the x-axis. At any such point, the y-value is 0, and that's why they're called zeroes.

When you see a function written out in algebraic form, you can "solve for x," and find out where the graph of the line or curve crosses the x-axis. Thus, you can match up a graph with a particular function on the SAT II Math test.

Suppose you get hit with the following function and need to recognize the graph of such function.

$$f(x) = x^3 - 3x^2 - x + 3$$

The first thing you should do is factor the function and see what you get.

$$f(x) = x^3 - 3x^2 - x + 3$$
$$f(x) = x^2(x - 3) - 1(x - 3)$$
$$f(x) = (x^2 - 1)(x - 3)$$
$$f(x) = (x - 1)(x + 1)(x - 3)$$

Now that your initial equation is factored into something more workable, you can more easily associate this function with the following graph.

See how the graph in Figure 5-17 of the preceding function intersects with the x-axis at three distinct points? Those points of intersection are the roots or zeroes of the function, that are, coincidentally, the solution you get when you "solve for x" in the equation. This means that the points of intersection with the x-axis give you the solution $y = 0$. As you can see, the curve crosses the x-axis at the points -1, 1, and 3, which are the *roots* of the function.

Figure 5-17:
Looking for distinct roots.

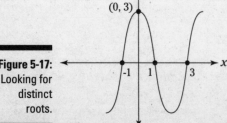

How many distinct roots does the function $f(x) = x^3 - 6x^2 + 32$ have?

(A) 0

(B) 1

(C) 2

(D) 3

(E) 4

Your first inclination may be to just say there are three distinct roots because this function is a polynomial of the third degree. Thus, you may choose D as your answer. Big mistake!

Your safest bet is to factor this polynomial into its component factors:

$$f(x) = x^3 - 6x^2 + 32$$

$$f(x) = (x^2 - 8x + 16)\,(x + 2)$$

$$f(x) = (x - 4)^2\,(x + 2)$$

This gives you two separate roots: –2 and 4. The correct answer is C, as there are two *distinct* roots. Choice E is impossible because a third degree function or polynomial may have a maximum of 3 roots, but it cannot have *more* than 3 roots. And if you chose A or B, you did not factor the polynomial down far enough.

Symmetry

You should also know that certain kinds of functions result in graphs that have symmetry in relation to the *x*- and *y*-axes. *Symmetry* means that the graph of a function has a mirror image of itself on either side of an axis. A parabola having its vertex on the *y*-axis is said to be symmetrical with the *y*-axis. Figure 5-18 shows these kinds of functions:

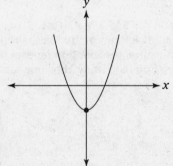
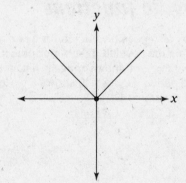

Figure 5-18: Symmetry.

A function that is symmetrical across the *y*-axis is an *even function*, because it has an even-numbered exponent. An example of an even function is $f(x) = x^2$ or $f(x) = \cos x$. An even function has the following property.

$$f(x) = f(-x)$$

If you plug any number into the *x* position of an even function, the *y*-value is going to come out the same regardless of whether you use a negative or positive number for your *x*-value.

An *odd function,* on the other hand, is not just some witty name you have for all the functions you've been studying. An odd function is one that has an odd-numbered exponent, such as $f(x) = x^3$ or $f(x) = \sin x$. An odd function has the following property:

$$f(x) = -f(x)$$

The bottom line of an odd function is that for any x-value put into the function, the result will be an opposite y-value. Odd functions have symmetry with the origin. Figure 5-19 illustrates some odd functions with origin symmetry:

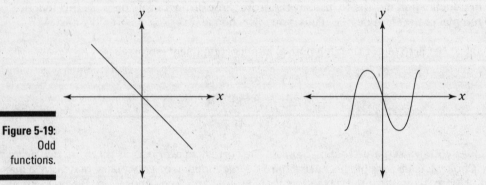

Figure 5-19:
Odd
functions.

As you can see the graph of these functions pictured in Figure 5-19 show a mirror image across the origin.

Occasionally, you may see a graph of an equation that appears symmetrical across the x-axis. This type of equation cannot be a function, because it would fail the vertical line test for functions. Any graph that has symmetry with the x-axis would have numerous places where a vertical line could intersect it at more than one point.

Periodic functions

You may also possibly encounter a *periodic function* on the SAT II Math test, but only if you are taking the Level IIC test. A periodic function is simply a function that keeps repeating the same values over and over again. The measure of how often the function repeats itself is called a *period.* Figure 5-20 shows a periodic function. Coincidentally, this function also has origin symmetry:

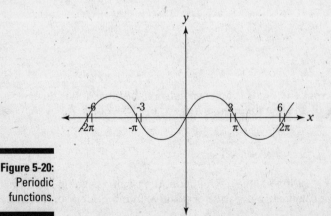

Figure 5-20:
Periodic
functions.

The equation of the periodic function in figure 5-20 is $y = \sin x$. It repeats itself in wave after wave, and the period of this function is 2π. On the Level IIC test, you should be able to spot the length of the period in a periodic function and address questions related to the period and its pattern.

Degrees of Functions

The *degree* of a function is the highest power (exponent) of any variable or term that occurs in the function. So for example, the function $f(x) = 2x^3 + 3x^2 + 6x - 2$ is a third-degree function. The function $g(x) = x^2 - 3x + 2$ is a polynomial function of degree 2.

The main thing you really need to know (and this is strictly for the Level IIC test), is that the *degree of the function* tells you *at most how many roots,* or zeroes, the function has. Keep in mind that the roots of the function can be distinct roots or they can be the same. To have distinct roots, the polynomial function needs to have separate and distinct solutions to the function.

The following polynomial function is a fifth-degree function having 5 roots, but it has only 3 *distinct* roots.

$$g(x) = 5x^5 - 10x^4 + 36x^3 - 54x^2 + 27x$$

That polynomial function looks a bit too large to attack all at once, so break it down into its component factors, like so:

$$g(x) = (x - 3)^3 (x - 1) x$$

In this more manageable state, you can see that there are a total of five roots to this function, but there are only three *separate and distinct* roots, while three of its roots equal 3:

$$g(x) = 0 \text{ when } x = 0, 1, 3, 3, \text{ or } 3$$

The graph of this function would look a lot different than it would if it had five distinct roots.

Another thing to be aware of is that the degree of a function tells you much about the shape of the function. A first-degree function; that is, a linear function, does not have any exponents greater than 1. Thus, a first-degree function is simply a straight line. It doesn't have a *minimum* or *maximum* value like, for example, a parabola. The graph of the first-degree function simply goes from one side of the graph to the other.

A parabola, on the other hand, is a second-degree function, based on a quadratic equation where the highest degree is the power of 2 in any one term. A parabola and other second-degree functions have a low point or high point. The *vertex* of the parabola is called the *extremum,* which is the minimum or maximum point for the function.

Figure 5-21 shows a parabola with an extremum that is a maximum value of the function. As you can also see, there is only one extremum with a parabola. This value is called the *global maximum* for the function. The graph of a second-degree function can have no more than one extremum. The graph of a third-degree extremum can have no more than two extrema. Two more extrema are often called *local extrema.* So, for example, the function in Figure 5-22 has a local minimum and a local maximum.

A *local maximum* is a point where the value of the function is higher than its immediate surrounding points. A *local minimum* is a point where the value of the function is lower than its surrounding points.

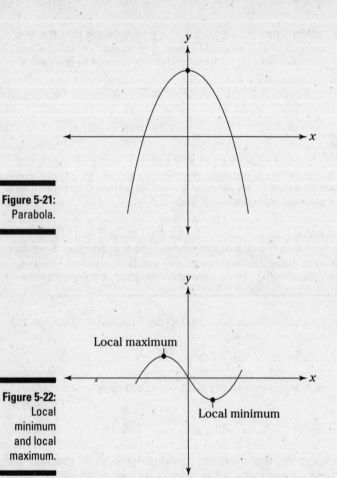

Figure 5-21:
Parabola.

Figure 5-22:
Local
minimum
and local
maximum.

A fourth-degree function can have up to, but no more than, three extreme values.

Figure 5-23 illustrates what could be at least a fourth-degree function. You can see that it has three extreme values, with two local minimums and a local maximum. You may also notice that this function has four distinct zeroes; that is, four points where the graph of the function crosses the *x*-axis.

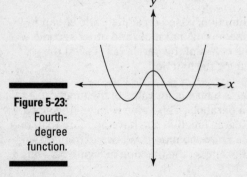

Figure 5-23:
Fourth-
degree
function.

Part III

It Has Curves, and You Have Angles: Geometry and Trigonometry

In this part . . .

Measuring lines, angles, and arcs and the shapes they create makes up almost 30 percent of the questions on the Level IIC test, and the Level IC test has even more questions in this area. The Level IIC test focuses on trigonometry and coordinate geometry, and the Level IC test concentrates on plane geometry. This part has it all: parabolas, Pythagorean triples, polar coordinates, spheres, trigonometric identities, and so much more. So, get out your graphing calculator and start punching the numbers.

Chapter 6

Angling Your Perspective: Plane Geometry

Plane geometry is the study of lines and shapes in two dimensions. Imagine a tool that could prove that the Earth is round and that the planets move around the sun in predictable orbits. Those are some of the wonders of geometry. It has been extremely important in the history of mathematical development. Using geometry, we can make models of the physical world and apply mathematical concepts to them. We make hypotheses and predictions about the real world and use geometry to proves that's really how the world goes 'round. Geometry starts with the basics, in this case plane geometry, and builds on that foundation to construct ever-increasing complex models to more accurately portray the real world.

The Straight and Narrow (and Sometimes Wide): Lines and Angles

The SAT II Math test spends about 20 percent of the Level IC test on plane geometry and measurement. Although the Level IIC portion does not test you on plane geometry per se, you're still expected to know the basic principles of plane geometry in order to work the more advanced level coordinate and solid (3-dimensional) geometry and to succeed on that test. This chapter focuses on the geometry of shapes and figures that can be found on a single plane; that is, a flat surface that extends infinitely in two dimensions and has no thickness.

Getting the skinny: Some basic definitions

The first thing you need to do in understanding geometry is to get to know the various terms for geometric shapes and forms. While you aren't tested on the definitions, it's important to understand their meaning to solve problems on the SAT II Math test. Here are the more common terms that will pop up at one time or anther on the test:

✔ **Plane:** A perfectly flat surface that has no thickness and extends forever in two directions.

✔ **Line:** A straight path of points that extends forever in two directions. A line does not have any width or thickness. Because a point is very, very tiny, a line is very, very thin. Arrows are used to show that the line goes on forever. See line *AB* in Figure 6-1. The word line is often used to indicate a line segment or a ray.

✔ **Line segment:** The set of points on a line between any two points on the line, basically just a piece of a line from one point to another that contains all the points in between. See line *CD* in Figure 6-1.

✔ **Ray:** A ray is like half of a line; it starts at an endpoint and extends forever in one direction. You can think of a ray as just like a ray extending from the sun (the endpoint) and shining as far as it can go. See line *EF* in Figure 6-1. While the sun's rays may eventually run out of energy on their path, a ray in geometry keeps going and going.

✔ **Midpoint:** The point halfway between two points on a line segment. If a point along a line segment is the same distance from each of the two ends of the line segment, that point is the midpoint on the line segment.

✔ **Bisect:** To cut something exactly in half, such as a line segment cutting another line segment or an angle or a polygon into two equal parts. A bisector is a line that divides the line segment, angle, or polygon into two equal parts.

✔ **Intersect:** Just like it sounds, it simply means to cross; that is, when one line or line segment crosses another line or line segment.

✔ **Collinear:** A set of points that lie on the same line.

✔ **Vertical:** Lines that run straight up and down.

✔ **Horizontal:** Lines that run straight across from right to left (or from left to right if you're holding your paper upside down).

✔ **Parallel:** Lines that run in the same direction always remaining the same distance apart. Parallel lines will never intersect with one another.

✔ **Perpendicular:** When two lines intersect to form a square corner. The intersection of two perpendicular lines forms a right angle or a 90° angle.

Where did geometry come from?

Good question! Plane geometry is also known as Euclidean geometry, so it's customary to start with Euclid and other Greek mathematicians. But many cultures developed the use of geometry on their own. The Ancient Egyptians made far-reaching use of geometry in the building of their empire. They were accomplished engineers who made use of mathematics in their surveying and construction projects. They were so successful that the Great Pyramid at Giza remained the tallest building in the world for 4,400 years! The Babylonians also left evidence of their understanding of the basic rules for geometry, although nothing as grand as the pyramids. From these two civilizations, the Greeks created the most extensive development of recorded geometry that humankind has found. Their use of mathematics ran the gamut from the seemingly simple — there can be a line between any two points — to problems that are elegantly complex, such as squaring the circle. You'll begin this study as mathematicians often do, by clearly stating the obvious and building from there.

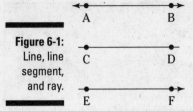

Figure 6-1:
Line, line
segment,
and ray.

- **Angle:** The intersection of two rays sharing a common endpoint. The common endpoint is called the vertex. The size of an angle depends on how much one side rotates away from the other side. An angle is usually measured in degrees or radians.

- **Acute angle:** Any angle measuring less than 90°. Like an acute or sharp pain, the acute angle has a sharp point. See Figure 6-2.

- **Right or perpendicular angle:** An angle measuring exactly 90°. It makes up a square corner. See Figure 6-3.

- **Obtuse angle:** An angle that measures more than 90° but less than 180°. While an acute angle can be quite sharp, an obtuse angle couldn't poke a hole in butter. An obtuse angle is actually quite dull or blunt. See Figure 6-4.

- **Straight angle:** An angle that measures exactly 180° is straight. A straight angle appears to be a straight line or line segment.

- **Complementary angles:** Angles that, when added together, total 90°. Together, they form a right angle, so just remember that it's the "right" thing to do to give an angle a complement.

- **Supplementary angles:** Angles whose measurements total 180° are supplementary. They form a straight line. Just remember that vitamin supplements can keep you on the straight and narrow.

- **Congruent:** Objects that are equal in size and shape are congruent. Two line segments having the same length are congruent. Two angles having the same measure are congruent. Two congruent triangles have their corresponding sides all the same length, and their corresponding angles are all the same measurement.

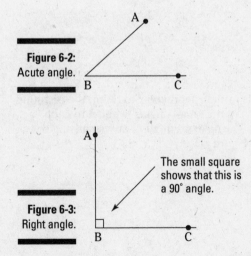

Figure 6-2:
Acute angle.

Figure 6-3:
Right angle.

The small square
shows that this is
a 90° angle.

An endless number of points

Did you know that each of these figures contains an infinite number of points? It's clear with the line and still fairly obvious with the ray, but it's also true for the segment that there is a limitless number of points on any one of these figures. It would take you forever to count an infinite number of points.

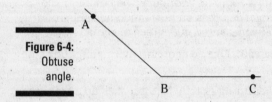

Figure 6-4:
Obtuse
angle.

Fishing for answers: Some rules for lines and angles

The rules for lines and angles are direct applications that arise from the basic definitions you've just studied.

When two lines intersect, the opposite angles are always congruent or equal, and the adjacent angles are always supplementary. Opposite angles are also known as *vertical* angles. *Adjacent* angles have a common side, so they are right next to each other.

In Figure 6-5, ∠ ABC and ∠ DBE are congruent. ∠ ABC and ∠CBD form a straight line and are, therefore, supplementary.

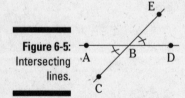

Figure 6-5:
Intersecting
lines.

When parallel lines are crossed by a third line that is not perpendicular to them, the resulting small and large angles share certain properties. Each of the small angles is equal to each other. The large angles are also equal to each other. The measurement of any small angle added to that of any large angle will equal 180°. See Figure 6-6.

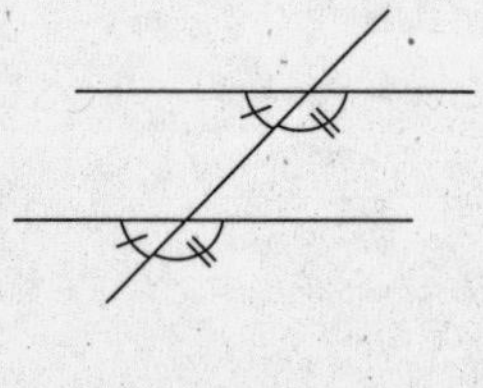

Figure 6-6:
Line inter-
secting par-
allel lines.

In the following figure, line *m* is parallel to line *n,* and line *t* is a transversal crossing both lines *m* and *n.* Given the information contained in this figure, what is the measure of ∠ *e*?

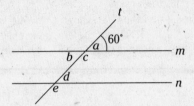

(A) 30°

(B) 60°

(C) 100°

(D) 120°

(E) Cannot be determined from the information provided

Because lines *m* and *n* are parallel, you know that ∠ *e* is equal to ∠ *c.* ∠ *c* lies along a straight line with ∠ *a,* so ∠ *a* + ∠ *c* = 180°.

Because ∠ *a* = 60°, ∠ *c* must equal 120°.

And because ∠ *c* = ∠ *e,* ∠ *e* must also equal 120°.

The correct answer is D.

A Sturdy Foundation: Triangles

The majority of questions about plane geometry found on the SAT II Math test involve triangles. What's more, the properties and rules involving triangles have been found to be useful in a variety of ways within and without geometric applications since the infancy of civilization.

A triangle is a three-sided polygon. Each point where two of the sides intersect is called a *vertex.* Triangles are generally identified by these vertices, so a triangle with vertices *A, B,* and *C* is referred to as △ *ABC.*

Thinking in threes: Properties of triangles

Just as lines and angles have rules that apply to lots of situations, triangles have rules that apply to all triangles. But some triangles are so special that some rules exist just for them.

- An *isosceles triangle* has two equal sides, and the angles opposite those two sides are also equal to each other.

- An *equilateral triangle* has three equal sides and three equal angles.

- A *right triangle* has one 90° angle. An interesting side note is that this triangle can have only one 90° angle, while the remaining two angles must add up to 90°. You should also know that the side opposite the 90° angle is called the *hypotenuse.*

The three angles of any triangle will always add up to 180°. See Figure 6-7.

Older than we thought

The Babylonians demonstrated that they had knowledge of Pythagoras' Theorem roughly 1,000 years before his name was attached to this famous mathematical relationship, but the Greeks get the credit. Probably because the Pythagorean theorem sounds cooler than the Babylonian theorem.

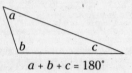

Figure 6-7:
Sum of the angles of a triangle.

$a + b + c = 180°$

In the following figure, line *AB* is congruent to line *CB*, and line *m* is parallel to line *n*. Given the information that is contained in the figure, what is the measure of ∠ *GBH*?

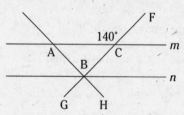

Note: Figure not drawn to scale.

(A) 40°

(B) 75°

(C) 100°

(D) 125°

(E) 140°

The figure in this question has two parallel lines traversed by two intersecting lines. You are to determine the measurement of ∠ *GBH*.

What do you know? You know that ∠ *FCA* = 140° and that it lies along line *FC* with ∠ *BCA*. Therefore, ∠ *BCA* = 40° (140 + ∠ *BCA* = 180°).

You also know that because line *AB* and line *BC* are congruent, triangle *ABC* is an isosceles triangle. Therefore, ∠ *BAC* also equals 40°. The angles of a triangle equal 180°, so ∠ *ABC* = 100°. (∠ *ABC* = 180 − 40 − 40.)

∠ *ABC* is directly opposite ∠ *GBH*, so the angles are equal. Because ∠ *ABC* is 100°, ∠ *GBH* is also 100°. The correct answer is, therefore, C.

EXAMPLE

In △ *ABC* in the following figure, the measure of ∠ *A* is 35°, and the measure of ∠ *B* is twice the measure of ∠ *A*. What is the measure of ∠ *C*?

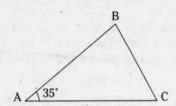

(A) 35°

(B) 70°

(C) 75°

(D) 90°

(E) 105°

This basic triangle problem tells you that ∠ *A* is 35° and that ∠ *B* is twice the measure of ∠ *A*.

Therefore, ∠ *B* = 2(35) or 70°.

Solve for C.

$$35 + 70 + C = 180$$
$$105 + C = 180$$
$$C = 75$$
$$\angle C = 75°.$$

Therefore, the correct answer is C.

EXAMPLE

In the following figure, line *SA* is parallel to line *TB*. If the measure of ∠ *BTU* is 60°, what is the measure of ∠ *SAT*?

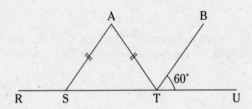

(A) 30°

(B) 40°

(C) 50°

(D) 60°

(E) 80°

Line *RU* traverses the parallel lines *SA* and *TB*. Therefore, ∠ *BTU* equals ∠ *AST*. Because ∠ *BTU* is 60°, ∠ *AST* must also be 60°.

It is important to recognize that because line *SA* equals line *TA*, △ *SAT* (what an appropriate name for a triangle on this test!) is an isosceles triangle. The angles opposite these two lines are equal.

One of these angles is ∠ *AST*. ∠ *AST* = 60°; ∠ *STA* = ∠ *AST*; therefore, ∠ *STA* also = 60°. You know that the angles of a triangle add up to 180°, so now you know that you are actually looking at an equilateral triangle. All angles = 60°. Therefore, ∠ *SAT* = 60°, and the correct answer is D.

As demonstrated in Figure 6-8, the side that is opposite of an angle in a triangle is proportional to that angle. So, the smallest angle will be facing the shortest side of the triangle. If two or more angles have the same measurement, their opposite sides will be equal lengths.

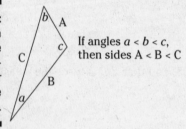

Figure 6-8: Angles of a triangle are in proportion to their opposite sides.

If angles $a < b < c$, then sides A < B < C

The *Third Side Rule* states that the length of any side of a triangle must be greater than the difference of the other two sides and less than their sum. So, in Figure 6-8, the length of A is greater than C – B, but smaller than C + B. Makes sense, doesn't it?

Finding the area of a triangle

The area of a triangle equals one half the base times the height (or altitude), where the base is one of the sides and the height or altitude is the distance that a perpendicular line runs from the base to the angle opposite the base. The formula for the area of a triangle can be expressed as A = ½ *bh*.

Notice that, as shown in Figure 6-9, the height or altitude is always perpendicular to the base, and that the height can be placed either inside or outside the triangle.

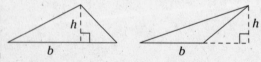

Figure 6-9: The area of a triangle.

Applying the Pythagorean theorem and other cool stuff about right triangles

The Pythagorean theorem simply states that the sum of the squares of the legs of a right triangle are equal to the square of the hypotenuse, or $a^2 + b^2 = c^2$, where a and b are the two sides or legs of the right triangle and c is the hypotenuse. The legs of a right triangle are the sides that form the right angle, and the hypotenuse is the side opposite it.

The hypotenuse is always the longest side of a right triangle because it is opposite the largest angle. Because the three angles in a triangle must equal 180° and you already have 90° with the right angle, the other two angles together must equal 90° total. In Figure 6-10, c is the hypotenuse; a and b are the legs.

Figure 6-10: A right triangle.

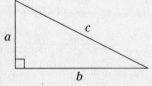

Keep in mind that the Pythagorean theorem works only with right triangles. You don't want to be fooled by look-alike triangles designed to make you think you've solved the problem. Look for that telltale square in the right angle of any figure. If the figure doesn't have the square in the angle or the question doesn't provide you with information that tells you a triangle is right, you can't assume that it's a right triangle even if it looks like one!

Which of the following is the length, in inches, of the remaining side of a right triangle if one side is 4 inches long and the hypotenuse is 11 inches long?

(A) 7

(B) $\sqrt{105}$

(C) 10

(D) 12

(E) 105

You may find it helpful to draw a right triangle on your paper to visualize the problem, but it's not necessary. If the hypotenuse is 11" and one side is 4," the measurement of the remaining side is figured as follows:

First, apply the formula

$$11^2 = 4^2 + a^2$$
$$121 = 16 + a^2$$
$$105 = a^2$$

You know that a^2 is 105, but the question asks for the value of a, not a^2. So the measurement of the remaining side is the square root of 105, which is answer B.

Common ratios of right triangles

You may find it handy to memorize some ratios based on the Pythagorean theorem. That way you don't have to work out the whole theorem every time you deal with a right triangle.

The most common ratio of the three sides of the right triangle is the 3:4:5 triangle (3 is the measure of the shorter leg; 4 is the measure of the longer leg; and 5 is the measure of the hypotenuse) and related multiples, for example, 6:8:10, 9:12:15, and so on. As soon as you recognize that two sides fit the 3:4:5 ratio or a multiple of the 3:4:5 ratio, you automatically know the length of the third side.

Other proportions of right triangles you should try to remember are 5:12:13, 8:15:17, and 7:24:25. Knowing these proportions may allow you to more quickly solve some of the problems on the SAT II Math test.

In the following figure, line *AB* is 6 units long, line *AC* is 8 units long, and line *BD* is 24 units long. How many units long is line *CD*?

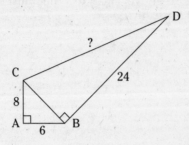

(A) 26

(B) 32

(C) 80

(D) 96

(E) 100

This problem would be very time consuming to solve if you didn't know the common ratios of right triangles. To determine the length of line *CD*, you must first know the length of line *CB*. You could use the Pythagorean theorem, but you know an easier and faster way. Because line *AB* = 6 and line *AC* = 8, △ *ABC* is a 3:4:5 triangle with twice those dimensions, a 6:8:10 triangle. Therefore, the length of the hypotenuse, line *BC* = 10.

This makes △ *BCD* a 5:12:13 triangle with twice those dimensions, a 10:24:26 triangle. Therefore, the length of line *CD* = 26, and the correct answer is A.

The 30:60:90 triangle

There are some other right triangles you should know. One is the 30:60:90. When you bisect any angle in an equilateral triangle, you get two right triangles with 30°, 60°, and 90° angles. In a 30:60:90 triangle, the hypotenuse is 2 times the length of the shorter leg, as shown in Figure 6-11. The ratio of the three sides is $s : s\sqrt{3} : 2s$, where s = the length of the shortest side.

Figure 6-11:
Bisected
equilateral
triangle with
two 30:60:90
triangles.

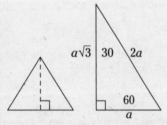

The new hypotenuse of one of the 30:60:90 triangles you've made was once one of the sides in the former equilateral triangle, and the short leg of the new triangle is one-half the length of one of the legs from the old equilateral triangle. The longer leg in the 30-60-90 triangle is $\sqrt{3}$ times the shorter leg.

In terms of *a*, what is the length of line *MN* in the following figure?

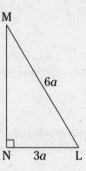

(A) ½*a*

(B) 2*a*

(C) $\sqrt{3a}$

(D) $(3a)\sqrt{3}$

(E) $\sqrt{45a}$

Because the hypotenuse of this triangle is twice the measure of the short side of the triangle, this is a 30:60:90 triangle.

To find the length of the long side, use the ratio shown in Figure 6-11. And don't let the *a* in the question worry you; all of the answer choices appear in terms of *a*. The measure of the long side is 3*a* times the square root of 3, which is the answer choice D.

The 45-45-90 Triangle

If you bisect a square with a diagonal line, you get two triangles that each have two 45° angles. Because the triangle has two equal angles (and, therefore, two equal sides), the resulting triangle is an isosceles right triangle, or 45:45:90 triangle. The length of the hypotenuse is equal to $\sqrt{2}$ times the length of a leg. It's important to recognize this also means that the length of a leg is equal to the length of the hypotenuse divided by $\sqrt{2}$. The ratio of sides in an isosceles right triangle is, therefore, $s : s : s\sqrt{2}$ (where *s* = the length of one of the legs) or $\frac{s}{\sqrt{2}} : \frac{s}{\sqrt{2}} : s$ (where *s* = the length of the hypotenuse). Figure 6-12 shows the formula.

Figure 6-12:
Bisected
square
with two
isosceles
right
triangles.

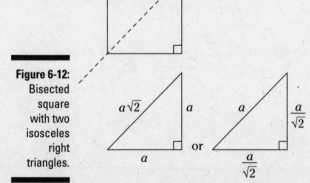

In △ STR, ∠ TSR measures 45° and ∠ SRT is a right angle. If line SR is 10 units long, how many units is line TR?

(A) 5

(B) 5√2

(C) 10

(D) 10√2

(E) 20

You could draw the triangle, but with what you know about 45:45:90 triangles, you probably don't need to.

Because ∠ SRT is a right angle, you know that the triangle in this question is a right triangle. If ∠ TSR = 45°, then ∠ RTS must also equal 45°. (180 = 90 + 45 + ∠ RTS.) Therefore, the triangle is also an isosceles triangle, and the sides opposite the equal angles are equal in length. Line SR must equal line TR. Line SR = 10, so line TR = 10. The correct answer is C.

Noticing a resemblance: Similar triangles

Triangles are similar when they have the same angle measures. Similar triangles have exactly the same shape as each other even though their sides have different lengths. So, the corresponding sides of similar triangles are in proportion to each other. The height or altitude of the two triangles is also proportional. Figure 6-13 illustrates the relationship between similar triangles.

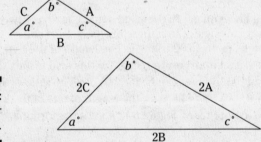

Figure 6-13:
Similar
triangles.

The similar right triangles in the following figure (△ RST ~ △ ABC) have sides of the indicated length. In square inches, what is the area of △ ABC?

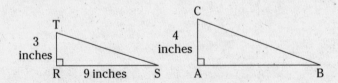

(A) 4

(B) 9

(C) 13

(D) 24

(E) 52

To find the area of △ ABC, you can determine the length of the base (line AB). The height is 4", but how do you know the base? You know that the two triangles are similar, which means that they are proportionate to each other. Therefore, line TR is proportionate to line CA, and line RS is proportionate to line AB. If x represents line AB, the following is true: ³⁄₉ = ¹⁄ₓ. Cross-multiply: 3x = 36. So, x (or line AB) = 12.

Now you can determine the area of △ ABC. If A = ½bh, then A = ½(12)(4), or 10 √2. At this point, you can multiply 12 and 4 and divide by 2 (A = ⁴⁸⁄₂), or you can divide the numerator and denominator by 2 and then multiply 12 and 2. Either way, A = 24, and the correct answer is D.

In the following figure, points S, U, and T are collinear, and line RU is perpendicular to line ST. Line RS is 10 units long, and line ST is 30 units long. Also, line RS is perpendicular to line RT. How many units long is line SU?

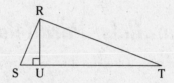

Note: Figure not drawn to scale.

(A) ³⁄₁₀

(B) 3

(C) ¹⁰⁄₃

(D) 10

(E) 30

Don't let the vocab fluster you. Remember, *collinear* means that S, U, and T are on the same line.

Because you are told that line RS is perpendicular to line RT, you know that △ RST is a right triangle bisected by line RU to form two 45° angles, ∠ SRU and ∠ URT. This means that every angle of each triangle other than the right angles measures 45 °. Therefore, you know that all three triangles, SRU, RUT, and RST, are proportional.

ST is the hypotenuse of △ RST, and it equals 30 units. SR is the hypotenuse of △ SRU, and it equals 10 units. So, the ratio of these two triangles is 30: 10 or 3: 1. Apply the ratio: $\frac{3}{1} = \frac{10}{SU}$. So, 3(SU) = 10; and SU = ¹⁰⁄₃.

Before you take the time to calculate ¹⁰⁄₃, check the answer choices. You'll see that this question does not require you to make the calculation. Answer C is ¹⁰⁄₃.

On All Fours: Quadrilaterals

A *quadrilateral* is a four-sided polygon, and a *polygon* is any closed figure made of line segments that intersect. The quadrilaterals you're most likely to see on the SAT II are parallelograms (which include rhombuses, rectangles, squares) and trapezoids. Your primary concern is to know how to find their areas and perimeters. The measure of their perimeters is always the sum of their sides.

TIP

The SAT II Math test may toss in a funky-looking quadrilateral that looks something like the shapes in Figure 6-14. Even though these strange shapes don't have a bunch of set properties to help you find their areas, don't give up. You may be able to divide them into shapes that you can find the area of and then add them together. One thing you can count on is that the sum of the angles of a quadrilateral is always 360°.

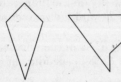

Figure 6-14:
Irregular
quadri-
laterals.

Drawing parallels: Parallelograms

Most of the quadrilaterals tested on the SAT II are parallelograms.

TIP

Parallelograms have very useful properties for solving SAT II Math problems.

- The opposite sides are parallel and equal in length.
- The opposite angles are equal to each other.
- The adjacent angles add up to 180°, so they are supplementary to each other.
- The diagonals of a parallelogram bisect each other. In other words, they cross at the midpoint of both diagonals.

Figure 16-15 provides a visual representation of the very important properties of parallelograms. All sides are equal in this parallelogram, which makes it a rhombus. Not all parallelograms have four equal sides, but all have equal opposite sides.

Figure 6-15:
A parallel-
ogram.

The area of any parallelogram is its base times its height ($A = bh$). The height is determined in very much the same way you find the height or altitude in a triangle, but you draw the perpendicular line from the base to the opposite side (instead of the opposite angle in the triangle). See Figure 6-16.

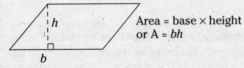

Figure 6-16:
Finding the
area of a
parallelo-
gram.

Area = base × height
or A = bh

You can use the Pythagorean theorem to help you find the height of a parallelogram. When you drop a perpendicular line from one corner to the base to create the height, it becomes the leg of a right triangle. If the problem gives you information to determine the length of other sides of the triangle, you can plug in the formula to find the length of the perpendicular line, or the height of the parallelogram.

Parallelograms come in a variety of types.

✔ A *rectangle* is a parallelogram with four right angles. Because all rectangles are parallelograms, they have all the properties of parallelograms, including the formula for finding the area. Only with a rectangle is the height or altitude the same as one of its sides.

✔ A *square* is a rectangle with four equal sides, so it also has four right angles, and its sides all have the same length. Because a square has four equal sides, you can easily find its area by knowing the length of only one side. The area of the square can be expressed as $A = s^2$ or $A = s \times s$, where s is the length of a side. The perimeter of a square is $4s$.

✔ The parallelogram in Figure 6-15 is also a *rhombus*. All four sides of a rhombus are equal in length. A square has equal sides. So, what makes a rhombus different from a square? A rhombus doesn't necessarily have right angles like a square does. You can find the area of a rhombus by multiplying the two diagonals d by each other and then dividing by 2, or $A = \frac{1}{2} d_1 d_2$

There's also another neat trick for finding the area of the square if the only measurement you know is the length of the diagonal. Where the diagonal $= d$, you can say $A = \frac{d^2}{2}$. Remember that the diagonal of a square is the hypotenuse of an isosceles right triangle, and it has some special formulas. When you think about it, this shortcut is just another way of using the Pythagorean theorem in reverse.

Raising the roof: Trapezoids

A trapezoid is a quadrilateral with two parallel sides and two nonparallel sides. The parallel sides are called the bases, and the other two sides are called the legs. It's a bit tricky to find the area of a trapezoid, but it can be done with an extra step, as long as you know the length of both bases along with the height or altitude. To find the area, you take the average of the two bases and multiply by the height or altitude. Thus, $A = (b_1 + b_2) \div (2 \times h)$. See Figure 6-17 for a visual.

An isosceles trapezoid is a special case where the legs of the quadrilateral are the same length. It looks kind of like an A-frame with the roof cut off.

Figure 6-17:
The area of
a trapezoid.

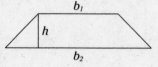

What is the measure, in square units, of the area of polygon *ABCD*, shown in the following figure?

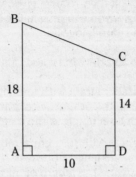

(A) 20

(B) 64

(C) 160

(D) 216

(E) 230

The fastest way to determine the area of this polygon is to recognize that it is a trapezoid. *CD* and *BA* are its bases, and *DA* is its height. You have all the information you need. The answer is C.

$$A = h\,[(b_1 + b_2) \div 2]$$
$$A = 10[(14 + 18) \div 2]$$
$$A = 10(16)$$
$$A = 160$$

If you forget the formula for the area of a trapezoid, you can also draw a line from *C* perpendicular to line *BA*. This forms a rectangle and a right triangle. From the information, you can find the area of the triangle and rectangle and add them together. This method takes a little longer than the first, however.

In the following figure, square *ABCD* has sides the length of 6 units, and *M* and *N* are the midpoints of line *AB* and line *CD*, respectively. What is the perimeter, in units, of *AMCN*?

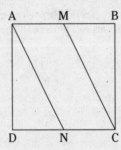

(A) 12

(B) $12\sqrt{5}$

(C) $6 + 6\sqrt{3}$

(D) $6 + 6\sqrt{5}$

(E) $6 + 12\sqrt{3}$

Even though this question asks you to determine the perimeter of parallelogram *AMCN*, it also incorporates what you know about triangles and simplifying radicals.

If *M* and *N* are the midpoints, then *AM* = 3(⅝) and NC = 3. Now you know the short sides of *AMCN* = 3. You can see that the long sides of the parallelogram are the hypotenuses of the right triangles within the square. The lengths of the legs of the right triangles are 3 and 6. These do not fit with any of the special ratios associated with right triangles, but you can use the Pythagorean theorem.

$$3^2 + 6^2 = c^2; 9 + 36 = c^2; 45 = c^2; c = \sqrt{45}$$

So, the perimeter = $(2 \times 3) + (2)\sqrt{45}$, or $6 + (2)\sqrt{45}$

This answer is not available, so you must simplify the radical. (If you need to review how to simplify radicals, read Chapter 3.)

$45 = 9 \times 5$, and the square root of 9 is 3. Multiply 3 and 2 to get 6. The answer is $6 + 6\sqrt{5}$, which is the choice provided by answer D.

Pirates without Parrots: More Polygons

The SAT II Math folks may throw in some of the other types of polygons to make things interesting. Some of the more common ones are shown in Figure 6-18 and are listed here for you:

- ✔ **Pentagon:** A five-sided figure
- ✔ **Hexagon:** A six-sided figure (the *x* makes it sound like *six*)
- ✔ **Heptagon:** A seven-sided figure
- ✔ **Octagon:** An eight-sided figure (like *oct*opus)
- ✔ **Nonagon:** An nine-sided figure
- ✔ **Decagon:** A ten-sided figure (like *deca*thlon)

Figure 6-18:
Regular
polygons.

Pentagon   Hexagon   Heptagon   Octagon   Decagon

In general, whenever you run across one of these polygons, the test will most likely tell you that they are *regular polygons*. This may be the only way to solve some of the problems because a regular polygon simply means that all of the sides are the same length and all of the angles are equal. The rules for similar triangles apply also to similar polygons. That is, if two polygons have exactly the same shape and the same angles, the length of their corresponding sides are proportional to one another.

There is no set formula for determining the area of a polygon. You need to create quadrilaterals and triangles within the polygon, find their areas, and add them together to get the total area of the polygon.

Top secret polygon

Did you know that the Pentagon building near Washington D.C. is just about as regular as a pentagon can get? While the interior angles are probably pretty close to 108° apiece, the length of each wall is 921 feet. That covers a lot of ground, with over 17 miles of corridors inside. Now don't go spreading state secrets.

You may remember how the sum of the angles of a triangle is 180° and the sum of the angles for a quadrilateral is 360°. Are you starting to detect a pattern here? Just add another 180° and you have the sum of the angles in a pentagon, that is 540°. But if you had to add the angles up like this, you'd soon run out of fingers to count on. So instead of adding up the number of extra sides in an awkward fashion, here is an easier formula for determining the sum of the interior angles for any polygon: Sum of the angles = $(n - 2) \times 180°$, where n is equal to the number of sides. Works every time. If it's a regular polygon, you can also determine the measure of the angles. You divide the sum of the angles by the total number of angles. Therefore, each angle in a regular pentagon measures $540 \div 5 = 108°$.

The formula for determining the measure of an angle in a polygon only works if the SAT II tells you it is a regular polygon.

Pieces of Pi: The Properties of Circles

A circle is a set of points in a plane that are at a fixed distance from a given point. That point is called the center. A circle is best drawn with the aid of a compass, but you can use your hands for the SAT II Math test.

Rounding up the basics: Radius, diameter, and circumference

Almost any SAT II Math problem regarding circles requires you to know about radius, diameter, and circumference and the formulas to determine their measurements.

The *radius* of a circle is the distance from the center of the circle to any point on the circle. Think of it as a *ray* going out from the center to the edge of the circle. The radius is usually indicated by the letter r as it is in Figure 6-19.

The *diameter* of a circle is the distance that a line goes from one side of the circle to the other and passes through the center. The diameter is twice the length of the radius, and it is the longest possible distance across the circle. The diameter is generally indicated by the letter d as shown in Figure 6-19.

Figure 6-19:
The radius and diameter of a circle.

The *circumference* of a circle is the distance around the circle. While not quite technically true, you can think of the circumference almost as if it were the perimeter of the circle. It's really more technically accurate to say you're trying to find the perimeter of a regular polygon with an infinite number of sides as it gets rounder and rounder. Rather than trying to figure out how many sides add up to infinity, it's better to use the following formula:

$$C = 2\pi r \text{ or (because } d = 2r) \ C = \pi d.$$

You can manipulate this formula to find the diameter or the radius of a circle as long as you know the circumference. For example:

- ✔ The formula for the radius is $r = C \div 2\pi$.
- ✔ The formula for the diameter is $d = C \div \pi$.

The area of a circle is $A = \pi r^2$.

You can manipulate this formula to find the diameter or the radius if all you know is the area.

For example:

- ✔ The formula for the radius is $r = \sqrt{(A \div \pi)}$
- ✔ The formula for the diameter is $d = 2\sqrt{(A \div \pi)}$

Assisting Noah: What you should know about arcs

You should know at least the concept of the following terms so you aren't running in circles on the SAT II Math test:

- ✔ An *arc* of a circle is a portion along the edge of the circle. Because it runs along the edge, an arc is actually a part of the circle. See Figure 6-20.

    - A *minor arc* is one of two arcs on a circle that is less than 180°.

    - A *major arc* is one that is greater than 180°. In the preceding example of the 90° central angle, the minor arc is 90°, while the major arc is 270°. You will most likely be working with minor arcs rather than major arcs on the SAT II Math test.

- ✔ A *central angle* of a circle is an angle that is formed by two radii; it's called a central angle because its vertex is in the center of the circle. The measurement of the central angle is the same as the arc that the radii on the central angle intercept. So a 90° central angle will intercept one-quarter of the circle, or a 90° arc.

Figure 6-20:
An arc and central angle.

Arc

Central angle

In the following figure, *A* and *B* lie on the circle centered at *C*. Line *CA* is 9 units long, and the measure of ∠ *ACB* is 40°. How many units long is minor arc *AB*?

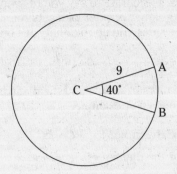

(A) π

(B) 2π

(C) 9π

(D) 18π

(E) 36π

First, determine how many degrees are in arc *AB*. Because *CA* and *CB* are radii of the circle, the degree measurement of the central angle *ACB* is the same measurement of the arc the ends of the radii form on the circle. Therefore, the minor arc *AB* is 40°. How does that help you determine the length of the arc? You know that a circle is 360° and 40° is ⅑ of 360°. Therefore, arc *AB* is ⅑ of the circumference of the circle. Determine the circumference:

$$C = 2\pi r; \ C = 2\pi 9; \ C = 18\pi; \ \text{⅑ of } 18\pi = 2\pi$$

Therefore, the correct answer is B.

Striking a chord: Chords, sectors, and segments

A *chord* is a line segment cutting across a circle that connects two points on the edge of a circle. Those two points at the end of the chord are also the endpoints of the arc. See Figure 6-21.

Figure 6-21:
A chord.

A *sector* of a circle is the area inside the circle that is carved out from the central angle all the way out to the arc. It's like a slice of pie that is defined by the central angle. See Figure 6-22. Think of the *c* in sector as standing for the center to distinguish sector from segment.

Figure 6-22:
A sector of
a circle.

A *segment* of a circle is the portion of the area inside a circle that includes a chord, an arc, and everything else in between. See Figure 6-23. Think of the *g* in segment as someone grabbing off a piece of pie toward the edge. Because the chord is a segment of a line, it lops off a segment of the circle.

Figure 6-23:
A segment
of a circle.

Staying in the lines: Inscribed and circumscribed figures, tangents, and inscribed angles

The SAT II may throw in some extra lines and figures when it questions you about circles. The extra features may appear within or outside the circle.

The ins and outs: Inscribed and circumscribed figures

An inscribed figure is any figure (angle, polygon, and so on) that is drawn inside another figure. For example, a triangle could be drawn inside another circle so that all its vertices touch at point on the circle. Figure 6-24 shows an inscribed triangle.

A *circumscribed* figure is one that is drawn around the outside of another shape, such as a circle drawn around a triangle so that all the vertices of the triangle touch the circle. The circle in Figure 6-24 is circumscribed around the triangle.

Figure 6-24:
Inscribed
and circum-
scribed
figures.

The difference between an inscribed and a circumscribed figure occurs in terms of reference. You refer to the figure on the outside of another figure as a circumscribed figure and the figure on the inside of another figure as an inscribed figure.

The SAT II Math test may see whether you can calculate the area of a figure that is shaded when using circumscribed and inscribed figures. When you get a "shaded area" problem, it's often best to calculate the area of both figures, and then subtract the area of one from another.

In the following figure, the circle centered at B is internally tangent to the circle centered at A. The smaller circle passes through the center of the larger circle and the length of line AB is 4 units. If the smaller circle is taken out of the larger circle, how much of the area in square units of the larger circle will remain?

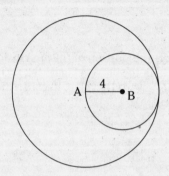

(A) 16π

(B) 36π

(C) 48π

(D) 64π

(E) 800π

Because the smaller circle passes through the center of the larger one, the radius of the larger circle is two times the radius of the smaller one. Therefore, the radius of the larger circle equals 8. To find the area in question, you must find the area of the larger circle and subtract the smaller from the larger. To determine the area of the larger circle, apply the area formula:

$$A = \pi(8^2)$$
$$A = 64\pi$$

The area of the smaller circle is determined in the same way:

$$A = \pi(4^2)$$
$$A = 16\pi$$

Now subtract the two areas:

$$64\pi - 16\pi = 48\pi$$

The correct answer is C.

Off on a tangent

A *tangent line* is one that intersects the circle at just one point. A good way to think of a tangent line in the real world is like a wheel rolling along a road. The road is tangent to the wheel. Figure 6-25 shows a line AB that is tangent to the circle, and that line is also perpendicular to the radius that touches the circle where the tangent intersects it. To use the wheel analogy again, if that wheel had an infinite number of spokes coming from its center, only one spoke would touch (be perpendicular to) the ground at any one time.

Rules for inscribed angles

You can get some valuable information about measurements from an angle inscribed within a circle. An *inscribed angle* is an angle inside a circle with its vertex and endpoints touching points on the circle. In Figure 6-26 the vertex of the inscribed angle touches the circle at a and its endpoints are b and c.

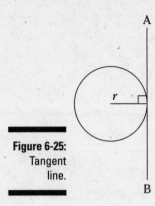

Figure 6-25:
Tangent
line.

Remember that a central angle has the same measure as the arc that it intercepts. Figure 6-26 shows a central angle of 90° intercepting an arc that is also 90°. If you take the endpoints of that arc and make them intercept an inscribed angle, your inscribed angle will be 45° or half the measure of the central angle.

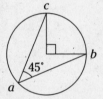

Figure 6-26:
Central
angle and
inscribed
angle inter-
cepting
an arc.

Notice in Figure 6-27 that if you leave in place the endpoints where the inscribed angle intercepts the circle, you can move the vertex of the inscribed angle anywhere along the circumference of the circle and still have the same angle measure. The angle with its vertex at *a* has the same measure as the angle with its vertex at *b* because the angles have the same endpoints.

After you know the measurements of the inscribed angles, you have a wealth of information. For example, you know that the two triangles formed in Figure 6-27 are similar triangles, because they have at least two angles with similar measurements. With knowledge such as this, you can solve the mysteries of plane geometry.

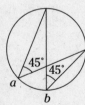

Figure 6-27:
Inscribed
angles
moving
along the
circumfer-
ence of
a circle.

Chapter 7

Getting Graceful: Coordinate Geometry

● ●

In This Chapter

▶ Exploring the coordinate plane

▶ Graphing linear equations and inequalities

▶ Graphing conic sections

▶ Exploring coordinate space: the final frontier

● ●

Coordinate geometry is the perfect marriage between the study of algebra, on the one hand, and plane and three-dimensional geometry, on the other. You can study those particular areas in more detail in other chapters in this book. This chapter shows you how those concepts are tied together; that is, you study how equations and numbers relate to geometric forms and shapes, such as a straight line or a parabola.

You can expect to encounter coordinate geometry questions on about 10 percent of the problems on the SAT II Math test, give or take one or two. You'll most likely find a higher concentration of these questions on the Level IIC test. And, as you may expect, the problems on the IIC test will probably be a bit more difficult.

Ready for Take-off: The Coordinate Plane

This plane does not have wings. Before you get started, take just a minute to refresh your memory of a few relevant terms that may pop up on the SAT II Math subject tests. While you won't be asked to define these terms, it's absolutely essential that you know what they mean. Otherwise, you'll have a hard time figuring out how to answer certain questions.

A well defined line: Some basic definitions

Here are some of the more common terms for coordinate geometry that show up from time to time on the SAT II Math subject tests.

✔ **Coordinate plane:** The coordinate or Cartesian plane is a perfectly flat surface that contains a system where points on the plane can be identified by their position using an ordered pair of numbers that represent their distance from an origin on perpendicular axes. The coordinate of any particular point is the set of numbers that identify the location of the point, such as (3, 4) or (x, y).

✔ **x-axis:** The x-axis is the horizontal axis on a coordinate plane, where values or numbers start at the origin that has a value of 0. Numbers increase in value to the right of the origin and decrease in value as they move to the left. The x-value of a point's coordinate is listed first.

Descartes before dehorse

Seventeenth century French philosopher and mathematician Rene Descartes, who coined the phrase "I think, therefore I am," is also credited with performing the wedding ceremony between algebra and geometry by devising the coordinate plane as we know it today. Although the concept actually came from the ancient Greek mathematician, Appolonius, Descartes perfected the idea, and that's why it's called the Cartesian plane. A little-known fact is that he entered a Jesuit college at the ripe young age of 8 and was allowed to sleep in every morning until 11:00 a.m.! With all that rest, he probably aced the SAT II. Take a cue from Descartes — get plenty of rest before you take the SAT II Math test, but set your alarm before 11:00.

- **y-axis:** The y-axis is the vertical axis on a coordinate plane, where the values run from the origin upward for greater values and downward for decreasing values. The value of y is listed second in a point's coordinate.

- **Origin:** The origin is the point (0, 0) on the coordinate plain where the x- and y-axes intersect.

- **Ordered pair:** Also known as a coordinate pair, this is the set of two numbers that show the distance of a point from the origin. The horizontal (x) coordinate is always listed first and the vertical coordinate (y) is listed second.

- **Abscissa:** This funky word is another name for the x-coordinate in an ordered pair. The SAT II probably won't refer to the x-coordinate as the abscissa, but you know what it is just in case.

- **Ordinate:** Ordinate is another way of referring to the y-coordinate in an ordered pair.

- **X-intercept:** This is the value of x where a line, curve, or some other function crosses the x-axis. The general convention is that y has a value of 0 at the x-intercept. The x-intercept is often the *solution* or *root* of an equation.

- **Y-intercept:** This is the value of y where a line, curve, or other function crosses the y-axis. The convention is that x has a value of 0 at the y-intercept.

- **Slope:** Slope measures how steep a line is and is commonly referred to as "the rise over the run."

Life in the flats: Defining the coordinate plane

The coordinate plane has no thickness and extends in two directions to infinity. That means it's a two-dimensional concept having length and width. Of course, you could never find such an infinite flat surface in the real world, even in the days of Christopher Columbus, when the earth was flat. But despite the fantasy, the coordinate plane is an extremely helpful way to graphically work with equations having two variables, usually x and y.

The coordinate plane has two intersecting number lines perpendicular to each other. The horizontal number line is called the x-axis. The vertical number line is called the y-axis. The point where the two axes intersect is called the *origin*. The arrows at the end of the axes show that they go on to infinity. You can identify any point on the coordinate plane by its location along the x- and y-axes, called its *coordinate*, which is also known as an *ordered pair*.

For example, the ordered pair (3, 5) means that the coordinate point is located three places to the right of the origin along the horizontal (x) number line, and five places upward on the vertical (y) number line. As you can see, the x-coordinate is listed first, and the y-coordinate shows up second. Pretty simple so far, huh? These Cartesian coordinates are also called rectangular coordinates, not to be confused with polar coordinates, which follow a different pattern. Don't be too concerned about polar coordinates, though. It's very unlikely they will ever show up on the SAT II Math subject test.

The intersection of the *x*- and *y*-axes forms four quadrants on the coordinate plane, oddly enough known as Quadrants I, II, III, and IV. See Figure 7-1. That's just so crazy, it may work!

- ✔ All points in Quadrant I have a positive *x*-value and a positive *y*-value.

- ✔ All points in Quadrant II have a negative *x*-value and a positive *y*-value.

- ✔ All points in Quadrant III have a negative *x*-value and a negative *y*-value.

- ✔ All points in Quadrant IV have a positive *x*-value and a negative *y*-value.

- ✔ All points along the *x*-axis have a *y*-value of 0.

- ✔ All points along the *y*-axis have an *x*-value of 0.

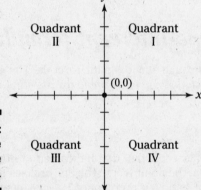

Figure 7-1:
The coordinate plane.

Quadrant I starts to the right of the *y*-axis and above the *x*-axis, the upper right portion of the coordinate plane. The other quadrants move counter-clockwise around the origin, as shown in Figures 7-1 and 7-2. Figure 7-2 also shows the location of coordinate points A, B, C, and D.

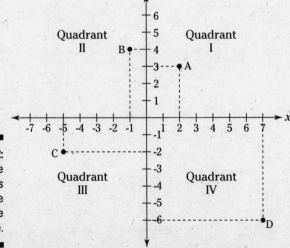

Figure 7-2:
Coordinate points on the coordinate plane.

✔ Point A is in Quadrant I and has coordinates (2, 3).

✔ Point B is in Quadrant II and has coordinates (–1, 4).

✔ Point C is in Quadrant III and has coordinates (–5, –2).

✔ Point D is in Quadrant IV and has coordinates (7, –6).

No, you will not be asked to pick your favorite quadrant on the SAT II Math subject test, but you may be asked to identify what points hang out in which quadrant.

A Slippery Slide: Slope and Linear Equations

One of the handiest things the coordinate plane can do is to graph the location of lines and linear equations. In fact, that's probably the most likely type of coordinate geometry question you'll see on the SAT II Math test, so try and get a firm understanding of lines on the plane, at least for the Level IC test. The Level IIC exam is likely to test you on some of the more elaborate concepts such as ellipses and hyperbolas.

Formulating lines: The slope-intercept formula

The characteristics of a line can be conveyed through a mathematic formula. The equation of a line generally shows y as a function of x. The *slope-intercept formula* for the equation of a line demonstrates this graphically as follows.

$$y = mx + b$$

In the slope-intercept formula, the coefficient m is a constant that indicates the slope of the line, and the constant b is the y-intercept; that is, the point where the line crosses the y-axis. So, a line with a formula $y = 4x + 1$ has a slope of 4 and a y-intercept of 1. You can see a graph of this line in Figure 7-3.

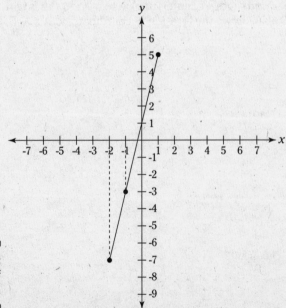

Figure 7-3:
The graph of
$y = 4x + 1$.

Whenever you get an equation for a line that doesn't neatly fit into the slope-intercept format, go ahead and play with the equation a little bit (sounds fun, doesn't it?) to get it into the $y = mx + b$ format. That way you can either solve the problem or get a visual idea of the graph of the line. For example, if you saw the equation $\frac{1}{3}y - 3 = x$, you would simply manipulate both sides of the equation until it looks like the slope-intercept form that you know and love.

$$\frac{1}{3}y - 3 = x$$

$$\frac{1}{3}y = x + 3$$

$$y = 3x + 9$$

Voilà! You now know the slope of the line (3) as well as the y-intercept (9) by using some basic algebra. Pretty tricky!

You can also just solve any equation like this for y.

Notice that if a line is not parallel to one of the coordinate axes, it either rises or falls from the left hand side of the coordinate plane to the right hand side. The steepness of the line's rising or falling is called the *slope*.

The slope of the line is just a number that measures how steep a line is. Think of the slope as a fraction. In Figure 7-3, a line having a slope of 4 is the same as if it had a slope of $\frac{4}{1}$. In other words, its points are a distance 4 units from each other vertically and 1 unit from each other horizontally.

As Figure 7-3 shows, a slope of 4 is very steep indeed. The greater the absolute value of the slope, the steeper it will be. Technically, the slope is a ratio of the vertical change from one point to another to the horizontal change between the same points, also known as "the rise over the run." The line in Figure 7-3 has a slope of 4 because it goes up 4 units (the rise) every time it goes 1 unit forward in horizontal distance (the run). Don't drive your car up a surface that steep, and if you park, you'd better turn your tires toward the curb.

A map for lost skiers and boarders: Finding the slope

The *formula for the slope* contains coordinate points (x_1, y_1) and (x_2, y_2) and is stated as a ratio of the change in the vertical height and the horizontal distance between the two points as follows.

$$Slope(m) = \frac{change\ in\ vertical\ coordinates}{change\ in\ horizontal\ coordinates} = \frac{y_2 - y_1}{x_2 - x_1}$$

Very important: When you subtract the values in the numerator and denominator, remember to subtract the x- and y-values of the first point from the respective x- and y-values of the second point. Don't fall for the trap of subtracting $x_2 - x_1$ to get your change in the run and then subtracting $y_1 - y_2$ for your change in the rise. That kind of backward math will mess up your calculations, and you'll very soon be on a slippery slope.

The graph in Figure 7-4 shows how important it is to perform these operations in the right order.

On the graph in Figure 7-4, use the coordinate point $(0, 2)$ as your (x_1, y_1), and the coordinate point $(4, 0)$ as (x_2, y_2). Now, it may be tempting to subtract the 0 in each coordinate point from the corresponding greater number in the other coordinate point, but doing that switches the order of how you subtract the x- and y-values in the two coordinate points.

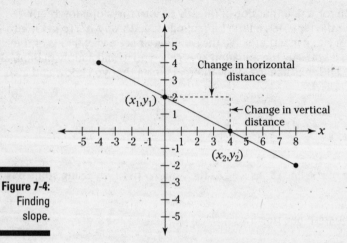

Figure 7-4:
Finding
slope.

For the slope formula to work, you need to take 0 minus 2 for your $y_2 - y_1$ operation (which equals –2), and then take 4 minus 0 as your $x_2 - x_1$ (which equals 4). The resulting ratio or fraction is –2⁄4 or –½. This gives you a slope of –½. You also know that the y-intercept is 2, and therefore, the formula for the line in Figure 7-4 is:

$$y = -\tfrac{1}{2}x + 2$$

You can see from the graph in Figure 7-4 that the line is falling from left to right. Aside from noticing the nice ski slope effect, that's your visual clue that the line has a negative slope. Figure 7-5 shows how you can eyeball a line to quickly get a good idea of whether the slope is positive or negative.

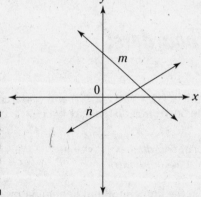

Figure 7-5:
Negative
and positive
slope.

In Figure 7-5, line *m* has a negative slope. Line *n* has a positive slope.

✔ A line with a negative slope falls from left to right (its left side is higher than its right), and its slope is less than 0.

✔ A line with a positive slope rises from left to right (its right side is higher than its left), and its slope is greater than 0.

Line *o* shown in Figure 7-6 has a slope of 0. Line *p* has a slope that is undefined; it has no slope.

TIP

✔ A horizontal line has a slope of 0; it neither rises nor falls and is parallel to the *x*-axis.

✔ The slope of a vertical line is undefined, because you don't know whether it's rising or falling; it has no slope and is parallel to the *y*-axis.

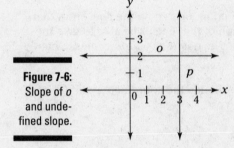

Figure 7-6:
Slope of *o* and undefined slope.

The equation for line *o* is *y* = 2. The equation for line *p* is *x* = 3.

In Figure 7-7, line *q* is negative and has a slope of –1. Line *r* is positive and has a slope of 1.

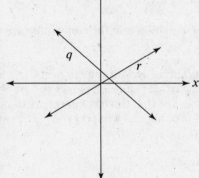

Figure 7-7:
The slopes of lines at 45-degree angle to the *x*- and *y*-axes.

TIP

To save time on the SAT II coordinate geometry questions, memorize the following rules about slope.

✔ A line having a 45-degree angle to the *x*-axis that rises from left to right has a slope of 1; it moves *up* one unit for every unit that it moves over. The converse is also true, that a line having a slope of 1 has a 45-degree upward angle to the *x*-axis.

✔ A line having a 45-degree angle to the *x*-axis that falls from left to right has a slope of –1; it moves *down* one unit for every unit that it moves over. The converse is also true, that a line having a slope of –1 has a 45-degree downward angle to the *x*-axis.

✔ A line having less than a 45-degree angle to the *x*-axis has a slope that is less than 1 and greater than –1, and vice versa.

✔ A line having greater than a 45-degree angle to the *x*-axis has a slope that is either greater than 1 or less than –1, and vice versa.

✔ Parallel lines have the same slope.

✔ The slopes of perpendicular lines are the negative reciprocal of each other (which means that if one line has a slope of 3, the other has a slope of $-\frac{1}{3}$), so long as neither line is vertical; the product of their slopes is -1.

Working backward: The point-slope form

Occasionally, you may be told the slope of a line and a point located on the line. From that information, you will be expected to recognize the graph of that line or be able to give the equation for the line using the slope-intercept formula. The *point-slope form* comes in handy for this type of question.

Suppose you are told that the slope is -3 and the line contains the coordinate point $(2, -2)$. The point-slope form of the equation for a line is stated as

$$y - b = m(x - a)$$

where m is the slope and (a, b) is the known point. In this case, plug the known variables into the point-slope form.

$$y - b = m(x - a)$$
$$y - (-2) = -3\,(x - 2)$$
$$y + 2 = -3x + 6$$
$$y = -3x + 4$$

You have now turned the point-slope form of the equation into the slope-intercept form! That's going to help you on the SAT II Math test.

You can also come up with this same equation graphically. You can connect the known point $(2, -2)$ to the y-axis by a line with the known slope of -3. Figure 7-8 shows the graph of this line. When you draw the line in Figure 7-8, you can see that the line intercepts the y-axis at $(0, 4)$. Plug that number into the slope-intercept formula and what do you get? Presto! The equation of the line in slope-intercept form.

$$y = -3x + 4$$

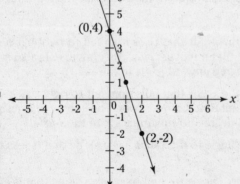

Figure 7-8:
Graphing to
find the
equation
of a line.

You could also determine the equation of the line if you are given only two points. In Figure 7-8, for example, you are told that points (2, –2) and (1, 1) are on the line. Knowing the location of two points, you can find the slope using the slope formula. Then you can draw the line that connects to the *y*-axis at the coordinate (0, 4). That's another way to figure out that the slope is –3. Again, it's a slam-dunk deal for you, and it just puts the SAT II folks in a quandary trying to figure out the next way to stump you.

Line *r* is perpendicular to line *s*, and line *s* passes through the origin. If the equation for line *r* is *y* = 2*x* – 3, then line *s* must pass through which of the following points?

(A) –1, 2

(B) –2, 1

(C) 2, 4

(D) 4, 2

(E) –1, –2

From the equation given for line *r*, you know that the slope of that line is 2. Because line *s* is perpendicular to line *r*, line *s* has a slope of –½, the negative reciprocal of line *r*. And, because line *s* passes through the origin, you can easily plot the points that line *s* will pass through. The correct answer is B, or (–2, 1). It's right next to the origin, and you get there by taking the slope of –½ and moving up 1 unit and backwards 2 units. If you selected A, you gave *s* a slope of –2 instead of –½. If you chose C, you gave line *s* a slope of ½ instead of the negative reciprocal –½. If you picked D, you gave line *s* a slope of 2 instead of –½. If you picked E, then you are still stuck on the positive slope ½.

Knowing that the line *s* has a negative slope and passes through the origin means that it travels through Quadrants II and IV. This should help you immediately eliminate choices C, D, and E. Feel better already?

Where does the function *f*(*x*) = –7*x* + 14 intersect the horizontal axis?

(A) 14

(B) –14

(C) 2

(D) –2

(E) 0

The question asks you to find the *x*-intercept. That is always the case when *y* = 0. The easiest way to tackle this one is to substitute 0 for *y*, or *f*(*x*):

$$0 = -7x + 14$$

$$7x = 14$$

$$x = 2$$

Therefore, C is the correct answer. If you chose A, you were looking at 7*x*, not *x*. If you chose B, then you were looking at –7*x*. D is the opposite of *x*, or –*x*. Finally, 0 is the value of *f*(*x*) at the *x*-intercept.

What is the equation of a line with a slope –⅔ and a *y*-intercept of 7?

(A) $3x + 2y = 21$

(B) $-2x + 3y = 14$

(C) $2x - 3y = 21$

(D) $2x + 3y = 14$

(E) $2x + 3y = 21$

When you look at the answer choices, you know that they all have the same format.

$$ax + by = c$$

You need to convert your equation to that format as well. Because the slope intercept formula for the line is $y = mx + b$, you know that *m* is the slope and *b* is the *y*-intercept. So what are you waiting for? Plug these values that you know into the slope intercept formula.

$$y = (-\tfrac{2}{3})x + 7$$
$$3y = -2x + 21$$
$$2x + 3y = 21$$

This means the correct answer is E. If you convert all the other choices to the slope intercept form, here's what you get:

A is equivalent to saying $y = (-\tfrac{3}{2})x + 2\tfrac{1}{2}$.

B is the same as $y = (\tfrac{2}{3})x + 1\tfrac{4}{5}$. This line has a slope of ⅔ and a *y*-intercept of 1⅘.

C in the slope-intercept form gives you $y = \tfrac{2}{3}(x - 7)$.

D converts to $y = (-\tfrac{2}{3})x + 1\tfrac{4}{5}$. The slope is right, but the *y*-intercept is wrong.

E is correct because this equation can be solved for *y* to find an equation with a slope of –⅔ and a *y*-intercept of 7.

What quadrants would the equation $y - 2x + 3 = 0$ for all real numbers pass through?

(A) I, II, and III

(B) III and IV only

(C) I, II, III, and IV

(D) I, III, and IV

(E) II and IV only

You need to know where the quadrants are. For reference, see Figure 7-1. You can immediately see that choice C is impossible. There is no straight line that can pass through all four quadrants, so cross out C.

The best way to start this one is to convert the linear equation into the slope-intercept form. The equation $y - 2x + 3 = 0$ can be converted to $y = 2x - 3$.

For this kind of question, you may want to draw on your scratch paper a coordinate plane graph and label the quadrants I, II, III, and IV. Nothing fancy, mind you, just enough to get your bearings. Now draw a point below the origin on the *y*-axis representing –3, the *y*-intercept.

Now, draw a line that travels upward from left to right rising two units toward the top of the paper for every one to the right. Your figure doesn't have to be perfect. From your drawing, you can immediately see that the line passes through quadrants I, III, and IV.

Therefore, D is your best choice. Choice A would be correct if you had a parallel line with a positive *y*-intercept. B is partially correct, but not complete. B would only be the case for a line parallel to the *x*-axis and a negative *y*-intercept. C is not possible for a straight line and is also incorrect. E would require a line with a negative slope passing through the origin, which is not the case with this equation.

Any line must travel through at least two quadrants unless the line runs directly on top of either the *x*- or *y*-axis. The lines that can travel through *only* two quadrants are the ones that pass through the origin or are parallel to either the *x* or the *y*-axis. A line that lies directly on top of an axis does not go *through* any quadrant. All other lines must eventually travel through three quadrants.

Midpoints and distances

Some of the questions on the SAT II Math test may ask you to identify the midpoint coordinate on a line segment or the distance between two points on a line. You can solve these problems with coordinate geometry!

A couple of formulas that can help you figure these questions out are the *midpoint formula* and the *distance formula*. They're really quite simple.

Assume you have two points A (x_1, y_1) and B (x_2, y_2) on a line:

- ✔ The formula to find the midpoint coordinate (M) between points A and B is

$$M = \left(\frac{x_2 + x_2}{2}, \frac{y_1 + y_2}{2} \right)$$

- ✔ The formula to find the distance between AB is

$$AB = \sqrt{(x_2 - x_1)^2 + (y_2 - y_1)^2}$$

You can see the midpoint formula at work in Figure 7-9.

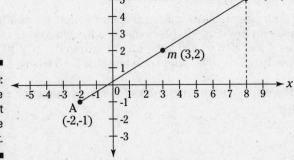

Figure 7-9:
Finding the midpoint of a line segment.

Notice that point A has coordinates (–2, –1) and point B has coordinates (8, 5). To get the coordinates for the midpoint (M), plug those numbers into the midpoint formula, and you get the following:

$$M = \left(\frac{x_2 + x_2}{2}, \frac{y_1 + y_2}{2}\right)$$

$$M = \left(\frac{-2 + 8}{2}, \frac{-1 + 5}{2}\right)$$

$$M = \left(\frac{6}{2}, \frac{4}{2}\right)$$

$$M = (3, 2)$$

The graph in Figure 7-10 shows how the distance formula actually works.

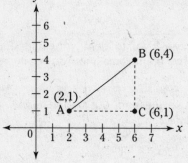

Figure 7-10: Finding the distance between two points.

Notice that point A has coordinates (2, 1) and point B has coordinates (6, 4). To find out the distance between these two points, you will want to plug these numbers into the distance formula as follows:

$$AB = \sqrt{(x_2 - x_1)^2 + (y_2 - y_1)^2}$$

$$AB = \sqrt{(4)^2 + (3)^2}$$

$$AB = \sqrt{16 + 9}$$

$$AB = \sqrt{25}$$

$$AB = 5$$

If you are thinking that this formula looks familiar, you're absolutely right! It's another use for your good buddy the Pythagorean theorem. (If this theorem is foreign to you, check out Chapter 6.) By connecting points A and B to a third point C as shown in Figure 7-10, you get a right triangle, which in this case happens to be your tried-and-true 3-4-5 triangle.

What is the distance between the origin and the midpoint of the line connecting coordinate points (–8, –4) and (4, –2)?

(A) 3.61

(B) 4.47

(C) 5

(D) 8.94

(E) 13.42

The first thing you need to do is to find the midpoint of the line connecting the two given coordinates. Use the midpoint formula.

$$M = \left(\frac{x_1 + x_2}{2}, \frac{y_1 + y_2}{2} \right)$$

$$M = \left(\frac{-8 + 4}{2}, \frac{-4 + (-2)}{2} \right)$$

$$M = \left(\frac{-4}{2}, \frac{-6}{2} \right)$$

$$M = (-2, -3)$$

Now that you have found the midpoint, you can figure out the distance between the coordinates of the origin (0, 0) and the midpoint (–2, –3):

$$AB = \sqrt{(x_2 - x_1)^2 + (y_2 - y_1)^2}$$

$$AB = \sqrt{(-2)^2 + (-3)^2}$$

$$AB = \sqrt{4 + 9}$$

$$AB = \sqrt{13}$$

$$AB = 3.61$$

Choice A is your answer. If you chose B, you were looking for the distance to the second coordinate point given (4, –2). If you chose C, you simply took the coordinates for the midpoint (–2, –3) and added them together for your distance. This does not work. Selection D is thrown in as the distance from the origin to the first point given. Finally, E is the distance between the two coordinate points given.

Notice that it doesn't matter in what order you subtract the *x*- and *y*-coordinate points from each other, because you end up squaring their difference, and your answer will always be a positive number.

Keep in mind that in the end, the distance between two points is always a positive number. If you ever see a negative number as an answer choice for a distance question, don't give that bubble the time of day with your number-2 pencil.

All Things Are Not Created Equal: Linear Inequalities

Graphing a linear inequality is almost exactly the same as graphing the equation for a line, except a linear inequality covers a lot more ground on the coordinate plane. While the graph of an equation for a line simply shows the actual line on the coordinate plane, the graph of a linear inequality shows everything either above or below the line on the plane. The graph appears as a shaded area to one side of the line or the other. Figure 7-11 shows variations on a theme of the equation $y = 2x - 4$.

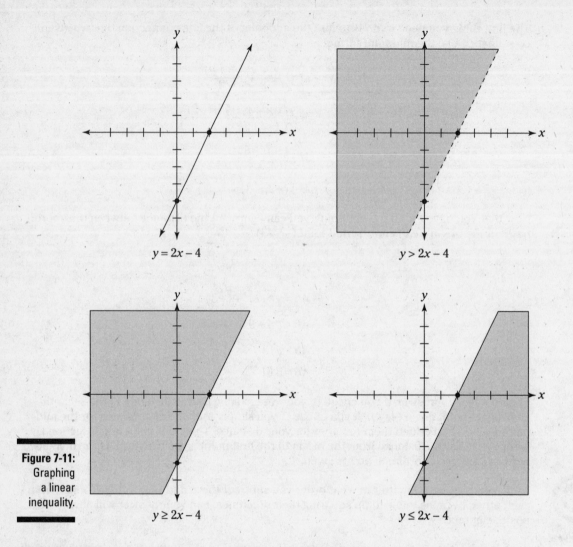

Figure 7-11:
Graphing
a linear
inequality.

Working on the Figure: Other Equations of Shape

The SAT II math exam may test your knowledge of equations for some of the conic sections such as the parabola, circle, hyperbola, and ellipse. They're called *conic sections,* oddly enough, because they are formed when a plane intersects with a cone and forms different shapes.

Passing parabolas

The parabola helps you visualize quadratic equations. When a plane cuts across a cone as shown in Figure 7-12, it forms a parabola.

A *parabola* is a figure on a plane where every point on the figure is the same distance from a fixed point called the *focus* and a fixed line called a *directrix.* If you can't make heads or tails of that definition of a parabola, don't sweat it. For purposes of the SAT II Math test, just know that the parabola is commonly shaped like a curve that opens either upward or downward as shown in Figure 7-13.

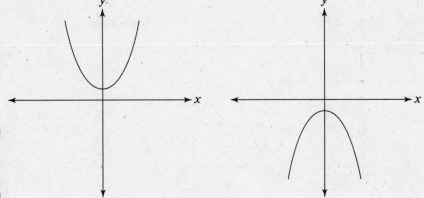

Figure 7-12:
When
a plane
cuts across
a cone.

Figure 7-13:
The
parabola.

There are two properties of the parabola you should know about.

- ✔ **The axis of symmetry:** The vertical line that bisects the parabola so that each side is a mirror image of the other is the axis of symmetry. In Figure 7-13, the *y*-axis is the axis of symmetry.

- ✔ **The vertex:** This is the rounded end of the parabola, which is the lowest point on a curve that opens upward and the highest point on a curve that opens downward. The *vertex* is the point where the parabola crosses the axis of symmetry.

You'll notice that the further the parabola travels away from the vertex, the steeper its sides get. And if you're keeping score, the vertex is the midpoint between the focus and the directrix.

The first formula for the equation of a parabola: Standard form

The most you'll really need to remember about the parabola for SAT II is summed up in two formulas for the equation of the parabola.

The first formula for the equation of a parabola is the *standard form.*

$$y = a(x - h)^2 + k \text{ (where } a \neq 0)$$

This is a nifty equation, and it tells you a few things about the graph of the parabola. The *a, h,* and *k* in the standard form equation are constants. Believe it or not, they all help define where the figure stands in relation to the *x*- and *y-axes.*

✔ The coordinate point (h, k) is the vertex of the parabola.

✔ The vertical line $x = h$ is the axis of symmetry of the parabola.

✔ If the coefficient a is a positive number, then the parabola opens upward.

✔ If a is negative, then the parabola opens downward.

The equation of the parabola in Figure 7-14 may be something very simple such as $y = x^2$. In this case, there is no h or k in the equation. More precisely, the values of h and k are 0. So the vertex (h, k) is right there at the origin making things neat and tidy. Oh, if they could all be this simple, and you could just be sipping a latte!

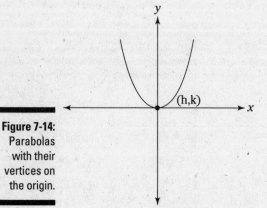

Figure 7-14: Parabolas with their vertices on the origin.

Figure 7-15 contains other parabolas with their equations.

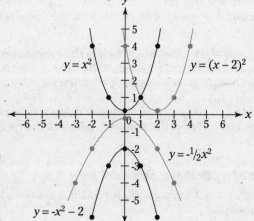

Figure 7-15: More parabolas.

You can visually see from Figure 7-15 that the following statements are true.

✔ If the coefficient a is positive, the parabola opens upward. If the coefficient a is a negative number, the parabola turns upside down; that is, the parabola opens downward.

✔ If the coefficient a is less than 1 but greater than –1, the curve gets fatter and less steep. Conversely, a coefficient a greater than 1 or less than –1 causes the curve to slim down and get steeper.

✔ A change in the constant h moves the parabola sideways along the x-axis. If there is no h in the equation (that is, if $h = 0$), then the vertex is on the y-axis.

✔ The constant k determines whether the vertex of the parabola moves up or down in relation to the y-axis. If there is no k in the equation (that is, if $k = 0$), the vertex of the parabola is on the x-axis.

The second formula for the equation of a parabola: General form

The second formula for the equation of a parabola is called the *general form*.

$$y = ax^2 + bx + c \text{ (where } a \neq 0)$$

If this second form looks familiar, that's because it's an awful lot like the quadratic equation discussed in Chapter 4. The only difference is that you've taken the 0 out of the old equation and substituted it with a y to get the equation of the parabola. When you solved the old quadratic equation, you would get you the *solutions, roots,* or *zeroes* of x. Here, when you solve this equation for y, you end up with the graph of the parabola. Coincidentally (but you probably may have guessed this one), the *zeroes* are the points where the parabola crosses the x-axis; that is, where y is equal to 0.

Just as with the quadratic equation, the $a, b,$ and c of the parabola's general form all represent constants. And these numbers tell you something about the shape and location of the parabola:

✔ If the coefficient a is a positive number, the parabola opens upward.

✔ If the coefficient a is a negative number, the parabola opens downward.

✔ The coordinate point $(0, c)$ represents the y-intercept of the parabola.

Also, you recall that the quadratic formula gives you certain information, and so does the general form of the equation of the parabola. As a refresher, here is the quadratic formula:

$$x = \frac{-b \pm \sqrt{b^2 - 4ac}}{2a}$$

The quadratic formula gives you more information about the location of the parabola:

✔ The axis of symmetry is the line $x = -\frac{b}{2a}$. This also tells you that the x-coordinate of the vertex is $-\frac{b}{2a}$.

✔ You can also find out the y-coordinate of the vertex by simply substituting the value of $-b2a$ in for x in the general form of the equation of the parabola.

The *discriminant* (the number under the radical sign in the quadratic formula) also tells you something about the parabola:

✔ If the discriminant is 0, there is only one point where the parabola touches the x-axis, and that also means that the vertex is on the x-axis.

✔ If the discriminant is positive, there are two points where the parabola intersects with the x-axis.

✔ If the discriminant is negative, the parabola does not intersect with the x-axis.

What is the vertex of the graph of the equation $y = -2(x + 3)^2 - 4$?

(A) $-3, -4$

(B) $3, 4$

(C) $3, -4$

(D) $-3, 4$

(E) $-6, -4$

Don't let those positive or negative signs push you around with this kind of question! Remember the standard form of the equation of a parabola: $y = a(x - h)^2 + k$. The equation in the question switches the positive and negative signs around from the standard form and takes you a bit outside your comfort zone.

You can immunize yourself to those kinds of traps, though. In this case, simply switch the signs back the way they should be. That means that your best choice here is D, because both signs got switched. The other choices are all variations on the theme, and choice E is just out in left field as it uses the product of a and h to come up with the off-the-wall x-coordinate as -6.

The best way to attack parabola problems is by knowing the general form of the equation.

What is the vertex of the parabolic function $f(x) = 3x^2 - 12x + 7$?

(A) $3, -12$

(B) $-12, 7$

(C) $12, -7$

(D) $4, 7$

(E) $2, -5$

This question asks you the same thing as the last one did, with a twist. Here, you have the equation of the parabola, only it's in the general form instead of the standard form. Just use some fancy footwork with the constants in the equation.

Compare: $ax^2 + bx + c$ with $3x^2 - 12x + 7$. Remember, you use one part of the quadratic formula, $-b/2a$, to get the axis of symmetry. Try it:

$$-b \div 2a$$
$$-(-12) \div (2 \times 3)$$
$$-(-12) \div 6$$
$$2$$

Not only do you have the axis of symmetry, but you also have the value of x at the vertex. This is just another way of saying that the equation of the axis of symmetry is $x = 2$. If x is equal to 2, you next need to plug that number into the equation in the original question to get the value for y, or in this case, $f(x)$ which is equal to y.

$$y = f(x) = 3x^2 - 12x + 7$$
$$y = 3(2) - 12(2) + 7$$
$$y = 6 - 24 + 7$$
$$y = -5$$

That leaves you with the coordinate point (2, –5) as the vertex of this parabola. Thus, your correct answer is E. All the other choices take only a superficial bite out of the general form of the equation without using that information to systematically find first the axis of symmetry and then the vertex.

Exacting the equation of a circle

The circle is just another conic section that is formed when a plane cuts across a cone at a 90° angle as shown in Figure 7-16. The circle really gets around!

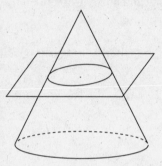

Figure 7-16:
When a plane cuts across a cone at a 90° angle, a circle appears.

A circle is simply a perfectly round shape where all the points are the same distance from a fixed point known as the center. That distance from the center to the points on the circle is called the radius, or *r*.

There's really only one equation for the circle you need to worry about for the coordinate geometry questions on the SAT II Math test. That's the standard form of the equation of a circle, which is

$$(x - h)^2 + (y - k)^2 = r^2$$

The radius of the circle is *r*, and the center of the circle is the coordinate point (*h*, *k*).

Suppose the center of the circle is the origin, that is, the coordinate point (0, 0). That makes the equation for the circle even simpler:

$$x^2 + y^2 = r^2$$

Try a couple of practice questions to check your understanding.

What is the equation of a circle with the coordinate point (5, –1) as its center and a radius of 6?

(A) $(x - 5)^2 + (y + 1)^2 = 6$

(B) $(x - 5)^2 + (y - 1)^2 = 36$

(C) $(x - 5)^2 + (y + 1)^2 = 36$

(D) $(x + 5)^2 + (y - 1)^2 = 12$

(E) $(x - 5)^2 - (y + 1)^2 = 36$

Your choices look so similar you definitely need to know your equation for the formula for a circle.

You know that the radius is 6, so the radius squared is 36. Thus you can easily eliminate choices A and D. Choice E is obviously incorrect because you know that the two terms must be added together in the formula for the circle, and choice E subtracts the two terms. That leaves you B and C. B would be okay if the center had the coordinate point of (5, 1), but in your question, the *y*-coordinate of the center is negative. Thus, you switch the sign around when you plug it into the standard form equation, and you get choice C.

Elucidating the ellipse

An ellipse is one of those conic sections that's formed when a plane cuts through a cone like a circle, only off-center as in Figure 7-17. You will not see an ellipse problem on the Level IC test.

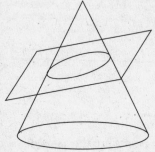

Figure 7-17:
When a plane cuts through a cone off of center, an ellipse appears.

The resulting two-dimensional figure is shown in Figure 7-18.

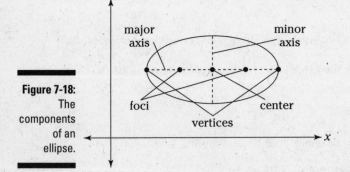

Figure 7-18:
The components of an ellipse.

An ellipse is a lot like a circle, only it's a bit flattened out, either from the topside downward or as if the sides are squished inward. Here's the technical low-down:

- An *ellipse* is all those points on a plane that are located the same distance from the sum of two fixed points.

- Those two fixed points are each called a *focus* (the plural is *foci* — that's Latin).

- The long distance across the ellipse is called the *major axis*. One-half that length is called the *semi-major axis*. The major axis is kind of like the diameter across the long part of the ellipse.

✔ The short distance across the ellipse is called the *minor axis.* One-half that length is called the *semi-minor axis.* The minor axis is kind of like the diameter across the short part of the ellipse.

✔ The *center* of the ellipse is the midpoint of the two foci.

✔ The *vertices* are the two points furthest from the center; that is, the points on the ellipse that intersect with the major axis.

The formula for the equation of an ellipse whose center is at the origin is $\frac{x^2}{a^2} + \frac{y^2}{b^2} = 1$, where

a and *b* are the lengths represented by the semi-major axis and the semi-minor axis.

✔ If the constant represented by *a* is greater than the constant represented by *b*, the ellipse will be parallel to the *x*-axis, and the axis parallel to the *x*-axis is the major axis.

✔ If the constant *b* is greater than *a*, the ellipses is flattened parallel to the *y*-axis, and the axis parallel to the *y*-axis is the major axis. (Okay, now you're catching on!)

✔ You can find the area of an ellipse with the following formula: $A = \pi ab$

That last part sure looks a lot like the formula for the area of a circle, doesn't it? That's because the constants *a* and *b* are a lot like the radius for the long part and the short part of the ellipse.

This is about as much as you'll ever need to know about ellipses if this kind of question even shows up on the Level IIC test (it doesn't appear on every administration of the test). But in case you're not asked the easy formula for the ellipse — that is, if the center of the ellipse is not at the origin — here is the standard formula for any other kind of ellipse they may throw at you.

$$\frac{(x-h)^2}{a^2} + \frac{(y-k)^2}{b^2} = 1$$

where *a* and *b* represent the respective length of the semi-major axis and the semi-minor axis, depending on which axis is greater.

What is the length of the minor axis of an ellipse with the equation $\frac{(x-2)^2}{30.25} + \frac{(y-3)^2}{6.25} = 1$?

(A) 2.5

(B) 4

(C) 5

(D) 6.25

(E) 11

Getting the right answer should be fairly easy as long as you know what you're looking for. The *semi-minor* axis is the square root of the smaller of the two denominators in the fractions in the equation for the ellipse. This means that the square root of the smaller denominator is one-half the length of the minor axis. That leaves choice 5, or choice C, which is exactly twice the length of the semi-minor axis and the answer you are looking for. Choice A is merely the length of the semi-minor axis without doubling it to get the correct answer. Choice B is the square of the number in the numerator and affects only the position of the ellipse along the horizontal axis, not its shape. Choice D is simply the smaller denominator of the two fractions in the equation, without getting the square root to get the length of the semi-minor axis. Finally, choice E is the length of the semi-major axis, which is also not what you're looking for.

Highlighting the hyperbola

The hyperbola is another conic section that you will not see on the Level IC test. However, it may be worth more than just a giggle or two to take a look at if you're taking the Level IIC test. The hyperbola is formed when a plane shaves across two portions of a cone as illustrated in Figure 7-19.

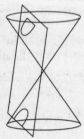

Figure 7-19:
When a plane shaves across two parts of a cone, a hyperbola appears.

A hyperbola is pretty much the opposite of the ellipse. If you're a technical geek, a *hyperbola* is all those points in a plane where the *difference* of the distance between two fixed points is constant. Figure 7-20 shows a hyperbola as basically two badly shaped parabolas that are mirror images of each other. One image is the positive branch, and the other is the negative branch.

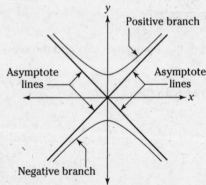
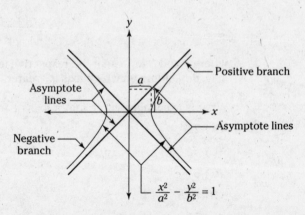

Figure 7-20:
A hyperbola.

The formula for a hyperbola is nearly the same as the one for the ellipse. But instead of *adding* the distances to the fixed points together, as was done with the ellipse, you are *subtracting* the distances from two fixed points to get the hyperbola.

Here is the formula for the equation of a hyperbola having the origin as its center: $\frac{x^2}{a^2} - \frac{y^2}{b^2} = 1$.

You also may have noticed a couple of imaginary lines that run alongside the two branches of the hyperbola. These *asymptote lines* are simply straight lines that the curve of the hyperbola comes very close to as it stretches out into infinity but never quite touches.

There is also a *standard form of equation* for a hyperbola in case the figure is not centered at the origin. The standard form goes as follows:

$$\frac{(x-h)^2}{a^2} - \frac{(y-k)^2}{b^2} = 1$$

✔ The standard form of the hyperbola is very much like the standard form for the ellipse; the one difference is that the equation for the hyperbola looks for the *difference* between, instead of the *sum* of, the distance from two fixed points.

✔ The center of the hyperbola is the coordinate point (h, k).

✔ The a and b represent the respective distances from the fixed points.

✔ If the term with the x is negative, then the hyperbola opens upward and downward. The branch opening upward is the positive branch; the branch opening downward is the negative branch.

✔ If the term with the y is negative, then the hyperbola opens to the left and right. The branch opening to the right is the *positive branch;* the branch opening to the left is the *negative branch.*

Which quadrants does the positive branch of the equation $\frac{x^2}{a^2} - 1 = \frac{y^2}{b^2}$ pass through?

(A) I and II

(B) II and III

(C) III and IV

(D) I and IV

(E) It cannot be determined with the information given.

This is a variation on some of the information you have just been given. At first glance, it looks harder than it is.

The equation given in the problem is simply the general form of the hyperbola with its center at the origin, except that it's just switched around a bit. To get this equation back into the general form, simply add 1 to each side of the equation, and then subtract y^2/b^2 from each side.

$$\frac{x^2}{a^2} - 1 = \frac{y^2}{b^2}$$

$$\frac{x^2}{a^2} - 1 + 1 = \frac{y^2}{b^2} + 1$$

$$\frac{x^2}{a^2} = \frac{y^2}{b^2} + 1$$

$$\frac{x^2}{a^2} - \frac{y^2}{b^2} = \frac{y^2}{b^2} + 1 - \frac{y^2}{b^2}$$

$$\frac{x^2}{a^2} - \frac{y^2}{b^2} = 1$$

Now you have the standard form of the equation for a hyperbola with the origin as its center. It's a very simple step to say that because the y term is negative, the curves of the hyperbola open to the right and left. In this case, the positive branch is on the right-hand side of the y-axis, because that branch opens up in a positive direction along the x-axis. Therefore, D is your correct answer. The quadrants I and IV encompass the positive branch of this hyperbola. Choice A is the positive branch when the x term is negative, so it's wrong. Choice B would be true if you were looking for the negative branch of the given equation, but you're not, so eliminate it. C would be the right answer if you were looking for the negative branch and if the x term were negative; this is not the case for this problem, so C is out. And you certainly have enough information to solve the problem, so E is out.

The Third Dimension: Triaxial Coordinates

The SAT II Math test may ask you to identify coordinates in space as opposed to coordinate points on a plane. You can construct a system for three-dimensional coordinate space by drawing a third axis, the z-axis, perpendicular to the x- and y-axes. One example is in Figure 7-21.

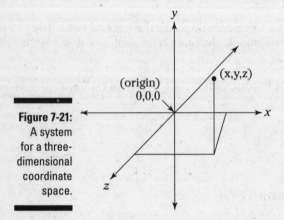

Figure 7-21: A system for a three-dimensional coordinate space.

The x-axis represents the horizontal dimension (length). The y-axis goes up and down in a vertical fashion and shows width. Finally, the z-axis goes back and forth and indicates height or depth, depending on how you want to look at it.

 For coordinate geometry purposes, the only real difference between the two-dimensional figures you worked with on the coordinate plane and three-dimensional objects on the triaxial coordinate system is that you now will need to label a point in space with its third dimensional coordinate. You give a coordinate point its position in space by adding the z-coordinate like so: (x, y, z). For example, the coordinate point $(-3, 5, 9)$ is located 3 places to the left of the origin on the x-axis, 5 places upward on the y-axis and 9 units in a positive direction on the z-axis.

 When you're drawing your picture of the triaxial coordinates, be sure and clearly label which is which so you know what you're trying to figure out! There is no clear convention here, and you don't want to be lost in space!

 You can find out the distance between two points in space by using a modified distance formula you studied earlier. For the distance between points A and B, having coordinates (x_1, y_1, z_1) and (x_2, y_2, z_2), simply use this distance formula that takes into account the third dimension.

$$AB = \sqrt{(x_2 - x_1)^2 + (y_2 - y_1)^2 + (z_2 - z_1)^2}$$

As you can see, this is just a modification of the Pythagorean theorem. You know the length of 3 "sides," and you are trying to find the length of the fourth side using right angles and the hypotenuse.

What is the distance between points A and B in space having coordinates $(-3, -2, 4)$ and $(3, 5, 8)$?

   (A) 7.14

   (B) 10.05

   (C) 12.37

   (D) 16

   (E) 17

This question is not too difficult as long as you remember to follow the distance formula for three-dimensional space. It doesn't matter whether you reverse the order and, for example, subtract x_1 from x_2 and then subtract y_2 from y_1, just as long as you remember to subtract x's from x's and so on. Because you are going to square your results, it doesn't matter if you get a negative number when you do your initial subtraction operations. Here is how to get the correct answer.

$$AB = \sqrt{(x_2 - x_1)^2 + (y_2 - y_1)^2 + (z_2 - z_1)^2}$$

$$AB = \sqrt{(3 - (-3))^2 + (5 - (-2))^2 + (8 - 4)^2}$$

$$AB = \sqrt{(6)^2 + (7)^2 + (4)^2}$$

$$AB = \sqrt{36 + 49 + 16}$$

$$AB = \sqrt{101}$$

$$AB = 10.05$$

The correct answer is B. Choice A is what you get if you multiply the corresponding coordinates, then add them up and take the square root. You get choice C if you *added* the corresponding coordinates in the first instance instead of *subtracting* them as the distance formula requires. Choice D is the result of adding the differences between the coordinate points before you get their squares. E is another variation on the theme from the wrong answer in D.

The nice thing about the distance formula in space is that it's pretty easy and makes logical sense. Also, any questions you may encounter on the SAT II Math test about 3-D coordinate space will generally be about as easy as this last question.

Chapter 8

Living in 3-D: Three-Dimensional Geometry

Three-dimensional geometry, or solid geometry, puts some depth to the plane geometrical figures you can read about in Chapter 6. 3-D geometry is just about as simple as plane geometry with one added dimension. You can use a lot of the same strategies to understand and measure the area of plane figures and apply your knowledge to solids. The trick is to add just a smidgen of common sense. You will most likely be asked no more than a handful of these questions on the SAT II Math test.

You will generally multiply the area of a two-dimensional object by the third dimension to get the volume, and you'll generally add up the area of all the sides of three-dimensional objects to get the total area.

A Chip Off the Old Block: Rectangular Solids

You can make a rectangular solid by taking a simple rectangle and adding a third dimension. A good example of a rectangular solid would be a brick, a cigar box, or something yummy like a box of your favorite cereal. A rectangular solid is also known as a *right rectangular prism* because it has 90-degree angles all around. Figure 8-1 shows some examples of rectangular solids.

A *prism* is more than a piece of glass that makes pretty colors when you shine light through it. A prism is simply two congruent polygons on parallel planes that are connected to each other by their corresponding points. The two polygons that are connected together are the *bases* of the prism.

A rectangular solid has three dimensions: length, height, and width. Beyond that, you really only need to worry about two basic measurements of the rectangular solid on the SAT II Math test: the total surface area and the volume.

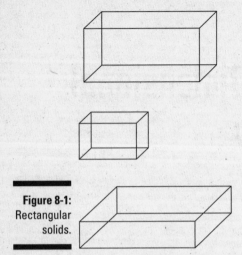

Figure 8-1:
Rectangular
solids.

The volume (*V*) of a rectangular solid is a measure of how much space it occupies, or for more fun, how much yummy cereal your cereal box holds inside. You measure the volume of an object in cubic units, but you would probably not use cubic spoonfuls in the cereal box example. The formula for the volume of a rectangular solid is simply its length (*l*) times width (*w*) times height (*h*).

$$V = lwh$$

Another way of saying this is the volume is equal to the base times the height,

$$V = Bh$$

where *B* is the area of the base. See Figure 8-2.

Figure 8-2:
Volume of a
rectangular
solid.

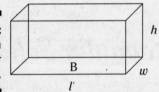

You can visualize the *surface area* of a rectangular solid, or any solid figure for that matter, if you flatten out all of the sides and put them next to each other (see Figure 8-3), sort of like taking apart a cardboard box to get it ready for recycling, only now you get to measure it. Lucky you!

You can find the surface area (*SA*) of a rectangular solid by simply taking the area of all six sides of the object and then adding them together.

First you take the length (*l*) times height (*h*), then the area of length (*l*) times width (*w*), and finally width (*w*) times height (*h*). See Figure 8-4. Now multiply each of those 3 area measurements times 2. Why? Because after you find the area of one side, you know that the opposite side is equal. Thus, the formula for the surface area of a rectangular solid is

$$SA = 2lh + 2lw + 2wh$$

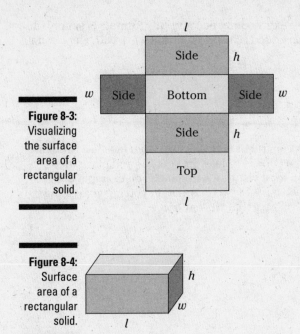

Figure 8-3:
Visualizing
the surface
area of a
rectangular
solid.

Figure 8-4:
Surface
area of a
rectangular
solid.

The line segments that connect the sides or surfaces of the rectangular solid to one another are called, amazingly enough, *edges*. Some problems on the SAT II Math test may involve finding the length of an edge, or you may be asked to find a diagonal on one of the faces of the solid.

Finding these diagonals, of course, would rely on your knowledge of triangles and rectangles from plane geometry, which Chapter 6 discusses. The Pythagorean theorem comes in quite handy in those instances.

Another measurement you may likely be asked to figure out on the SAT II Math test is the length of the long diagonal of a rectangular solid. The long diagonal is the line that runs from the upper corner of one side of the box through the center and all the way down to the lower corner on the very opposite side. See Figure 8-5.

Figure 8-5:
The length
of the long
diagonal
of a
rectan-
gular solid.

You can find the length of the long diagonal of a rectangular solid if you know the length, width, and height of the rectangular solid. The formula for this measurement is

$$a^2 + b^2 + c^2 = d^2$$

If this formula looks vaguely familiar, you get extra points because this formula is simply the Pythagorean theorem with a third dimension added. This formula will work with all rectangular solids, including cubes.

A Perfect Square: Cubes

A *cube* is simply a rectangular solid with six congruent faces; that is, each face has the same length and width. Another way of looking at a cube is to take six squares and place them together to form a solid figure. See Figure 8-6 for a visual. If a domino is an example of a rectangular solid, then a pair of dice is a good example of two cubes. Can you imagine trying to roll a couple of dominos on a gameboard? Don't try that trick at home.

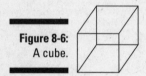

Figure 8-6:
A cube.

You can use the same formulas you use with rectangular solids to find the area and volume of a cube. But because all the faces on a cube are perfect squares, it's a lot easier to find these measurements with some simple formulas. Figure 8-7 shows sides *a* as represented in the following formulas.

The volume of a cube with an edge having a length *a* is a^3.

$$V = a^3$$

The surface area of a cube is simply the area of one side times 6.

$$SA = 6a^2$$

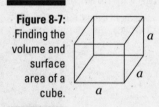

Figure 8-7:
Finding the
volume and
surface
area of a
cube.

You can also use some simple formulas for finding the diagonal of a face on a cube and the long diagonal of the cube as well. Figure 8-8 shows these diagonals.

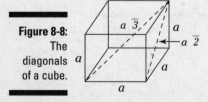

Figure 8-8:
The
diagonals
of a cube.

Using the same formula for finding the diagonal of a square that is provided in Chapter 6, you can find the diagonal of a face on a cube with an edge having a length of a.

$$\text{Face diagonal} = a\sqrt{2}$$

Additionally, to find the length of the long diagonal on a cube, you can either use the enhanced Pythagorean theorem like you do with the rectangular solid, or you can use an even easier formula made especially for squares.

$$\text{Long diagonal} = a\sqrt{3}$$

The converse is not necessarily true. Just because these formulas work for cubes does not automatically mean you can use these shortcuts on rectangular solids. Stick with the tried-and-true cereal box method for measuring rectangular solids.

The following figure shows a cube with an edge having a length of 10 inches. If points B and D are midpoints of the edges of the cube, what is the area of the region ABCD?

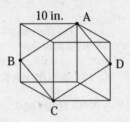

(A) 11.18 square inches

(B) 22.36 square inches

(C) 44.72 square inches

(D) 125 square inches

(E) 15,625 square inches

To find the area of the polygon ABCD, first find the length of each side. If you look closely enough at the figure, you notice that each side of the the polygon ABCD is actually the hypotenuse of right triangles. The short leg of the right triangle is 5 inches, because it is half the length of an edge of the cube, while the long leg (an edge of the cube) is 10 inches. When you use the Pythagorean theorem, you find:

$$5^2 + 10^2 = c^2$$
$$25 + 100 = 125$$
$$\sqrt{125} = 11.18$$

Now, you have the length of one side of the region. Because all four sides are equal, your region is a square, and the area of this region is 11.18^2 or simply 125. Expressed in units, the area is 125 square inches.

Beware of really simple answers, such as simply finding one side of the region, and then choosing that as your answer. That's what happens if you chose A. Answer B is simply multiplying the one side times 2. You get choice C if you multiply one side times 4, which would be correct if you were looking for the perimeter. That's why your correct choice is D, which is the length times the width. The formula for finding the area of a square is simple with your calculator. It's simply the length of a side (c) squared, or c^2. Choice E is taking the correct answer and squaring it again.

The volume of the cube in the following figure is 64. What is the length of diagonal AB?

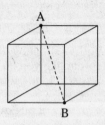

(A) 5.04

(B) 5.66

(C) 6.93

(D) 11.31

(E) 13.86

For this problem, you first must find the length of one of the sides. Don't fall into the trap of going for only the *square* root of 64 and come up with 8. When the volume of the cube is 64, the easy way to get this (if you're not able to do it in your head) is to use your trusty calculator and find the *cube* root of 64 to get the length of the side.

The cube root of 64 is 4. But don't stop there. Memorize the formula for the long diagonal of the square, which is

$$a \times \sqrt{3}$$

$$4 \times \sqrt{3} = 6.93, \text{ which is choice C.}$$

The wrong choices are simply the result of either using the wrong formula for the long diagonal or of not finding the correct length of the edge of the cube. Your first choice A is just $4\sqrt[3]{2}$. Choice B is almost there with $4\sqrt{2}$, and this would be okay if you were looking for the diagonal of one side of the cube. D is the result if you made the mistake of coming up with 8 instead of 4 for the length of the edge and then applying the formula $8\sqrt{2}$. Finally, E is the answer you'd get if you used the correct formula but again had the incorrect length for an edge, or $8\sqrt{3}$.

If you forget the short-cut formula for the long diagonal of the cube, you can also use the variation on the Pythagorean theorem as it applies to rectangular solids. While it may take a bit more time, you'll at least have the correct answer in a pinch.

$$a^2 + b^2 + c^2 = d^2$$

$$4^2 + 4^2 + 4^2 = d^2$$

$$16 + 16 + 16 = 48$$

$$\sqrt{48} = 6.93$$

Take a square piece of cardboard, each side measuring 10 inches, and cut off a 1.5 inch square in each corner as shown in the figure below. Fold up the remaining sides that are shortened along the dotted lines to form a box without a top. What is the volume of the box that is formed? (See the following figure.)

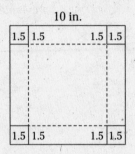

10 in.

1.5 | 1.5   1.5 | 1.5

1.5 | 1.5   1.5 | 1.5

(A) 22.5 cubic inches

(B) 49 cubic inches

(C) 73.5 cubic inches

(D) 100 cubic inches

(E) 150 cubic inches

You can determine the volume in many cases by simply multiplying the length times width times height. So just multiply the sides together, and then multiply the height, right? Maybe so, but not so fast.

First, you have to determine the dimensions by subtracting the 1.5 from each side at the corners. This leaves you with 7 inches remaining on each side. Multiply the sides to get the base of 49, and then multiply the base times the height (1.5) to get your volume of 73.5, which is choice C. Answer A is way too low, because it only multiplies the area of the original cut out corners times the length of the original side before the corners are cut. The number in B would be okay if you were looking for the area of the base, but it doesn't give you the volume. You're still one step shy of the area. D simply multiplies the length of the original sides together and gives you the area of the square before the corners are cut so you would not have any height. Finally, E gives you the volume if you used your original sides of 10 and multiplied by a height of 1.5, but it does not account for subtracting the 1.5 inches from each side when you cut the corners.

Soda Cans and Other Cylinders

A *cylinder* is a circle that grows straight up into the third dimension, as if it became a can of soda. Like prisms, the cylinder has two congruent circles on different planes as its *bases,* and all the corresponding points on the circles are joined together by line segments. The line segment connecting the center of one circle to the center of the opposite circle is called the *axis.* You will most likely be dealing with right circular cylinders rather than the more complicated oblique circular cylinders, but Figure 8-9 shows you a picture of both.

A cylinder has the same measurements as a circle. That is, a cylinder has a radius, diameter, and circumference, but in addition, it also has a third dimension of height or altitude. Check out Figure 8-10 to see where these measurements lie.

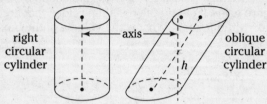

Figure 8-9:
A right circular cylinder and an oblique circular cylinder.

Figure 8-10:
The dimensions of a cylinder.

Finding the volume of the cylinder is rather intuitive. To get the volume, take the area of the base of the cylinder (a circle), that is πr^2, and multiply it by the height (h) of the cylinder. Thus, the formula for a cylinder's volume is, you guessed it:

$$V = \pi r^2 h$$

If you want to find the total surface area of a cylinder, you have to add up the areas of all the surfaces. Imagine taking a soda can and cutting off the top and bottom sections, and slicing it down one side. Then you spread out the various parts of the can. See Figure 8-11 for a visual. You can measure each one of these sections and you'll get the total surface area.

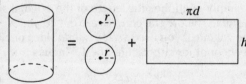

Figure 8-11:
Surface area of a cylinder.

When measuring surface area of a cylinder, don't forget to include the top and bottom of the can in your calculation!

The formula for the total surface area (SA) of a cylinder is

$$SA = \pi dh + 2\pi r^2$$

where the diameter d is 2 times the radius (r).

If you simply want to know the *lateral area* (LA) of the cylinder, find the area of the rectangle portion of the preceding formula, which is the circumference times the height, or:

$$LA = \pi dh$$

You can use another time-honored formula from plane geometry to help figure out the length of a diagonal line from the top of one side of the cylinder to the bottom of the opposite side. See Figure 8-12. You simply use the Pythagorean theorem to figure it out.

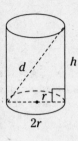

Figure 8-12:
Finding the length of the diagonal line from the top of one side of a cylinder to the bottom of the opposite side.

The formula for this long diagonal is

$$(2r)^2 + h^2 = d^2$$

where the diameter of the cylinder is $2r$, the height is h and the diagonal is d.

If an aluminum can which is a perfect rectangular cylinder contains 250 cm³ of soda, what is the diameter of the can in centimeters if it is 10 cm tall?

(A) $\dfrac{5}{\sqrt{\pi}}$

(B) $\dfrac{10}{\pi}$

(C) $\dfrac{10}{\sqrt{\pi}}$

(D) 10π

(E) $5\sqrt{\pi}$

Start with the formula for the volume of the rectangular cylinder, that is the area of the base times the height of the can

$$V = \pi r^2 h$$

where r is the radius of the base and h is the height of the can. You know that the volume is 250 cm³.

$$V = 250 \text{ cm}^3$$

Now plug in the number 10 for h in the volume equation.

$$250 \text{ cm}^3 = \pi\, 10\text{cm} \times r^2 \text{ cm}^2$$
$$25 \text{ cm}^2 = \pi r^2 \text{ cm}^2$$
$$25 \text{ cm}^2 \div \pi = r^2 \text{ cm}^2$$
$$r = \sqrt{\frac{25}{\pi}} \text{ cm}$$
$$r = \frac{5}{\sqrt{\pi}}$$

WARNING!

You're not done yet! You found the radius, but the question asks for the diameter, so in a sense, you're only halfway there.

$$2r = 2\left(\frac{5}{\sqrt{\pi}}\right) = \frac{10}{\sqrt{\pi}}$$

Look down the list of answers and you can see that the correct choice is C. A is the radius, not the diameter. B is 10 divided by π rather than the square root of π. D is ten times π rather than ten divided by the square root of π. Finally, while E resembles the radius, the square root of π appears in the numerator instead of the denominator.

Not Just for Holding Ice Cream: Cones

You can think of a *cone* as a circle (the *base*) that extends from all of its points upward into a single point, called a vertex. This is known as a circular cone and is the one you'll most likely be asked about on the SAT II Math test. You'll be expected to work with the various dimensions of the cone including the radius, the height (or altitude), and the slant height (length of its side). A cone may look like an ice cream cone or the region covered by a spotlight on a stage. You usually see a cone in the inverted position with the point at the top, but it can show up in any orientation on the test. See Figure 8-13. Remember, an ice cream cone doesn't change its shape just because you turn it upside down to eat it!

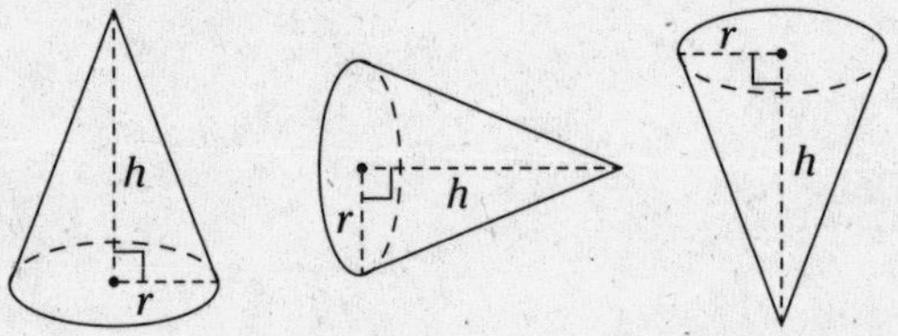

Figure 8-13:
Cones.

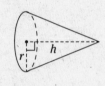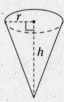

Notice that the the angle formed by the height (or altitude) and the radius is a right angle. This, of course, means that you can use that ubiquitous Pythagorean theorem to determine the slant height if you know the radius and the height. You'll be given the formula for the volume and surface area of a cone at the beginning of each SAT II Math test, but for the sake of time, you should know them extremely well.

The volume of a cone is ⅓ of the area of the base times the height (or altitude), or

TRAPS & TRICKS

$$V = \frac{1}{3}\pi r^2 h$$

The total surface area of the cone is not the same as the lateral area of the cone.

The total surface area of a cone is the sum of the area of the base (πr^2, the part the ice cream plops into) plus the lateral surface area (the area that goes around the sides and keeps your ice cream from falling on the floor).

$$SA = \pi r l + \pi r^2$$

where *r* is the radius and *l* is the slant height of the cone. See Figure 8-14.

Figure 8-14:
The meas-
urements of
a cone.

A space capsule has the shape and dimensions of the following figure. What is the total volume of this capsule in cubic feet?

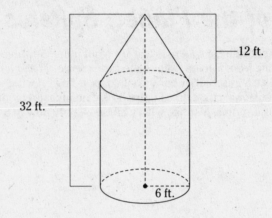

—12 ft.

32 ft. —

6 ft.

(A) 1,006.94

(B) 2,261.95

(C) 2,714.34

(D) 2,767.87

(E) 3,619.12

This is really a combination of a cone and a cylinder, and the answer is simply the sum of those two volumes. First, figure out the volume of the cone. Use the volume formula.

$$V = \tfrac{1}{3}\pi r^2 h$$

$$V = \tfrac{1}{3}\pi\,(36)\,(12)$$

$$V = \pi\,(12)\,(12)$$

$$V = 144\,\pi$$

$$V = 452.39$$

Now, figure out the volume of the cylinder portion of the space capsule. Start with the formula.

$$V = \pi r^2 h$$

$$V = \pi\,(36)\,(20)$$

$$V = (720)\,\pi$$

$$V = 2{,}261.95$$

Add the two volumes of the cone and the cylinder, and what do you get? If you came up with 2,714 cubic feet, or answer C, NASA would be proud of you.

If you chose A, you found only the surface area of the cone and cylinder without the bases. The question asked for volume, and this doesn't leave much room for astronauts to float around in. If you chose B, you found only the volume of the cylinder portion of the capsule. This puts the astronauts in the "coach" section. If you chose D, you miscalculated the volume of the cone by multiplying by the slant height instead of by the height. A bit more room for the space travelers, but still incorrect. Finally, choice E is another miscue on the cone because this answer does not factor the ⅓ part of the cone's volume into the equation. Now they may be flying first class, but the space capsule would no longer have the nose cone!

The Shape of the Planet: Spheres

Technically speaking, a *sphere* is the set of all points in three-dimensional space that are the same distance from a given point called the center. That's a fancy way of saying a sphere is something like a baseball or basketball, depending on what's in season. A sphere is basically a three-dimensional circle, like Figure 8-15. The measurement from the center to any point on the sphere is a radius, just like with a circle — only now in 3-D!

Figure 8-15: A sphere.

The volume of a sphere is ⅔ of π times the radius (*r*) cubed (see Figure 8-16), as stated in the following formula.

$$V = \tfrac{4}{3}\pi r^3$$

The surface area of a sphere is 4 times π times the radius (*r*) squared, or 4 times the area of the big circle that cuts through the center of the sphere. The formula for the surface area is stated as follows.

$$SA = 4\pi r^2$$

Figure 8-16: Finding the measurements of a sphere.

The SAT II Math folks are also gracious enough to provide these formulas to you at the beginning of both the Level I and II tests.

If any plane passes through more than one point on a sphere, you get a circle, and if the plane intersects with the sphere right through the center of it, you get what is called a *great circle*. See Figure 8-17. The circumference of a great circle is the same as the circumference of the sphere, and obviously they each have the same radius.

Figure 8-17:
A great circle.

If a plane intersects at only one point on a circle, then the plane is tangent to the circle, as shown in Figure 8-18.

Figure 8-18:
A tangent plane.

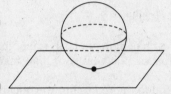

A ball is placed into a 10-inch diameter aluminum can holding 6 inches of water. When the ball is completely submerged, the water level rises 0.4267 inches. What is the diameter of the ball to the nearest half inch?

(A) 1

(B) 1.5

(C) 2

(D) 3

(E) 4

This is a bit tricky and involves what you have learned about cylinders as well as spheres. But take a deep breath and take the steps in logical order, and you should be able to figure it out. It's just a matter of taking the amount of displaced water in the can and converting it to volume in the form of a sphere. You can determine the volume of water in the can in the first instance by using the formula for volume in a cylinder. Remember, the diameter of the can is twice the radius!

$$V = \pi r^2 h$$

$$V = \pi (5)^2 (6)$$

$$V = \pi (25) (6)$$

$$V = 150 \pi$$

$$V = 471.24$$

So you know the volume of water in the can at a level of 6 inches is 471.24 cubic inches. If it rises 0.4267 inches with the ball submerged, you need to work backward to find out what that number means in terms of additional volume. Let 0.4267 represent the additional height in the new equation.

$$V = \pi r^2 h$$

$$V = \pi (5)^2 (0.4267)$$

$$V = \pi (25) (0.4267)$$

$$V = \pi (10.6675)$$

$$V = 33.51$$

Now you have the volume that is displaced by the ball in the can of water. Believe it or not, that's the exact same volume as the ball. Now plug that number into the formula for the volume of the sphere, and see what you get:

$$V = \tfrac{4}{3}\pi r^3$$

$$33.51 = \tfrac{4}{3}\pi r^3$$

$$33.51 = 4.1888\, r^3$$

$$33.51 \div 4.1888 = r^3$$

$$r^3 = 7.9999$$

$$r = 1.99999 \approx 2$$

Because the question wants your answer to the nearest half-inch, you can round just a tad to get 2 as your answer for the radius. So you're ready to pick B. Is that your final answer?!

So close, yet so wrong! You're still not done yet. You found the *radius* but the question asks for the *diameter*. With that last caveat, you're ready to proudly proclaim that the diameter of the ball is 4 inches. The answer is E. The other answers are too close for comfort here. Thus, you need to do the math to get the right answer. You may just as easily plug in answers here to see which one works. Try starting with choice C as your first try, and then work up to the right answer.

Thanks to the Egyptians: Pyramids

A *pyramid* is the three-dimensional object formed when you take a polygon and extend all of its points around its perimeter upward, turning them inward to a single point or *vertex*. A pyramid is kind of like a cone, only it's not nearly as tasty. And because the base of a pyramid is a polygon instead of a circle, it's more angular rather than the rounded surface of a cone. Check out Figure 8-19.

Figure 8-19:
Pyramids.

The widest part of the pyramid — the bottom — is called the *base* (funny how the base keeps reappearing in its supporting role with these solid objects). The distance from the vertex to the base is called the altitude or height. See Figure 8-20. The SAT II Math test generally asks questions based on the length of the altitude (or height) and the area of the base.

The greatest of pyramids

While the base of a pyramid can be any type of polygon from a triangle on up, the bases of the great pyramids of Egypt were squares. The biggest one of all is the Pyramid of Cheops, and the edge of its base is 252 yards long. Its original height was about 160 yards. That's one huge brickhouse! Isn't it better to be studying for the SAT II rather than slaving away at building one of those monstrosities?

Figure 8-20:
The base and altitude of pyramids.

The volume of a pyramid is ⅓ of the area of the base (*B*) multiplied by the height (or altitude). This formula is expressed as follows:

$$V = \tfrac{1}{3}Bh$$

Notice the similarity of the formula for the volume of the pyramid to that of the cone. Almost uncanny!

The sides of a pyramid are always triangles. Pyramids get their classification from the number of sides on the polygon forming the base. So for example, a pyramid with four faces (including the base) is called a *tetrahedron*. You may have guessed that the word *tetra* is not just a video game and actually means "four."

You may be asked to find the surface area of a pyramid, but it's not likely. There is no general formula for finding the surface area of a pyramid. In case you are asked to find it, you will be given enough information from the other measurements in the pyramid to figure it out based on what you know about the area of triangles and polygons from plane geometry in Chapter 6. After you figure out the area of each face, just add 'em up!

Captured: Solids within Solids

Just as with inscribed polygons covered in Chapter 6, the SAT II Math test may ask you about solids inscribed within one another.

Remember, the *inscribed* solid is the solid *inside* the other solid, and it's the largest one that can fit inside the other solid.

Figure 8-21 shows some common examples of inscribed solids:

Figure 8-21:
Inscribed solids.

SAT II Math questions usually ask you some variation on the difference between the measurements of the two solids. The key is to find the measurements of both solids first, and then you can answer the question asked, usually by subtracting one from the other.

Here are some other basic tips on how to find certain dimensions with inscribed solids.

Whenever you have a sphere or cylinder inscribed inside a cube, the diameter of the sphere or cylinder is the same as the length of an edge of the cube. See Figure 8-22.

Figure 8-22:
The diameter of a sphere or cylinder inscribed in a cube equals the length of the edge of the cube.

Whenever you have a rectangular solid inscribed inside a sphere, the length of the long diagonal of the solid is the same as the diameter of the sphere. See Figure 8-23.

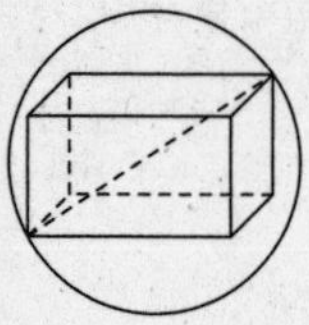

Figure 8-23:
The length of the long diagonal equals the diameter of the sphere.

Whenever you have a sphere inscribed inside a cylinder, they each have the same diameter. See Figure 8-24.

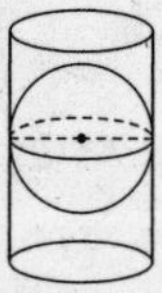

Figure 8-24:
A sphere inscribed in a cylinder has the same diameter as the cylinder.

Whenever you have a cylinder inscribed inside a sphere, the long diagonal of the cylinder is the same as the diameter of the sphere. See Figure 8-25.

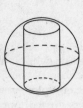

Figure 8-25:
A cylinder inscribed in a sphere has the same diameter as the sphere.

Assume that tennis balls are 2.5 inches in diameter and can be packed into a cylindrical tube so that their edges fit snugly against each other and against the inside of the tube. If three balls are packed in the tube in this manner, how much extra space, to the nearest hundredth of a cubic inch, is there inside the tube that is not taken up by the balls?

(A) 11.72

(B) 12.27

(C) 24.54

(D) 28.63

(E) 36.82

This problem is not too difficult as long as you realize that you are dealing with inscribed spheres inside a cylinder. Most likely, you would never in 1×10^6 years figure out the volume of the air pockets inside the tube directly. Instead, do it the indirect way and it's a piece of cake. You need to find out the volume of each item so that you can subtract the volume of the balls from the volume of the tube.

Because you know the diameter of the balls and the tube, you can figure out the volume of each. Simply divide the diameter by 2 so you now have the radius, 1.25, and then go ahead and plug it into the volume formulas. First, find the volume of one of the balls.

$$V = \tfrac{4}{3}\pi r^3$$

$$V = \tfrac{4}{3}\pi (1.25)^3$$

$$V = \tfrac{4}{3}\pi (1.9531)$$

$$V = 2.6041\pi$$

$$V = 8.1812$$

You multiply by 3 and quickly get the volume of all three balls.

$$V = 24.5437$$

Next, find out the volume of the tube, which can go a bit more quickly.

$$V = \pi r^2 h$$

$$V = \pi (1.25)^2 (7.5)$$

$$V = \pi (1.5625) (7.5)$$

$$V = 11.7188\pi$$

$$V = 36.8155$$

The only thing left to do now is subtract the volume of the balls from the volume of the tube.

$$36.8155 - 24.5437 = 12.2718$$

Your closest answer to the nearest hundredth of a cubic inch is 12.27, or B. As you can see, choice A is the volume of the tube if you don't multiply by π. Choice C is simply the volume of the balls without subtracting from the volume of the tube. If you chose D, you only subtracted the volume of one ball from the tube. And E, of course, is the complete volume of the tube without taking away anything. You can't have much of a tennis game without the balls!

Measuring Up: Volume and Dimensions

Suppose the SAT II Math test people do something sneaky and ask you to increase or decrease each side of a three-dimensional object in all three directions by a certain factor. Sounds simple, eh?

Just because you double the length, width, and height of, say, a cube, doesn't simply mean you double the area and volume of the same object. Au contraire — it's a bit more complicated than that. But don't fret because there is a simple trick for this kind of trap.

When you multiply or divide each dimension of a cube or rectangular solid — or the diameter of a sphere — by a certain factor, you must do the following.

- Correspondingly multiply or divide the total *surface area* of the cube, rectangular solid, or sphere by the *square* of that factor.

- Correspondingly multiply or divide the *volume* of the cube, rectangular solid, or sphere by the *cube* of that same factor.

A baseball has ⅛ the volume of a bowling ball. How many times greater is the radius of the bowling ball than that of the baseball?

(A) ¼

(B) ½

(C) 2

(D) 4

(E) 8

The information in the question mentions the smaller object first, and then asks the question in terms of how much the larger object is in relation to the smaller one. This effectively turns the equation around. Don't be fooled into thinking you are looking for how much smaller the baseball's dimensions are because of the way the question is set up. Always keep in mind what they are asking. As the professionals would say, stay on the ball (so to speak) when they phrase the question like this.

The question asks you to determine the relationship of the radius of two spheres based on the relationship of their volumes. If you decrease the volume of the larger object by a factor of 8 times to get the volume of the smaller object, you must consequently decrease the radius of the larger object by the cube root of that factor. The cube root of 8 is 2, so if you take ⅛ the volume of the bowling ball to get the volume of the baseball, you have to take ½ the radius of the bowling ball to get the radius of the baseball. But don't choose B, because the question asks you how many times greater the bowling ball's radius is. This makes C the correct choice. The radius of the bowling ball is twice that of the baseball.

Creative Motion: Rotation

You can take three types of two-dimensional (plane) geometric shapes and rotate them around an edge or axis to make them become three-dimensional (solid) figures. Here are some examples of this neat little trick.

Taking a circle and rotating it around the axis of its diameter, whether horizonally, vertically, or any which way, makes it a sphere. See Figure 8-26.

Figure 8-26: A circle becomes a sphere.

Rotate on axis   on vertical axis

Taking a square or rectangle and rotating it around its central axis or one of its edges gives you a cylinder. See Figure 8-27.

Figure 8-27: Rotating a square or rectangle.

Rotate on central axis   Rotate around left edge

Taking a right triangle and rotating it around one of its legs, or taking an isosceles triangle and rotating it around its axis of symmetry gives you a cone. See Figure 8-28.

Figure 8-28: Rotating a triangle.

Rotate on left leg   Rotate around axis of symmetry

The SAT II Math test generally asks you to find out how the area of a two-dimensional figure relates to the volume of the three-dimensional object that is formed as a result of rotation.

Assume that the following figure is a semicircle on top of a rectangle. What is the volume in cubic centimeters of the resulting object that is produced by rotating the figure around the central axis indicated?

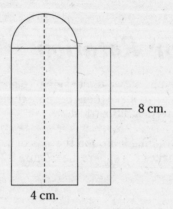

8 cm.

4 cm.

(A) 16.76

(B) 100.53

(C) 117.29

(D) 134.04

(E) 536.16

This is a cute combination of a semicircle and rectangle that, when rotated around the central axis, becomes a hemisphere on top of a cylinder — sort of like a missle silo. Just find the volume of each by plugging in the right formulas for those objects, and you're a rocket scientist. Start with the rectangle that forms that cylinder:

$$V = \pi r^2 h$$
$$V = \pi (2)^2 (8)$$
$$V = \pi (4) (8)$$
$$V = 32 \pi$$
$$V = 100.53$$

Now that you have the volume of the cylinder, you need to find the volume of the semicircle turned hemisphere. Because the hemisphere is only half the volume of a sphere, you can cut the formula in half. Don't you wish you could cut your work in half the same way? So, instead of the conventional formula for the volume of the sphere, you can use the following.

$$V = \tfrac{2}{3}\pi r^3$$
$$V = \tfrac{2}{3}\pi (2)^3$$
$$V = \tfrac{2}{3}\pi (8)$$
$$V = 16.76$$

Now add the two volumes together and you get 117.29, which is coincidentally choice C.

If you chose A, you stopped when you got the volume of the hemisphere. Keep working! If you chose B, you only found the volume of the cylinder. Again, your job is not over. If D was your final answer, you calculated the area of a *sphere* instead of the *hemisphere* that the problem asked for. Finally, if you chose E, you plugged in the diameter when you should have used the radius in the formulas for volume, and this answer would be off the chart.

Chapter 9

Give Me a Sine: Trigonometry

In This Chapter

▶ Understanding basic trigonometric functions

▶ Exploring trigonometric identities

▶ Graphing trigonometric functions

▶ Applying trigonometric functions to oblique triangles

▶ Learning polar coordinates

You're likely to encounter trigonometry questions on both the Level IC and Level IIC versions of the SAT II Math test. Trigonometry is simply the study and measurement of triangles. Trigonometry helps you find out the unknown values of the sides and angles of a triangle when you know only some of those sides and/or angles. The Level IC is more likely to cover the basic trigonometric functions and ratios found in right triangles. The Level IIC exam is more likely to throw in questions related to graphs of trig functions, inverse trig functions, the Law of Sines, Law of Cosines, and polar coordinates. Trigonometry questions may show up in just over 5 percent of the questions on the Level IC test, and they may make up about 15 percent of your Level IIC test. So study up, and you'll be chanting the Greek cheer in no time. What is the Greek cheer?

Cosine, secant, tangent, sine

Three point one four one five nine

Go team!

Examining Basic Trigonometric Functions

You generally work with right triangles to find your basic trigonometric functions. You should also know the measurement of one of the acute angles of the right triangle. An *acute angle* is one that measures less than 90 degrees. Armed with this knowledge, you can use basic trigonometric functions to compare the length of the sides of a right triangle with each other, and then determine the basic shape of the right triangle.

The angle that you start with is generally referred to by the Greek letter θ (theta). Figure 9-1 shows the basic setup.

Figure 9-1 shows a right triangle ABC. You are given angle A as θ. You generally want to designate the sides opposite the angles with the lowercase letter of the angle designation as shown. You know that a triangle has a total of 180 degrees. You know that the right angle at ∠ C is 90 degrees. Then if you know, for example, that the angle θ is 20 degrees, you can easily figure out that ∠ B is going to be 70 degrees, because 20 + 70 + 90 = 180.

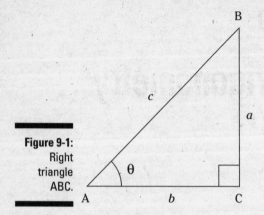

Figure 9-1:
Right
triangle
ABC.

As a result of knowing one of the acute angles of a right triangle, you can figure out the shape of the triangle, and you also know the proportions of the sides to each other. Consider the triangles in Figure 9-2.

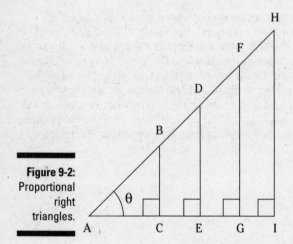

Figure 9-2:
Proportional
right
triangles.

Figure 9-2 contains several triangles with a right triangle: $\triangle$ ABC, $\triangle$ ADE, $\triangle$ AFG and $\triangle$ AHI. Each triangle also has as one of its angles the $\angle \theta$. Notice that each one of these is a similar triangle, because they share the same three angles. Also, the sides of each of these triangles are proportional and, thus, they all have the same shape.

Sine, cosine, and tangent

This is where trigonometry comes in handy. After you know the measure of one of the acute angles of a right triangle, you can figure out the proportions of the sides to one another. The three basic trigonometric functions are the *sine, cosine,* and *tangent.* Figure 9-3 shows how you arrive at these three basic functions.

For basic trig functions, the right triangle has, in relation to the angle θ, an opposite side (side *a*), an adjacent side (side *b*), and its *hypotenuse* (side *c*).

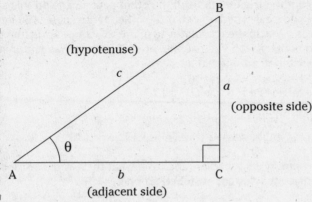

Figure 9-3:
A right triangle showing the opposite side, hypotenuse, and adjacent side

✔ The *sine* is the ratio of the length of the *opposite* side (side *a*) over the *hypotenuse* (side *c*), or *a/c*. You refer to the sine of ∠ θ as sin θ.

✔ The *cosine* is the ratio of the length of the *adjacent* side (side *b*) over the *hypotenuse* (side *c*), or *b/c*. Refer to the cosine of ∠ θ as cos θ.

✔ The *tangent* is the ratio of the length of the *opposite* side (side *a*) over the *adjacent* side (side *b*), or *a/b*. Refer to the tangent of ∠ θ as tan θ.

A convenient mnemonic device can help you remember these basic trig functions: the word *SOHCAHTOA*. Each letter in this word stands for a basic trig function along with the way you compare the sides of the triangle. Thus:

✔ *SOH* stands for **s**ine = **o**pposite over **h**ypotenuse

✔ *CAH* stands for **c**osine = **a**djacent over **h**ypotenuse

✔ *TOA* stands for **t**angent = **o**pposite over **a**djacent

That's really about all you need to know about trigonometry. Go ahead and skip the rest of this chapter. Just kidding! Frankly, though, if you're not taking the Level IIC test, these basic trig functions and their formulas comprise most of the trigonometry on the SAT II Math test. For Level IIC, the rest of trig simply builds on this foundation of knowledge.

The trig book you studied in high school may have whole pages of tables devoted to sines, cosines, tangents, and so on for all angles from 0 to 90 degrees. Unfortunately, the SAT II Math test is not an open-book test, so you can't bring your math book in with you. However, there is the next best, or even better, thing.

Your calculator will figure out sines, cosines, and tangents for all the angles in a triangle, and more. And you don't even have to memorize what page number they're on. Just push a button or two and it's right there in the palm of your hand. So, to figure out the trig functions of, for example, a 40-degree angle just hit the right keys on the calculator:

✔ The sine of a 40-degree angle is 0.643 (sin 40° = 0.643). You find this by pushing the sine key and 40.

✔ The cosine of a 40-degree angle is 0.766 (cos 40° = 0.766). You find this by pushing the cosine key and 40.

✔ The tangent of a 40-degree angle is 0.839 (tan 40° = 0.839). You find this by pushing the tangent key and 40.

Make sure your calculator is in the degree mode, and not the radian mode, for calculating trig functions.

You can just as easily figure out the trig functions of the other acute angle on this same triangle. If one acute angle is 40 degrees, you know that the other acute angle is 50 degrees (90 − 40 = 50). This means you can use the calculator to figure this angle's trig functions pretty quickly. You may also notice that if you use the other acute angle as your new θ, the sine and cosine are simply switched around. So now:

✔ The sine of a 50-degree angle is 0.766 (sin 50° = 0.766).

✔ The cosine of a 50-degree angle is 0.643 (sin 40° = 0.643).

✔ The tangent of a 50-degree angle is 1.192 (tan 50° = 1.192).

The new sine and cosine are simply switched around from what they were before because the opposite side and the adjacent sides got switched around.

The trig function of a positive acute angle is equal to the cofunction of its complementary angle. (The sine and cosine are cofunctions of each other.)

Arcsin, arccos, and arctan

Not only can you find the value of the trigonometric function of an angle with your calculator, but you can also find the measurement of the *angle* when you already have the value of a trig function. Your calculator should have separate keys for $\sin^{-1}$, $\cos^{-1}$, and $\tan^{-1}$ (also known as the arcsin, arccos, and arctan). These keys help to calculate the angle measure from the trig function value. So, for example, if you hit the $\sin^{-1}$ key, and then punch in a trig function value of 0.643, your result will be approximately 40 and, therefore, your answer is 40°. Another way to state this is

$$\text{arcsin } (0.643) = 40°$$

The *arcsin* is the inverse trig function of the sine, where the result is the degree measure of the angle when you know the sine.

Suppose you were to hit $\tan^{-1}$, and then punch in 1.192. Your answer is a nifty 50, or 50°. You can also state this as

$$\text{arctan } (1.192) = 50°$$

We're sure you've already guessed that the *arctan* is the inverse trig function of the tangent, which results in the degree of the *angle* where you already have the trig function measurement of the tangent.

You may see trig questions on the SAT II test that ask you to find out the trig functions of right triangles with special ratios (or Pythagorean triples), like the 3:4:5 triangle (see Chapter 6).

See if you can find out the sine, cosine and tangent of the ∠ θ of the 3:4:5 right triangle in Figure 9-4 and as the measure of its interior angles.

Even if you knew the measure of only two out of the three sides of the triangle in Figure 9-4, you could immediately recognize the length of the remaining side, because you know your Pythagorean triples from Chapter 6. Finding the sine, cosine, and tangent of ∠ θ should be a snap for this triangle

$$\sin θ = \tfrac{4}{5} = 0.8$$

$$\cos θ = \tfrac{3}{5} = 0.6$$

$$\tan θ = \tfrac{4}{3} = 1.333$$

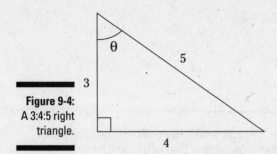

Figure 9-4:
A 3:4:5 right triangle.

Can you guess the measure of ∠ θ in Figure 9-4? Of course you can! Simply take any one of the trig functions you just came up with and plug it into your calculator using the arc- or inverse trig function key. So you could plug in the $\sin^{-1}$ key and hit 0.8, which gives you the angle measure of θ as 53.13°. Or, you could enter the $\cos^{-1}$ key and punch in 0.6. This also rings up the number 53.13; there are so many ways to find the angle from the trig functions just by going for the inverse.

Now see if you can use trig functions to help you find the length of the sides of a right triangle if you know the measurement of one angle and the length of one side. Determine the lengths of the two legs of a right triangle having a hypotenuse of 3.5 inches and a base angle that measures 37°. See if you can also find out the measurement of the other acute angle.

The best way to approach this problem is to draw a picture. Try to draw something like the triangle in Figure 9-5, and label the given measurements accordingly.

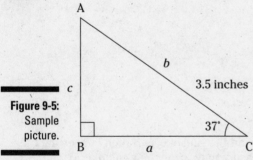

Figure 9-5:
Sample picture.

You can use your calculator to find out the sine of 37° = 0.6018. You also know from SOHCAHTOA that the sine is also equal to the opposite side over the hypotenuse.

$$\sin \angle C = c \div 3.5$$
$$\sin 37° = c \div 3.5$$
$$0.6018 = c \div 3.5$$
$$3.5 \, (0.6018) = c$$
$$2.106 = c$$

Now you know that the length of one leg of the triangle is 2.106 inches. You can find the length of the adjacent side of the triangle from the cosine of ∠ C, which has a value of 0.7986. Again, using SOHCAHTOA, you can compare the cosine value with the fact that you know that the cosine is equal to the adjacent side divided by the hypotenuse.

$$\cos \angle C = a \div 3.5$$

$$\cos 37° = a \div 3.5$$

$$0.7986 = a \div 3.5$$

$$3.5\,(0.7986) = a$$

$$2.795 = a$$

The length of the adjacent side of the triangle is 2.795 inches. You could just as easily have found the length of the adjacent side by using the tangent function of 37° and by comparing that value with the equation that divides the length of the known opposite side by the unknown adjacent side. The answer is the same.

The last thing you're asked is to determine the value of the other acute angle in the triangle. This one is pretty easy, and you probably already did it in your head. It's a simple case of subtracting 37 from 90 to get 53° for the remaining interior angle.

Here's another one that asks you to find the measure of an acute angle of a right triangle where you know only the length of two sides of the triangle.

Find the measure of $\angle \theta$ in the right triangle having two sides measuring 5 and 7 as shown in Figure 9-6.

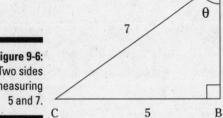

Figure 9-6: Two sides measuring 5 and 7.

This is a simple matter of using SOHCAHTOA to find the sine of θ, and then converting the value of the sine to its corresponding degree measurement. Grab your calculator!

$$\sin \theta = \text{opposite side} \div \text{hypotenuse}$$

$$\sin \theta = 5 \div 7$$

$$\sin \theta = 0.7143$$

Now that you have the value for the sine of θ, go ahead and plug in the calculator keys to find the arcsin, or $\sin^{-1}$, of that value.

$$\text{arcsin } 0.7143 = \theta$$

$$\text{arcsin } 0.7143 = 45.58°$$

$$\theta = 45.58°$$

That's about as hard as they get for the SAT Math Level IC test. There are a couple more things you need to know, but you get the picture. It should be a piece of cake for you if you just remember these few simple rules and know how to use your calculator. Try at least one sample problem that you may encounter:

From a point P on the ground the angle of elevation to a ledge on a building is 27°, and the distance to the base of the building is 45 feet from P. How many feet high is the ledge?

(A) $\dfrac{45}{\sin 27°}$

(B) $\dfrac{45}{\tan 27°}$

(C) 45 sin 27°

(D) 45 cos 27°

(E) 45 tan 27°

Make sure you set up the equation the right way. In this problem, you're asked to find the height of the ledge. Set up a drawing like the one in Figure 9-7.

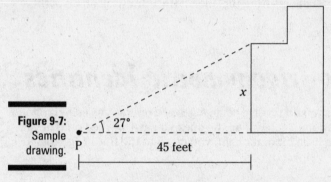

Figure 9-7:
Sample
drawing.

P   45 feet   27°   x

Think of the height of the ledge as the "opposite side" (x) of the triangle from an angle measuring 27°. The key here is to use your known information (the angle measurement and the length of the adjacent side) to find out the unknown, that is, the opposite side. If you divide the opposite side by the adjacent side, you get the tangent. Here's how it plays out.

$$\tan 27° = \frac{x}{45}$$
$$x = 45\tan 27°$$

The easy answer is, therefore, E. You can twist and turn all you want on trying to figure out why the other answers don't work, but the simplest way is to set your problem up in a form that forces you to concentrate on what you know and what you need to find out. In this case, you don't even want to bother with the sine or cosine, because they won't help you solve the problem. That throws out choices A, D, and E. If you chose B, you used some funny division when you did the math! Try another.

If cos θ = 0.745, what is the value of tan (θ × 2)?

(A) 0.009

(B) 0.035

(C) 0.382

(D) 1.791

(E) 9.032

You need to do a couple of operations to solve this one. First thing you do is find out what the angle θ is. If the cosine of θ is 0.745, then you need to find the arccos, or plug in the cos⁻¹ key on your calculator for 0.745. This gives you the angle measurement of 41.841°. Next, you multiply the angle measure by 2, giving you 83.682°. It's very simple, then, to hit the tangent key and come up with 9.032, or E, as your answer.

Be very careful when executing the operations and make sure you do them in the correct order. Don't fall for getting the tangent of 41.841 *first* and only *then* multiplying your result by 2. If you did that, you would have ended up with D as your answer, which is obviously not correct. If you *divided* instead of multiplied by 2 *before* getting the tangent, you would have gotten 0.382 as your result, and chosen C as your answer. If you chose A, you took the cosine of 0.745 as your angle instead of first finding the angle that you know the cosine of already. Then, you divided by 2 and found the tangent. That's the right answer for something, but not the question asked here. Finally, if you picked B, you did just about the same thing as for A — you took the cosine of 0.745 as your angle, and then multiplied your result by 2 and got the tangent. Again, you would be closer to doing the operations in the proper order, but you would have neglected to get the correct angle from the outset, thus throwing off all of the rest of your work.

Finding Oneself: Basic Trigonometric Identities

Trig functions on the SAT II Math test could involve some algebraic operations called trigonometric identities that are a bit more complicated than the ones you just worked on. They're actually not that difficult, and with a little bit of thought, you can find the third trigonometric function if you know the other two. Just watch!

Using the triangle in Figure 9-8, you know that the basic trig functions are as follows:

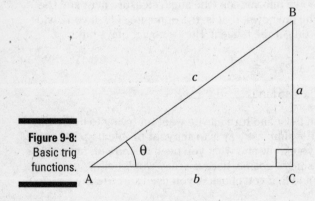

Figure 9-8:
Basic trig
functions.

$$\sin\theta = \frac{opposite\ side}{hypotenuse} = \frac{a}{c}$$

$$\cos\theta = \frac{adjacent\ side}{hypotenuse} = \frac{b}{c}$$

$$\tan\theta = \frac{opposite\ side}{adjacent\ side} = \frac{a}{b}$$

Suppose you know the sine and cosine and want to find the tangent. Try this one on for size.

$$\frac{\sin\theta}{\cos\theta} = \frac{a}{c} \div \frac{b}{c}$$

$$= \frac{a}{c} \times \frac{c}{b} = \frac{a}{b}$$

$$\frac{\sin\theta}{\cos\theta} = \tan\theta$$

This tells you the relationship between the sine and cosine. The sine divided by the cosine gives you the tangent. From that information, you can find out any one of these three functions as long as you know the other two. Just manipulate them accordingly.

$$\frac{\sin\theta}{\cos\theta} = \tan\theta$$

$$\frac{\sin\theta}{\tan\theta} = \cos\theta$$

$$\sin\theta = \tan\theta\cos\theta$$

And if that doesn't just blow your mind, here's another interesting relationship between the sine and cosine functions that may come in handy on the SAT II Math test. Using the triangle in Figure 9-8, start with the good old Pythagorean theorem.

$$a^2 + b^2 = c^2$$

$$\frac{a^2}{c^2} + \frac{b^2}{c^2} = \frac{c^2}{c^2}$$

$$\left(\frac{a}{c}\right)^2 + \left(\frac{b}{c}\right)^2 = 1$$

$$\frac{a}{c} = \sin\theta$$

Remember the formula for the sine and cosine.

$$\frac{b}{c} = \cos\theta$$

Therefore,

$$\sin^2\theta + \cos^2\theta = 1$$

Knowing the trigonometric identity that sine squared plus the cosine squared equals 1, you can also manipulate this relationship to form two other interesting relationships.

$$\sin\theta = \sqrt{1 - \cos^2\theta}$$

$$\cos\theta = \sqrt{1 - \sin^2\theta}$$

All this talk about relationships may give you the urge to pick up the phone and call someone you love, but don't take a break quite yet. There's more fun on deck!

Getting Tricky: More Complex Functions and Identities

You thought you knew all the functions you needed to know for trig, right? Well, if you're taking the Level IIC test, you'd best get familiar with the other trig functions: the secant, cosecant, and cotangent. The main thing to know about these remaining functions is that they are merely the reciprocals of the three primary functions you already know. That means that the sine is the reciprocal of the cosecant, the cosine is the reciprocal of the secant, and the tangent is the reciprocal of the cotangent. You would abbreviate these functions as follows: cosecant = csc; secant = sec; cotangent = cot.

I'll bet you're wondering why these are called the *reciprocal trigonometric functions.* Or maybe not. You would notice that the various relationships between any two sides of a triangle can be flipped around to get their reciprocal, and that's the whole idea behind reciprocal trig functions.

Thus, using Figure 9-9, the reciprocal functions go just like so:

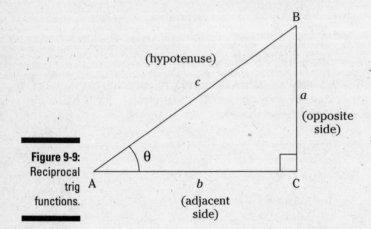

Figure 9-9:
Reciprocal
trig
functions.

$$\csc\theta = \frac{hypotenuse}{opposite\ side} = \frac{c}{a} = \frac{1}{\sin\theta}$$

$$\sec\theta = \frac{hypotenuse}{adjacent\ side} = \frac{c}{b} = \frac{1}{\cos\theta}$$

$$\cot\theta = \frac{adjacent\ side}{opposite\ side} = \frac{b}{a} = \frac{1}{\tan\theta} = \frac{\cos\theta}{\sin\theta}$$

That last one is quite a mouthful, but it shows several relationships among all three of the basic trig functions. Here are two additional trigonometric identities that stem from all of the trig functions:

✔ $\tan^2\theta + 1 = \sec^2\theta$

✔ $\cot^2\theta + 1 = \csc^2\theta$

Another interesting tidbit, in addition to these *reciprocal* functions, is that you have what are called *cofunctions* as well. Cofunctions come in pairs: sine and cosine; tangent and cotangent; and secant and cosecant. Any trig function of an angle in a triangle is equal to the cofunction of the complementary angle in that triangle. Looking at Figure 9-9 again, angle C is the right angle and adds up to 90 degrees. Angles A and B are *complementary* angles, and they add up to the remaining 90 degrees in the triangle. Thus, the *sine* of angle A is equal to the *cosine* of angle B, and vice versa. The same holds true for the tangent and cotangent of A and B respectively, as well as the secant and cosecant.

Most of the time on the SAT II Math test, you'll get some complicated looking algebraic questions that manipulate trig functions. Your mission is to try and transform the functions into their simplest form. Take a look at the relationships among trig functions and how they may be the foundation of something you'll see on the SAT II Math test:

Which of the following is equivalent to $\dfrac{1 - \cos^2 \theta}{\cos^2 \theta}$?

(A) $\sec^2 \theta$

(B) $(\csc^2 \theta) - 1$

(C) $\tan^2 \theta$

(D) $\sin^2 \theta$

(E) $-\dfrac{1}{\sin^2 \theta}$

This one's a bit of a warm-up for figuring out how trig functions interact. First, tackle the numerator. Going back to your equivalent functions, recall that sine squared plus the cosine squared is equal to 1. Therefore

$$\sin^2 \theta + \cos^2 \theta = 1$$
$$1 - \cos^2 \theta = \sin^2 \theta$$

You can see that the expression in the numerator is just another (more complicated) way of saying the same thing as the "sine of the angle theta." Now, what do you get when you take the square root of the terms in the numerator and denominator, and then divide them as the fraction that they are? Isn't that the same as dividing the sine by the cosine? And for your answer, you get your good old buddy, the tangent. So if you divide the sine by the cosine to get the tangent, the square of those terms divided by each other gives you the tangent squared, and C is your correct answer. Fill that one in. If you picked one of the other choices, you got carried away with the calculation and made this one more complicated than necessary.

Which of the following is equivalent to $\sec \theta \sin \theta$?

(A) $\cos \theta$

(B) $\csc \theta$

(C) $\tan \theta$

(D) $\cot \theta$

(E) $\dfrac{\sec \theta}{\cos \theta}$

Because the problem asks you to multiply the secant of θ times the sine of θ, give yourself a visual picture of what the cofunctions and the reciprocal functions that are being asked for look like. Ask yourself, how are the secant and sine related to each other? You know that the secant's reciprocal function is the cosine, and the cosine is the cofunction of the sine. Because the secant is the reciprocal of the cosine, you can express the problem like so:

$$\frac{1}{\cos \theta} \sin \theta =$$
$$\frac{\sin \theta}{\cos \theta} = \tan \theta$$

And what do you know! The answer was right there in front of you; that is, if you happened to choose C as the correct one. But unfortunately, the incorrect answers are also right there in front of you. The bad ones are thrown in just to give you extra work to do. For example, choice E is simply one reciprocal function divided by another, resulting in the number 1. You can certainly ignore those red herrings that pose as answers if you keep your wits about you and calmly keep the trig identities in mind. The best way to eliminate wrong answers is to find the right one in cases like this.

Picture This: Graphing

A picture is worth a thousand words, and trigonometric functions are no exception. After you see how trigonometric functions show up on a graph, you have a far better understanding of the repeating nature of these functions. They're called repeating functions, or *periodic* functions, because they repeat themselves in cycles over and over again.

The unit circle

One way to visualize this periodic trait is by looking at a unit circle, which is just another way to define the sine, cosine, and all the other trig functions. The *unit circle* is simply a circle on the coordinate plane with its center at the origin and having a radius of 1. The unit circle is a great way to define and graph the sine and cosine functions in terms of that radius. Figure 9-10 shows a bit how this circle works to graphically show these two functions.

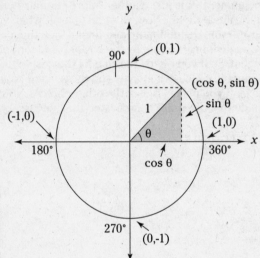

Figure 9-10: The unit circle.

As you can see, the center of the circle is at the origin on the coordinate plane. The angle θ is measured counter-clockwise from the x-axis. A ray from this angle extends to the circle and forms a right triangle. The radius itself is the hypotenuse of the triangle. One leg of the right triangle runs horizontally along the x-axis. Its measurement is cos θ, because it's also the adjacent leg in the right triangle used to calculate the cosine. Because the hypotenuse has a measurement of 1, the adjacent leg divided by the hypotenuse becomes the same measure as the cosine.

The third leg of the triangle is a vertical line starting from where the radius from the angle θ hits the circle and extends down to the x-axis. The vertical leg is exactly the same measurement as the sine, because if you divide this leg (the opposite side) by the radius (the hypotenuse), which has a length of 1, you end up with the sine measurement. The exact point where the hypotenuse lands on the circle has a coordinate position of (cos θ, sin θ).

From the vantage point of the unit circle, you can see how the sine and cosine are related to one another. You know that these two functions are the measurements of complementary angles. When one of the acute angles in the triangle gets bigger, the other acute angle gets smaller, and vice versa. Thus, the sine and cosine of an angle vary directly with each other. Take a look at Figure 9-11 to see how this works.

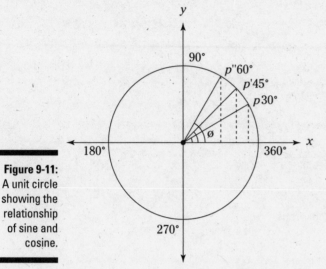

Figure 9-11:
A unit circle showing the relationship of sine and cosine.

Notice that as the angle increases and the radius shifts counterclockwise away from the *x*-axis, the sine increases, while the cosine decreases proportionately. The right triangle having point *p* as the endpoint of its hypotenuse has an ∠ θ that measures 30°. The sine and cosine are 0.5 and 0.866, respectively. Both of these functions are positive numbers.

The cosine should not gloat about being bigger than the sine at this point. As they say, what goes around comes around. The second triangle pictured, having point *p'* as the endpoint of its hypotenuse, has an ∠ θ measuring 45°. Its sine and cosine each measure 0.707. Thus, sine meets cosine at last, and their wits are matched.

The third triangle illustrated in Figure 9-11, with the endpoint of its hypotenuse at point *p''*, has ∠ θ at 60°. The tables have now turned in the battle between the sine and cosine for supremacy, as the sine moves ahead of the cosine. Their measurements are 0.866 and 0.5, respectively. So as the sine increases, the cosine decreases proportionately. They vary inversely with one another.

As the radius begins its journey from the 0° mark, the cosine is at its maximum of 1. At that point, the sine is equal to 0. That's because there is no triangle when the radius hits the circle at 0°. The adjacent side and the hypotenuse are one and the same, and there is no opposite side of a triangle to speak of. The same sort of thing happens when the angle widens and the radius hits the 180 ° mark. Again, there is no triangle, and the cosine at this point is at its opposite extreme of –1, while the sine measures 0. The cosine now has a negative measurement because the value of *x* is to the left of the origin.

The sine and cosine have some important properties about their positive and negative signs when the ∠ θ increases on the unit circle in a counterclockwise manner.

✔ Any angle between 0° and 90° has a positive sine and a positive cosine.

✔ Any angle between 90° and 180° has a positive sine and a negative cosine.

✔ Any angle between 180° and 270° has a negative sine and a negative cosine.

✔ Any angle between 270° and 360° has a negative sine and a positive cosine.

You also know this intuitively because the four quadrants on the coordinate plane determine whether the sign is positive or negative, as shown in Figure 9-12. By convention, the ordered pair for a coordinate point lists the cosine value (x) first, and the sine value (y) second.

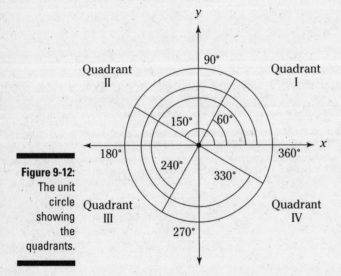

Figure 9-12: The unit circle showing the quadrants.

✔ In quadrant I, the sine and cosine are both positive. The 60° angle has a sine of 0.866 and a cosine of 0.5. The angle intersects the circle at the point (0.5, 0.866)

✔ In quadrant II, the sine is positive and the cosine is negative. The 150° angle has a sine of 0.5 and a cosine of –0.866. The angle intersects the circle at the point (–0.866, 0.5)

✔ In quadrant III, the sine and cosine are both negative. The 240° angle has a sine of –0.866 and a cosine of –0.5. The angle intersects the circle at the point (–0.5, –0.866)

✔ In quadrant IV, the sine is negative and the cosine is positive. The 330° angle has a sine of –0.5 and a cosine of 0.866. The angle intersects the circle at the point (0.866, –0.5)

The unit circle shows the repeating nature of trig functions. In other words, the measure of the sine at 0° is equal to 0 and the cosine measures 1. If you were to move the angle around the circle and come back to the beginning, the measure of the angle is now 360°, but the sine and cosine have the same measurement at this new angle as they had when the angle was 0°. In other words, the sine and cosine vary inversely proportionately with each other as the angle θ rotates around the origin in a counterclockwise manner, but both functions come back to their original reading when the angle hits 360°. The angle looks the same even though it has a difference of 360°. All that work for nothing!

Go ahead and use your calculator to double-check. You know that when ∠ θ measures 60° as in Figure 9-12, the sine measures 0.866 and the cosine is 0.5. Go ahead and plug in some new numbers that add 360° to your 60° angle. The new angle measures 420°, but the sine and cosine are the same as they were at 60°. Add another 360°. Your new angle is 780°, but your sine and cosine remain unchanged. What if you *subtracted* 360° from your initial 60° angle? Your new angle is 300°, but the sine and cosine are the same as ever. Do those functions ever get bored repeating themselves? Actually, not.

Try out a sample question.

If sin 25° = cos θ, then θ could equal

(A) –335°

(B) –295°

(C) –115°

(D) 25°

(E) 395°

Remember that the sine of 25° is the cofunction of its complementary angle. The complementary angle is simply that angle that adds up to 90 with your original angle. So if the original angle is 25°, the complementary angle is 65° because together they add up to 90°. You can check with a calculator (but you shouldn't have to) that the sine of 25° is 0.4226. Coincidentally, this is the same number as the cosine of 65° (the complementary angle). Now, it's child's play to find any angle in your choices that result when you add or subtract a multiple of 360° with your theta angle; that is, 65°. The obvious (and correct) choice here is B, or –295°.

You would have chosen A as your answer if you subtracted only 180° and if you took it away from the wrong angle of 25° instead of subtracting from the complementary angle 65°. If you chose C, you subtracted from the correct angle, but you again subtracted only 180°. If you chose D, you simply picked the wrong angle to begin with and didn't add or subtract anything from it. For shame! And if you ended up with E, you added 360° to the wrong angle. Not much better for your result.

You may be asked a question that tests your knowledge of *radians* instead of degrees as a way of measuring angles. You know that there are 360° in a full circle, so one degree is ⅟₃₆₀ of a circle. There's another way of measuring fractions of a circle besides degrees. Think of a pie where the radius is 1. You cut out a piece where the arc that is intercepted by the angle of that piece of pie is the same length as the radius (see Figure 9-13).

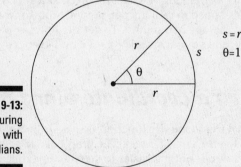

Figure 9-13:
Measuring
with
radians.

The *radian measure* of that angle is the ratio of the length of the arc to the radius. If the intercepted arc and the radius are the same length, the central angle = 1 radian, which is written 1^R. The radian measure is unit-free; that is, it doesn't matter whether you measure the radius of the circle in inches or miles; it's still simply a proportionate measure of the arc to the angle.

You know that the circumference of a circle is equal to $2\pi r$, so the total radian measurement of a circle is 2π radians. The radian measurement of a semicircle is π radians. You can have a little over 6 radians in a circle. This kind of measurement is just a wee bit awkward. You could say there are approximately 6.283185307 radians in a circle (or however many digits you want to carry pi out to). It's much easier (and takes far less paper) to say you have 2π radians in the circle. You can see from Figure 9-12 that an angle measuring 1 radian is about 60°. Figure 9-14 compares radian and degree measurement of angles in a circle.

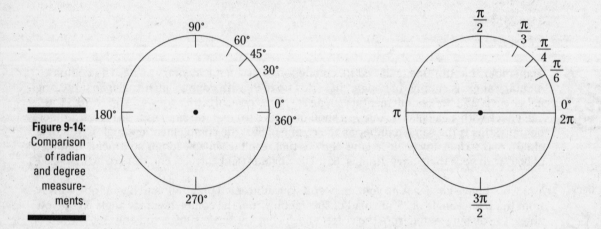

Figure 9-14: Comparison of radian and degree measurements.

Because there is a total of 2π radians in a circle, each increment of 30° is equal to ⅙ π radians. A 90° angle is ½ π radians, and it's written $\pi \div 2$ radians.

You can figure the rest of the radian measures out on your own if you need to. For example, if a 60° angle has a radian measurement of $\pi/3$, what is the radian measure of a 120° angle? That's about as easy as they come. Because 120° is twice as large an angle as 60°, the radian measure will also be twice the original angle. Multiply 2 times $\pi \div 3$, and you get $2\pi \div 3$. It's that simple. A 240° angle is twice again as big as the 120° angle, so its radian measure is $4\pi \div 3$.

See whether you can convert the following measurement. A wheel revolves at a speed of 80-rpm (revolutions per minute). Express this speed in radians per second. You know that 80-rpm = 80 (2π) radians per minute, or 160π radians per minute. To get the number of radians per second, you simply divide 160π by 60, and your answer is $8\pi \div 3 \pi$ radians per second.

Graphing trig functions on a coordinate plane

You may also have to recognize the graph of a trigonometric function on a coordinate plane on the SAT II Math test, but only if you're taking the Level IIC test. The graph of a trig function on a coordinate plane looks a bit like the unit circle, but it's easier to see the repeating nature of trig functions on the coordinate plane. The difference between the two graphs is this. On the unit circle, you measure the angle by how many degrees or radians a radius revolves around the origin at the center of the circle. On the typical coordinate plane, on the other hand, the x-axis itself becomes a numerical measure of the angle. The angle is usually measured in radians. Here are some graphs of trig functions on the coordinate plane.

Figure 9-15 shows the graph of the function $y = \sin x$. The degrees of the angle on this graph are shown by radians running along the x-axis. The measure of the angle starts at 0° or 0 radians. At that angle, the sine is equal to 0 as well. As the angle increases to 90° (that is $x = \pi \div 2$ radians), the sine of x is equal to 1. From its maximum, the sine starts to decrease. The sine becomes 0 where $x = \pi$ radians (that, is, the angle is 180°). The sine reaches its minimum value of –1 when $x = 3\pi/2$ radians, and the angle is 270°. Finally, the function bends back upward as it goes forward, reaching an x-value of 2π radians. The angle at that point is 360°, and the sine has come right back around again to 0, ready to start the whole cycle all over again. Obviously, then, it's a periodic function as it repeats itself and goes on forever in both directions. This function has a period of 2π radians, because it travels a distance of 2π radians and repeats itself again. The sine function has amplitude of 1, because it varies above the y-axis by 1, and goes below the y-axis by 1 as well.

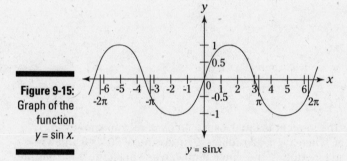

Figure 9-15: Graph of the function $y = \sin x$.

$y = \sin x$

The formula for the amplitude of a periodic function is: $\frac{M - m}{2}$, where M is the y-value at its maximum, and m is the minimum y-value. Thus, to calculate the amplitude of the sine, you plug your minimum and maximum values into this equation: $\frac{1 - (-1)}{2} = 1$

Figure 9-16 shows the graph of a cosine, $y = \cos x$. Notice that it also repeats and has the same period as the cosine. (Remember, the sine and cosine are cofunctions, and they vary indirectly and proportionately with each other.) You can see that the period of the cosine is also 2π radians, just like the sine graph. The graph of $y = \cos x$ starts at a y-value of 1, when $x = 0$. This means that when the angle is 0° or 0 radians, the cosine is equal to 1. Again, just like the sine, the graph of the cosine moves to the right and to the left and it goes on forever, swinging back and forth between its minimum and maximum values of 1 and –1 for the cosine, depending on what the angle, or x-value, is. The cosine function also has an amplitude of 1.

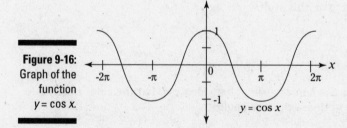

Figure 9-16: Graph of the function $y = \cos x$.

$y = \cos x$

Figure 9-17 shows the graph of a tangent function. This is also a periodic function, but it has a different shape than the sine or cosine graph. Why is that? First of all, the tangent function has a period of only π radians, that is, it repeats itself twice as often as the sine and cosine. Also, you'll notice it goes on forever upward and downward, but each cycle of the tangent has a limit on the x-axis. In the figure, the dotted vertical lines are *asymptotes*, that is, they are values of x where the function does not exist. The asymptote lines represent values that the tangent function approaches, but never quite touches.

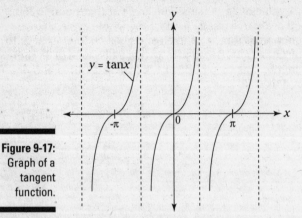

Figure 9-17:
Graph of a
tangent
function.

You may recall that the formula for the tangent is tan θ = sin θ ÷ cos θ. As you know, the cosine has a value of 0 at 90°; that is, when $x = π ÷ 2$ radians. The cosine of 90° is 0 and when you have a 0 in the denominator, the function is undefined. You can see in Figure 9-17 that the tangent function is undefined each time the x-value hits $π ÷ 2$, $3π ÷ 2$, and so on, as well as the negative of those values. The asymptotes are at 90°, 270°, and every 180° after, as well as the negatives of those values.

The main thing to remember about these graphs is that you should be able to recognize the function when you see it for the Level IIC test. If you're taking the Level IC test, fuhgetaboutit!

Above and Beyond: Trigonometry in Other Triangles

So far, you've played with trig functions on right triangles. That's pretty simple for the most part, right? Of course! But you can also use these trig functions to help you figure out measurements of sides and angles on irregular, or *oblique,* triangles as well. Take heart, though, if you're taking the Level IC test, because you don't need to worry about these new rules. Let the Level IIC folks do the heavy lifting on this stuff.

The Law of Sines

The first of these additional rules is the *Law of Sines.* That rule says that the sides of any triangle are proportional to the sines of their opposite angles.

Look at the triangle in Figure 9-18. The formula for the Law of Sines says that for any triangle *ABC*, where *a*, *b*, and *c* are the sides opposite of their corresponding angles, the following is true.

$$\frac{\sin A}{a} = \frac{\sin B}{b} = \frac{\sin C}{c}$$

You can also say

$$\frac{a}{b} = \frac{\sin A}{\sin B}, \frac{b}{c} = \frac{\sin B}{\sin C}, \frac{a}{c} = \frac{\sin A}{\sin C}$$

Figure 9-18:
Triangle
demonstrat-
ing the Law
of Sines

The Law of Sines comes in handy when you want to find out measurements of a triangle and you already know the following:

✔ One side and two angles, or

✔ Two sides and an angle opposite one of them

Try the following exercise to see how it works: In Figure 9-19, triangle *ABC* has $\angle A = 26°$, $\angle B = 87°$, and side $a = 275$. To find the missing angle and sides, go in two steps:

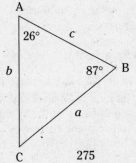

Figure 9-19:
Triangle
ABC.

First, find out the missing angle. That's simple enough. Add the two known angles and subtract from 180.

$$180 - (26 + 87) = 67$$

Next, use just part of the Law of Sines:

$$\frac{\sin A}{a} = \frac{\sin B}{b}$$

$$\frac{\sin 26°}{275} = \frac{0.9986}{b}$$

$$(0.4384)\,b = (0.9986)(275)$$

$$(0.4384)\,b = 274.6$$

$$b = \frac{274.6}{0.4384}$$

$$b = 626.4$$

You now know that side $b = 626.4$. You can use the rest of the Law of Sines to include the relationship between the sine of $\angle C$ and side c. Here you go.

$$\frac{\sin A}{a} = \frac{\sin C}{c}$$

$$\frac{\sin 26°}{275} = \frac{\sin 67°}{c}$$

$$\frac{0.4384}{275} = \frac{0.9205}{c}$$

$$(0.4384)\,c = (0.9205)(275)$$

$$(0.4384)\,c = 253.1$$

$$c = \frac{253.1}{0.4384}$$

$$c = 577.4$$

And now you have your final side: $c = 577.4$.

The riddle is solved. Maybe this stuff isn't so tricky after all.

The Law of Cosines

If they were all that easy, you could take a nap right now. But sometimes you don't have all the 411 to invoke the Law of Sines. Related to the Law of Sines is the Law of Cosines. Why does that sound so coincidental? And if you're perfectly happy taking only the Level IC exam, you can go and get that cup of coffee and skip over this material.

You know that the Pythagorean theorem gives the formula for a right triangle: $a^2 + b^2 = c^2$.

If you don't have a right triangle, though, don't fret. You can always use the Law of Cosines, which is really quite similar to the Pythagorean theorem. The *Law of Cosines* says that for every triangle, the square of the length of any one side equals the sum of the squares of the lengths of the other two sides decreased by twice the product of these two sides and the cosine of their included angle. That's a fancy way of saying the following formula.

$$c^2 = a^2 + b^2 - 2ab \cos C$$

If you had a right triangle, the cosine of $\angle C$ would be equal to 0 and the last term in this equation would simply cancel itself out to match the Pythagorean theorem.

Figure 9-20 labels the parts of the triangle for the Law of Cosines equation. The angles are *A*, *B*, and *C*. The sides are labeled *a*, *b*, and *c*. You can use the Law of Cosines if you know three of the variables in the formula. From a practical standpoint, the Law of Cosines comes in handy when:

 ✔ You know three sides of the triangle, or

 ✔ You know two sides of the triangle and their included angle

See whether you can find the length of side *c* in Figure 9-20. You know the length of two sides *a* and *b*, as well as the angle *C*, so it's just a matter of finding the cosine of *C*, and then plugging in the numbers from the preceding formula.

$$c^2 = a^2 + b^2 - 2ab \cos C$$

$$c^2 = 8^2 + 13^2 - 2\,(8)\,(13) \cos 65°$$

$$c^2 = 64 + 169 - 208\,(0.4226)$$

$$c^2 = 233 - 87.9$$

$$c^2 = 145.1$$

$$c = 12.05$$

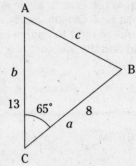

Figure 9-20:
Triangle
demonstrat-
ing the Law
of Cosines.

Now doesn't that third side look like it's just about 12.05 long? Of course it does, especially when you compare it with the other two sides.

Try a sample from the Level IIC test and use the Law of Sines and the Law of Cosines.

Determine the measure of the ∠ *B* in △ *ABC* where sides *a* = 5, *b* = 7, and *c* = 8 are the sides opposite ∠ *A*, *B*, and *C* respectively.

(A) 38

(B) 60

(C) 82

(D) 88

(E) It cannot be determined from the information given.

First, draw a picture and label the parts that you know, something like Figure 9-21.

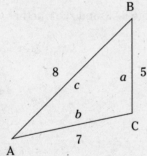

Figure 9-21:
Sample
picture.

Use the Law of Sines to determine the largest angle, namely *C*. (You know it's the largest angle because it's opposite the longest side.)

$$2ab\cos C = a^2 + b^2 - c^2$$
$$\cos C = \frac{a^2 + b^2 - c^2}{2ab}$$
$$\cos C = \frac{5^2 + 7^2 - 8^2}{2(5)(7)}$$
$$\cos C = \frac{25 + 49 - 64}{70}$$
$$\cos C = \frac{10}{70} = \frac{1}{7} = 0.142857$$
$$C = 81.8$$

You have found the measure of $\angle C$. Go ahead and round up to the nearest angle and you get 82°. Because C is the largest angle, the remaining angles must be acute angles, right? Use the Law of Sines to find at least one of them. Of course, you could go right to finding out what $\angle B$ is equal to by plugging it into that rule. Why not try and find the measure of $\angle A$ first?

$$\frac{\sin A}{a} = \frac{\sin C}{c}$$
$$\sin A = \frac{a \sin C}{c}$$
$$\sin A = \frac{5(0.990268)}{8} = \frac{4.9513}{8} = 0.6189$$
$$A = 38.2$$

You've figured out $\angle A = 38.2°$, which you can round down to a reasonable 38°. Now, it's a simple matter of subtracting $180 - (82 + 38) = 60$. Piece of cake when it comes right to getting this stuff, right? These kinds of problems will be a snap for you if you follow the few simple rules for the Law of Sines and the Law of Cosines. Make sure your calculator has good batteries.

Matching Snow Suits: Polar Coordinates

There's just one other little bit of trigonometry you should get comfortable with before moving on. If you cherished the coordinate plane in Chapter 7, you'll adore reviewing the concept of polar coordinates. It's very similar to the coordinate plane with just a couple of extra twists. Also, if you're not going to sit through the Level IIC test, you can skip around to some other stuff that's more relevant to your needs.

Labeling polar coordinates

Take a look at the point P on the coordinate plane in Figure 9-22, and see whether you can determine what the coordinate position is for this point.

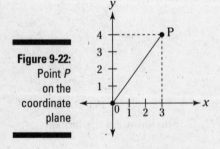

Figure 9-22:
Point P
on the
coordinate
plane

The conventional wisdom says you should label this point according to standard ordered pair notation (3, 4), listing the x-coordinate first and the y-coordinate second. If you were on the rectangular coordinate plane, you would be going to the head of the class. The polar coordinate is slightly different. Any point on a plane can be identified by its distance from the origin (r) and its angle of inclination (θ). The *polar coordinate* lists as the number in the first coordinate position the *distance* that point P is from the origin. The second coordinate number in the polar coordinate is the *angle* of a ray that would travel to that point with respect to the x-axis.

In Figure 9-22, you can use the Pythagorean theorem to figure out rather quickly that the distance from the origin to point P is 5. You get this, of course, from the equation $a^2 + b^2 = c^2$

$$3^2 + 4^2 = 5^2$$

Now, figure out the angle that points sits at in relation to the x-axis. You know the sine of that angle is $\dfrac{opposite\ side}{hypotenuse} = \dfrac{4}{5} = 0.8$.

With a sine of 0.8, you use the $\sin^{-1}$ key on your trusty calculator to figure out that the actual angle of the ray going from the origin to the point is approximately 53.13°. You can round this number to the nearest degree for purposes of this section, unless you are told otherwise. With an angle of 53°, your polar coordinate for point P is (5, 53).

Expressing polar coordinates in radians

Polar coordinates can express the angle measurement in degrees or radians. Look at another polar coordinate to see how the latter designation works.

In Figure 9-23, you can see that the rectangular coordinates for point P are (5, 5). The Cartesian (rectangular) coordinates can be converted to polar coordinates by first listing the length of the ray that goes to the point, and then the angle of that ray second. You can easily determine that the *angle* is 45°, or $\pi/4$ radians. But what is the length? Go with the Pythagorean theorem on the x- and y-values to find out the length of the hypotenuse of the triangle.

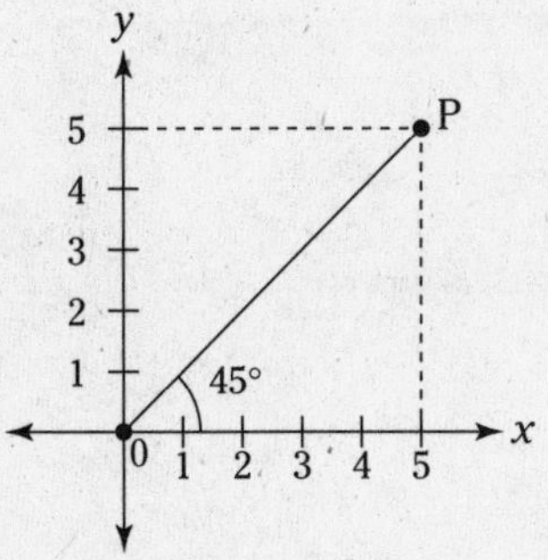

Figure 9-23:
Point with rectangular coordinates of (5, 5).

$$5^2 + 5^2 = c^2$$
$$25 + 25 = c^2$$
$$50 = c^2$$
$$c = 5\sqrt{2} = 7.07$$

Of course, because the x- and y-values are equal, you could have gone with the formula for the diagonal of a square with its side having a length of s. The diagonal is $s\sqrt{2}$. This gives you a polar coordinate of (7.07, 45). If you use radian measure for the angle, you end up with a polar coordinate of (7.07, $\pi \div 4$).

Suppose you have an angle that is less than 0°. Take a look at Figure 9-24 to see how that plays out.

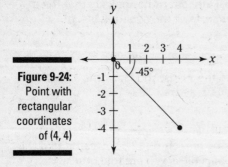

Figure 9-24:
Point with rectangular coordinates of (4, 4)

In Figure 9-24, you have an angle of –45°, with rectangular coordinates of (4, –4). The polar coordinate is therefore a bit different from the last one. The distance from the origin to the point *P* is 4 times the square root of 2, or 5.66. Your polar coordinate is (5.66, –45). In radians, it's (5.66, –π/4).

Suppose you find a point that is the same distance from the origin as that in Figure 9-24, but instead has a positive angle that is 180° larger than the angle shown. Figure 9-25 illustrates this scenario.

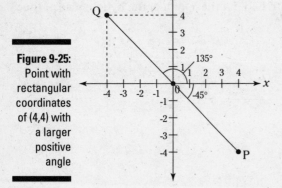

Figure 9-25:
Point with rectangular coordinates of (4,4) with a larger positive angle

Now, you have two coordinate points, *P* and *Q*. The rectangular coordinates for *P*, you recall, are (4, –4), while *Q* has coordinates of (–4, 4). You can't call these points anything else in the Cartesian scheme of things. Notice that both points are the same distance from the origin. *P* has polar coordinates of (5.66, –45), and you could say that *Q* has polar coordinates of (5.66, 135). But imagine you could call these points something else in the coordinate system. Don't go too far, because you can actually assign other values to these two points on the polar coordinate plane and still keep them right where they are.

Because these two points *P* and *Q* are 180° apart and lie the same distance from the origin, you can think of them as negatives of each other. Point *P* can be thought of as (5.66, –45), or (5.66, –π/4) in radian measure. But point *P* can also be (–5.66, 135). This means that it goes a *distance* of 5.66 in the *opposite* direction from where it would go if it were at a positive angle of 135° from the *x*-axis. You have essentially given the distance a negative value away by going in the opposite direction. The radian measure of this alternative polar coordinate designation is (–5.66, 3π ÷ 4).

Likewise, point *Q* has its polar coordinate quirks. You can call it (5.66, 135), or (5.66, 3π ÷ 4) in radian measure. You can also use the alternate designation of (–5.66, –45), or (–5.66, –π ÷ 4) radian measure. Either way, you end up with the same point *Q*.

Just as the unit circle allows you to add or subtract 360 degrees (or 2π radians) to or from an angle and wind up in the same place, you can do the same thing with the polar coordinates. So point P in the preceding example can also have polar coordinates (5.66, 315) or (5.66, 7π ÷ 4).

You can add 360° (or 2π radians) to point Q's angle measure of 135° and give it polar coordinates (5.66, 495), or (5.66, 11π/4) in radians. You could even give point Q the polar coordinates that assign its location in relation to point P. That is to say, you could say point Q has polar coordinates of (–5.66, 315), or (–5.66, 7π ÷ 4) in radians. The possibilities are limitless.

Converting polar coordinates

You may also want to know how to convert values of polar coordinates to rectangular coordinates and vice versa. Just use your basic trig functions, and you should be able to figure these out. Remember to draw a picture whenever possible and delegate the x-value in the coordinate plane as your *adjacent* side, and call the y-value your *opposite* side for trig function and conversion purposes.

Use a drawing such as in Figure 9-26 and fill in the values where you can. The trig functions you will use most often are $\sin\theta = \frac{y}{r}$, $\cos\theta = \frac{x}{r}$, $\tan\theta = \frac{y}{x}$.

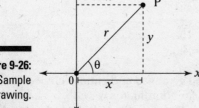

Figure 9-26: Sample drawing.

To convert from polar coordinates into rectangular coordinates, manipulate your trig functions accordingly:

$x = r\cos\theta$ $y = r\sin\theta$

To convert from rectangular to polar coordinates, liberally use the Pythagorean theorem and your $\tan^{-1}$ key. The following formulas should work for these problems

$r = \sqrt{x^2 + y^2}$, $\theta = \arctan\frac{y}{x}$

You may want to try a sample question to whet your appetite for the Level IIC test.

Convert polar coordinate (–5, π ÷ 6) to rectangular coordinates.

(A) (–4.3, –2.5)

(B) (–2.5, 4.3)

(C) (–3.5, –3.5)

(D) (2.5, 4.3)

(E) (4.3, 2.5)

It's always best to draw a picture when working with graphs. It's so much easier to visualize these problems on paper than in your head. There's an old saying: A picture is worth a thousand words.

Draw something similar to Figure 9-27. This problem really isn't too bad after you get the picture down on paper. You know the length of *r*, and you know θ is π/6 radians, which is another way of saying the angle measures 30°. Now go ahead and plug in values for the sine and cosine to convert the polar coordinates.

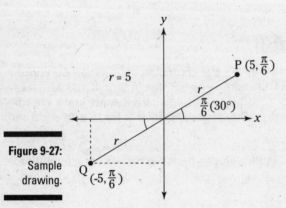

Figure 9-27: Sample drawing.

$$\sin\theta = \frac{y}{r} \qquad \cos\theta = \frac{x}{r}$$

$$\sin 30 = \frac{y}{5} \qquad \cos 30 = \frac{x}{5}$$

$$0.5 = \frac{y}{5} \qquad 0.866 = \frac{x}{5}$$

$$y = 2.5 \qquad x = 4.33$$

Armed with your knowledge of the *x*- and *y*-values, you go ahead and choose choice E, correct? Wrong! This would be an okay answer if your polar coordinates were (5, π/6), but the problem asks you to convert the coordinates (–5, π/6). The –5 in the polar coordinate tells you that you want to go 180° away from point *P* in Figure 9-27. You can figure out the rectangular coordinates for *P*, but realize that point *P* is in the first quadrant, and your answer is 180° in the opposite direction. If you turn tail and move 5 units in the opposite direction, you arrive at point *Q*. Now you're in Quadrant III, and the rectangular coordinate has negative values for *x* and *y*. Your correct answer is A.

If you chose B, you got your sine and cosine functions backward. If you drew a picture to help you out, you would know that choice B is wrong because the ordered pair doesn't add up to where your graph ends up. If you wound up with choice C as your answer, you used the wrong angle for your calculations. You would have based your sine and cosine on a 45° angle (π/4 radians) instead of the 30° angle the question asked about. Answer D is even worse than answer E. Not only is this answer choice in the wrong quadrant, but it also has the sine and cosine functions backward. Drawing a picture also helps you check your work.

Part IV
Highly Unlikely: Statistics, Probability, and Sets

The 5th Wave   By Rich Tennant

"I wish you'd practice for the SAT II Math test on your own time..."

In this part . . .

It's decidedly improbable that more than 10 percent of your Level IC or Level IIC test will cover statistics and probability. But you want to get every possible point, so this part reviews the highlights of data interpretation, probability, and sets, from the essential concepts of mean and mode to more complex calculations of standard deviation and probability. Our computations predict that you will have a high degree of success on these questions after you study Part IV.

Chapter 10

Playing the Numbers: Statistics and Probability

. . . .

In This Chapter

▶ Apprehending averages

▶ Perceiving probability

▶ Scrutinizing standard deviation

. . . .

The SAT II Math test is bound to have at least one or two questions related to probability and statistics. The science of statistics involves organizing, analyzing, and interpreting data in order to reach reasonable conclusions and to help in decision-making. While you will most likely encounter very few questions involving statistics, you no doubt may run across the occasional query asking you to determine probability or some type of statistical average or variation from the average. The questions you'll encounter in this area are not too difficult, but it pays to give this subject a cursory review.

The Level IC test is more likely to include questions about probability, mean, median, and mode. Level IIC is more likely to test you on standard deviation.

Finding a Middle Ground: Mean, Median, and Mode

To evaluate data correctly, you should know the central tendency of numbers and the dispersion of their values. Mean and mode give you tendency and range provides you with dispersion information.

Much of statistics involves measurement of the *central tendency* of numbers. A measurement of *central tendency* is a value that is typical, or representative, of a group of numbers or other information.

Common tools for describing a central tendency follow.

✔ Mean (the arithmetic mean)

✔ Median

✔ Mode

✔ Weighted mean

✔ Geometric mean

Analyzing the arithmetic mean

The most common of all of these is the arithmetic mean, or simply the *mean*. This is what people are talking about when they say the *average*.

The important concept to remember about mean is the following formula: Mean average = the sum of all numbers divided by the amount of numbers that make up the sum.

While people commonly refer to the mean as the *mean average,* it does not "mean" the SAT II Math testers are *mean* by any means.

Take a peek at this example.

Sara tried to compute the mean average of her 8 test scores. She mistakenly divided the correct sum of all her test scores by 7, which yields 96. What is Sara's mean test score?

(A) 80

(B) 84

(C) 96

(D) 100

(E) 108

Because you know Sara's mean test score when divided by 7, you can determine the sum of Sara's scores. This information will then allow you to determine her mean average over 8 tests.

Apply the average formula to Sara's mistaken calculation. (To review the average formula, check out Chapter 3.)

$$96 = \text{sum of scores} \div 7$$

$$96 \times 7 = \text{sum of scores}$$

$$672 = \text{sum of scores}$$

Now that you know Sara's test score sum, you can figure her true mean average.

$$\text{Mean average} = 672 \div 8$$

$$\text{Mean average} = 84$$

You know that her average must be less than 96 because you are dividing by a larger number, so you can automatically eliminate C, D, and E. The correct answer is B.

Mastering medians

The *median* is another type of average you may see on the SAT II Math test. The *median* is a value among a list of several values or numbers that falls exactly in the middle. To find out the median, put the values or numbers in order, usually from low to high. You can think of the median income of a community as, for example, $48,000. This means that half the people's income is below that amount, and half is above. For an odd number of values, just select the middle value. If there is an even number of values, just arrange the numbers as before and determine the value halfway between the two middle values.

Manipulating modes

The *mode* is the other common type of average you may encounter on the SAT II test. The *mode* is the value that occurs most often in a set of values. Questions about mode use a word such as *frequency* or "how often" something happens. An example of mode could be something like income (again), where you look to the amount of income that occurs most frequently in a given population or sample. Maybe more people in the population or sample have an income of $30,000 than any other amount of income by others. In that case, you'd say the mode is $30,000.

The mean, median, and mode are usually different numbers. If they were all the same, the numbers would fall into a bell curve pattern as shown in Figure 10-1.

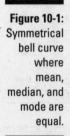

Figure 10-1: Symmetrical bell curve where mean, median, and mode are equal.

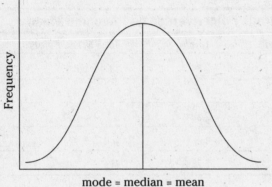

mode = median = mean

As you can see, the mean average of the bell curve pictured falls right in the middle. If you were to add up all the values and divide by the total number of values, the mean average would be right the in the middle of the curve. This amount is also the mode because the greatest number of any one value falls directly in the middle. It's also the median because there is an equal amount of numbers on either side of the center of the curve. This bell curve is *symmetrical,* meaning it's the same shape on both sides, and there is an even distribution of numbers throughout the entire curve.

If the different measures of central tendency — mean, median, and mode — were spread out unevenly, then you may say the results are *skewed,* as in Figure 10-2.

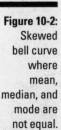

Figure 10-2: Skewed bell curve where mean, median, and mode are not equal.

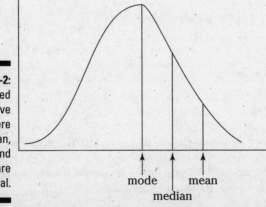

mode

median

mean

Figure 10-2 is a bell curve where the results are *skewed to the right,* or *positively skewed.* It's skewed to the right, because it trails off to the right. Guess what a curve would be called if the results were in the other direction. If you said it was *skewed to the left*, or *negatively skewed*, then you would get a gold star! It's important to be able to recognize these relationships between numbers for statistics on the test.

Wondering about weighted means

A *weighted mean* is where each individual value is multiplied by the number of times it occurs, and then the sum of these products is divided by the total number of times they all occur. In plain English, you may look at an example of grade point average (GPA). Suppose Becky's grades in classes with the amount of credits specified appear in Table 10-1:

Table 10-1	Grade Point Average As Weighted Mean		
Class	*Number of Credits*	*Grade*	*Total Grade Points*
Statistics	5	3.8	19
English	5	1.9	9.5
Speech	4	2.3	9.2
Bowling	1	4.0	4
Total	15	2.78 GPA	41.7

Like any weighted mean, in this example, you multiply the individual values (grades) by the number of times they each occur (credits) to get total grade points. Then you divide the total of these products (total grade points: 41.7) by the total number of times they all occur (total credits: 15). This gives you

$$41.7 \div 15 = 2.78 \text{ GPA}$$

Getting the geometric mean

The *geometric mean* is simply the average of relative numbers, percents, index numbers, growth rates, and so on. You should know what a geometric mean is, but you won't need to know how to calculate it. If you must *know* how to calculate it, however, the formula for a geometric mean is:

the *n*th root of the product of *n* values

All Have Strayed: Range and Standard Deviation

Besides knowing the main concepts of central tendency, you should also know about *variation* or *dispersion* of values in statistics. While the mean, median, and mode can tell you something about central tendency, *dispersion* tells you how spread out the values are from the center. If dispersion is small, the values are clustered around the mean. But a wide dispersion of values tells you that the mean average is not a reliable representative of all the values.

Scouting out the range

The easiest measure of dispersion is the *range*. The *range* is simply the biggest value minus the smallest value in the data. You can say that the range is the *difference* between the highest and lowest values in the set of data. For example, if the highest test score in the math class was 94% and the lowest was 59%, subtract the low from the high to get the range.

$$94 - 59 = 35$$

The range of test scores is 35. Simple as that!

The range of values in statistics can either come from a *population* or a *sample*. The *population* is the set of all objects or things; that is, the total amount of all data considered. A *sample* is just a part of the population.

From the set of numbers 47, 63, 53, 39, 72, 53, 54, and 57, what is the range?

(A) 8

(B) 53

(C) 39

(D) 72

(E) 33

The range of any set of numbers is the value of the greatest element minus the value of the least element. The biggest number here is 72, and the smallest is 39. The difference is 72 – 39 = 33.

Your correct answer for the range is E. A is the number of the elements in the given set, not the range. B is the mode, not the range. C is the least element, not the range. D is the greatest element, not the range.

Watching out for wanderers: Standard deviation

Another form of variation or dispersion you need to know about on the SAT II Math test is *standard deviation*. The standard deviation expresses variation by measuring how spread out the distribution is from the mean. While the range gives you an idea of the total spread, the standard deviation is a more reliable indicator of dispersion as it considers all the data, and not just the two on each end. The standard deviation is the most widely used number for expressing how much the data is dispersed from the mean.

As an example, suppose you get a grade of 75 on a test where the mean grade is 70 and the vast majority of all the other grades fall between 60 and 80. That's one heckuva lot better than if you get a 75 on the same test with the mean grade being 70 and most of the grades falling between 45 and 95. In the first situation, the grades are more tightly clustered around the central tendency. A standard deviation in this case is a small number. It's harder to get a score that's a standard deviation away from the mean. Your grade is higher compared to all the other test takers in the first group than your grade would be in the second scenario. In the second scenario, the standard deviation is a bigger number than it is in the first group, and a grade of 75 is not so hot.

Finding variance

Finding out the standard deviation is a two-step process, and one that you probably won't need to actually calculate for the SAT II Math test. The first step is to find the *variance*. The *variance* is the mean (average) of the squared deviations between each value in the group and the mean. You see, it's already starting to sound Greek. That's why the variance is expressed with the Greek letter sigma squared, or σ^2. Take heart, because you don't need to know the formula for finding the variance for the SAT II Math test. It involves way too many weird symbols, anyway.

Calculating the deviation

After you find out the variance, though, if you really *had* to, you calculate the *standard deviation* as the positive square root of the variance. If the standard deviation for a population is σ (the Greek lowercase letter sigma), the formula for the standard deviation is

$$\sigma = \sqrt{\frac{\sum(X-\mu)^2}{N}} \quad \text{or} \quad \sigma = \sqrt{\frac{\sum X^2}{N} - \left(\frac{\sum X}{N}\right)^2}$$

Well? We told you you wouldn't like this. You can always bone up on your stats book if you want to find out more about calculating the variance and standard deviation. But because you'll see so few questions on this subject at all on the SAT II test, you may want to concentrate on other areas of the test instead.

However, it's a good idea to be able to recognize that with a symmetrical bell curve, the relationship between the standard deviation and the mean is pretty scientific.

Figure 10-3 shows a bell curve and the distribution of the standard deviations away from the mean. The mean appears as an *X* with a line over it, and it shows the average value is right in the middle of all the values. If you stray 1 standard deviation in either direction from the mean, you'll have netted 68% of all the values. Going another standard deviation away from the center, you pick up another 32% of all values, giving you about 95% of all values. Finally, when you go ± 3 standard deviations from the mean, you now have about 99.7% of all the values in your population or sample.

Figure 10-3:
Distribution
of the
standard
deviations
from the
mean.

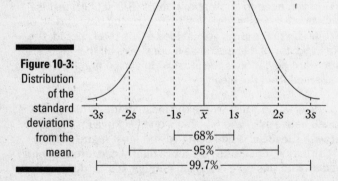

If this were a group of test scores, *over a majority* of test takers scored right within 1 standard deviation of the mean. The *vast majority* score within 2 standard deviations, and *virtually everyone* scores within 3 standard deviations. Say that the mean test score is 80, and one standard deviation may be 10 points on either side. This means that 68% of the students score between 70 and 90. If the second standard deviation was another 5 points in either direction, then you could say that 95% of the students score between 65 and 95 on the test. Finally, you could say that the third standard deviation is another 4 points away from the mean. This means that 99.7% of the students score between 61 and 99. That about sums it up for standard deviations.

A small value for the standard deviation means that the values of the group are more tightly clustered around the mean. A greater standard deviation means that the numbers are more scattered away from the mean. The more standard deviations there are in a group, the easier it is to deviate from the center. The less standard deviation there is, the harder it is to deviate from the center.

There's a formula to determine what percentage of a group lies within a particular standard deviation of the mean. For any set of observations, the minimum proportion of the values that lie within k standard deviations of the mean is at least $1 - \frac{1}{k^2}$.

At least 75% of the total number of values in a set lie within 2 standard deviations. Try the formula out with 2 standard deviations (that is, 1 standard deviation on either side of the mean).

$$1 - \frac{1}{k^2} = 1 - \frac{1}{2^2} = 1 - \frac{1}{4} = \frac{3}{4} = 75\%$$

Try this with 4 standard deviations (2 on either side of the mean).

$$1 - \frac{1}{k^2} = 1 - \frac{1}{4^2} = 1 - \frac{1}{16} = \frac{15}{16} = 93.75\%$$

It Could Happen: Probability

Probability is the measure of how likely it is that a particular event will occur. It's a bit more scientific than fortune telling and reading tarot cards. You express probability as a percent or decimal. Thus, you'd say that the probability of an event's occurring falls between 0% and 100%, or between 0 and 1. If the probability of an event's occurrence is 0 or 0%, it's impossible for the event to occur. If the probability is 1 or 100%, the event is certain to occur. Few things in life are as certain as death and taxes. It's also rare to assume that an event is impossible as well. While it may not be *likely* or *probable* for a certain baseball team to win the World Series, it's not necessarily *impossible*. The probability of the occurrence of an event is usually somewhere between 0 and 1.

Finding the probability of one event

Probability deals with *outcomes* and *events*. For situations where all possible outcomes are equally likely, the probability (P) that an event (E) occurs, represented by "P (E)," is defined as

$$P(E) = \frac{\textit{The number of outcomes involving the occurrence of } E}{\textit{The total possible number of outcomes}}$$

Because probability is expressed as a fraction, it can never be less than 0 or greater than 1.

What is the probability of getting both heads and tails with the flip of a coin? Because that's an impossible outcome, the probability of that particular event's occurring is 0. Now suppose you flip a coin with heads on both sides. What is the probability of getting heads on one flip? The probability is 1, because the number of possible outcomes is exactly the same as the number of outcomes that will occur.

Finding the probability of many events

There are rules of probability for *multiple events* happening. Consider a case where there are two possible outcomes of events, A and B. You can represent the probability of *one* of the events, A or B, occurring as

$$P (A \text{ or } B)$$

What is the probability that (A or B) will occur? There are two ways to state this.

- **P (A or B) = P (A) + P (B):** This is the *special rule of addition*. In order for this rule to work, two events have to be mutually exclusive.

- **P (A or B) = P (A) + P (B) – P (A and B):** The *general rule of addition*, on the other hand, assumes that the two events are not mutually exclusive.

The special rule of addition

An example of the special rule of addition is rolling a die and coming up with either a 1 or a 2. You can't get both on one roll. Thus, the probability of rolling a 1 or a 2 in one roll is: $\frac{1}{6} + \frac{1}{6} = \frac{2}{6} = \frac{1}{3}$.

The general rule of addition

For the general rule of addition, imagine there are three types of sodas in a cooler. Colas are numbered consecutively 1 through 5, orange sodas are numbered 1 through 7, and grape sodas are numbered 1 through 8. Let event A stand for when a cola is taken out, and let event B represent when a can with a number 2 is withdrawn. You want to know the probability of picking out *either* a cola *or* a can with the number 2 on it, but *not* a cola can with the number 2 on it. The probabilities would be

P (A) = 5 ÷ 20 (5 of the 20 cans are colas)

P (B) = 3 ÷ 20 (3 of the 20 cans are numbered 2)

P (A and B) = 1 ÷ 20 (only 1 of the 20 cans is a cola can numbered 2)

P (A or B) = $\frac{5}{20} + \frac{3}{20} - \frac{1}{20}$ = 7 ÷ 20

You can express this probability as 0.35 or 35%. The part of the above scenario that shows the situation of *both* A's and B's occurring is as follows: P (A and B)

The probability of both events happening is called *joint probability*.

The special rule of multiplication

The probability of multiple events occurring together is the product of the probabilities of the events occurring individually. This is the *special rule of multiplication*. This rule assumes that the events are independent of one another.

So, if you're rolling an individual die, what is the probability of rolling a 1 *and* 2 in two separate rolls?

$$\frac{1}{6} \times \frac{1}{6} = \frac{1}{36}$$

The general rule of multiplication

Suppose the outcome of the second situation depends on the outcome of the first event. You then invoke the *general rule of multiplication:*

$$P \text{ (A and B)} = P \text{ (A)} \times P \text{ (B/A)}$$

In this case, the term P (B/A) is a *conditional* probability, where the likelihood of the second event depends on the fact that A has already occurred. For example, what are the odds of drawing the Ace of Spades from a deck of 52 cards on one try and then, while holding the ace, pulling the King of Spades on the second draw? Here is how it would play out.

$$P \text{ (A and B)} = P \text{ (A)} \times P \text{ (B/A)}$$
$$P \text{ (A and B)} = \tfrac{1}{52} \times \tfrac{1}{51}$$

You have to slightly increase the probability of drawing the King of Spades on the second draw because you have already removed one card from the deck on the first draw. Your answer is

$$\tfrac{1}{52} \times \tfrac{1}{51} = \tfrac{1}{2,652}$$

Don't bet against the house on that one! Try a sample problem from the SAT II Math test.

A sack of marbles contains 3 blue, 2 red, 7 yellow, and 1 purple marbles. Two children blindly take 2 marbles each out of the sack. The first child returns her marbles to the sack before the second child draws. What is the chance that the second child draws both red marbles?

(A) $\tfrac{2}{169}$

(B) $\tfrac{1}{13}$

(C) $\tfrac{2}{13}$

(D) $\tfrac{1}{156}$

(E) $\tfrac{1}{78}$

You need to treat the drawing 2 red marbles as 2 events: drawing the first red marble, and then drawing the second red marble. The chances of drawing 1 red marble is 2 (the number of red marbles) divided by 13 (the total number of marbles in the sack), or $\tfrac{2}{13}$. When the child draws the second marble, one red marble is already gone. Now there is only 1 red marble in a sack of 12 marbles, so the chance of drawing it is $\tfrac{1}{12}$. The probability of both events happening is the product of the probability of each event occurring.

$$\tfrac{2}{13} \times \tfrac{1}{12} = \tfrac{2}{156} = \tfrac{1}{78}$$

Your correct answer choice is E. Choice A is $\tfrac{2}{13} \times \tfrac{1}{13}$ and reflects a failure to subtract the withdrawn red marble from the total number on the second draw. Choice B is the chance of drawing one red marble from a sack of 13 marbles containing only 1 red marble. In this question, it would also be the chance of drawing the purple marble. If you picked C, you were going for the chance of drawing the first red marble. Choice D is what you get if you came up with only $\tfrac{1}{156}$ instead of $\tfrac{2}{156}$ or $\tfrac{1}{78}$, and it's also incorrect.

Remember not to get too hung up on statistics. You'll probably have no more than a handful of statistical questions on either the Level IC or IIC tests.

Chapter 11

Sticking Together: Sets

• •

In This Chapter
▶ Grasping group problems
▶ Understanding unions and intersections
▶ Perceiving permutations
▶ Comprehending combinations
▶ Facing factorials

• •

*F*rom the time you mastered the ability to tie your shoes in kindergarten, you had to learn to work and play together in groups. That's a good thing! In this chapter, you understand how important it is to get a grip on groups. The techniques are not really earth-shattering. In fact, you could probably figure out the answer to most of these questions on your own given enough time. But of course, you don't have all the time in the world on the SAT II Math test. On the actual test, you may not encounter even a handful of problems that relate to the concepts in this chapter. But you either learn or review some of the wonderful shortcuts that give you enough of an edge to grab at least a few extra points on the test for those set questions you do come across.

Numerical Socialization: Grouping

When you see a question about groups on the SAT II Math test, don't flinch. Or, worse yet, don't become antisocial with the group by fleeing in the other direction. Look that group problem right smack in the eye and say, "Eureka!" or whatever else you may say in case you're not fluent in Greek.

The key to finding the answer to group problems on the SAT is just about the same as if you were counting out the answer. But counting is time consuming, and you want to be working smarter, not harder, to solve these questions. It all comes down to simple arithmetic and nothing else.

Group problems come in the form of populations of persons or objects. In the course of the problem, different parts of the population are grouped together in various categories. The questions generally ask you either to find the total or determine how many people or objects make up one of the subgroups.

TIP

The formula for solving group problems is

> Group 1 + Group 2 − Both Groups + Neither Group = Grand Total

You may see a problem something like the following.

The charter academy has exactly 110 students. 47 students are enrolled in Physical Education, while 56 students are enrolled in Driver's Training. 33 students are enrolled in both Physical Education and Driver's Training classes. How many students are not enrolled in either Physical Education or Driver's Training classes?

(A) 24

(B) 28

(C) 30

(D) 33

(E) 40

This question really isn't that tough, but it can throw you off guard if you aren't ready. In a pinch, you can try counting on your fingers, but it begins to get more difficult when you have to start using your toes for backup. After all, they may not let you remove your shoes when you're sitting there taking the SAT. There is an easier way, though, and you may not even need a calculator.

Go ahead and plug your numbers into this formula to set up your equation. Your variable (unknown) will be the group that doesn't take either Physical Education or Driver's Training classes.

$$\text{Group 1} + \text{Group 2} - \text{Both Groups} + \text{Neither Group} = \text{Grand Total}$$

$$47 + 56 + \text{Neither Group} - 33 = 110$$

The equation tells you that there are 47 in group 1 (taking Physical Education), 56 in group 2 (taking Driver's Training), and 33 in the group taking both classes. You also know there are 110 students total, so you need to find out how many students aren't taking either class.

$$47 + 56 + x - 33 = 110$$

$$70 + x = 110$$

$$x = 40$$

Now that wasn't so bad, was it? Try a couple more.

Of all taxpayers in the United States, one-third of them can deduct charitable contributions on their federal income tax returns, while 40% of all taxpayers are able to deduct state income taxes that they pay from their federal returns. If 55% of all taxpayers can deduct neither charitable contributions nor state sales tax, what portion of all taxpayers can claim both types of deductions?

(A) 15%

(B) 18%

(C) 20%

(D) 28%

(E) $\frac{17}{60}$

This is a bit of a tricky wicket. Don't be fooled into thinking that the one-third of the taxpayers who can claim charitable contributions is the same amount as 33% of those same taxpayers. The trouble is, one-third is very close to 33%, but it isn't exactly that amount. If you fell for that, you may have chosen answer D, because that answer is pretty close to the percentage of taxpayers who can claim both deductions. You would get that answer by setting up the equation as follows:

$$33\% + 40\% + 55\% - x = 100\%$$

$$128\% - x = 100\%$$

$$x = 28\%$$

Looks okay, but this short-cut doesn't pan out. A quick check on your calculator on choice E would tell you that $17 \div 60$ is equal to 28.333%. That's too close to give you confidence that D is your best choice. Remember, if you convert from one-third to a very close 33%, you're sacrificing accuracy to save time. There's a better way.

Convert *all* of your percentages to fractions, along with the ⅓ fraction you are given for those who deduct charitable contributions. Your lowest common denominator is 60.

$$\tfrac{1}{3} + 40\% + 55\% - x = 100\%$$

$$\tfrac{1}{3} + \tfrac{4}{10} + \tfrac{11}{20} - x = 1$$

$$\tfrac{20}{60} + \tfrac{24}{60} + \tfrac{33}{60} - x = 1$$

$$\tfrac{77}{60} - x = 1$$

$$x = \tfrac{17}{60}$$

Again, these group problems are not very difficult. You just need to think about what you're doing and the answer will materialize before you can say, "Where's my calculator?"

Try one more before you move on to the next set of problems.

A high school cafeteria can serve 340 students at one time. 178 students prefer to have hamburgers for lunch. 98 students like to eat macaroni and cheese. 127 students like both items for lunch. How many students do not like either type of meal?

(A) 64

(B) 148

(C) 162

(D) 191

(E) 242

Again, not a particularly difficult question to answer if you keep your operations straight. The simple way to do this is to set up the mixed group formula.

$$\text{Total} = \text{Group 1} + \text{Group 2} - \text{Both Groups} + \text{Neither Group}$$

$$340 = 178 + 98 - 127 + x$$

$$340 = 149 + x$$

$$x = 191$$

It's just a simple case of plugging the right numbers into the equation. The correct answer is D. You would have chosen A if you subtracted group 1 and group 2 from the total without accounting for the students who loved and hated both kinds of meals. You would have chosen B if you didn't do proper addition and subtraction correctly. You would have gotten C if you only subtracted the meat lovers from the total. And finally, you would have landed on E if you subtracted the mac-and-cheese eaters from the total.

Crossing Paths: Union and Intersection

A *set* is a collection of objects, numbers, or values. Those objects that are included in a set are called the *elements* or *members* of the set. Those elements or members of a set are said to belong to the set. The symbol ∈ means "is an element of," while ∉ means an object or thing or value "is not an element of" a set.

When dealing with ways to combine sets, the terms *union* and *intersection* are simply ways of stating how two or more sets relate to one another by the elements they contain. An *empty set* or *null set* is indicated by the symbol Ø and simply means that there is nothing in that set.

All together now: Unions

The union is the most inclusive. The *union* of two sets *A* and *B* (written as $A \cup B$) is the set of all values or numbers (elements) that are members of either set *A* or set *B* or both.

For example, the union of sets *A* = {0, 1, 2, 3, 4, 5, 6, 7, 8, 9} and *B* = {2, 4, 6, 8} is the set $A \cup B$ = {0, 1, 2, 3, 4, 5, 6, 7, 8, 9}.

Looking both ways: Intersections

The intersection is less inclusive. The *intersection* of two sets *A* and *B* (written as $A \cap B$) is the set of all numbers or values (elements) that must appear in both sets, not just in either one.

For example, the intersection of sets *A* = {0, 1, 2, 3, 5, 6, 7, 8, 9} and *B* = {2, 4, 6, 8} is the set $A \cap B$ = {2, 6, 8}

The preceding examples show how sets can be defined by using algebra. You can also illustrate the concept of sets with graphic diagrams. The classic Venn diagrams in Figure 11-1 give examples of how sets are related to one another using set terminology.

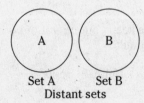

Set A   Set B
Distant sets

A∩B
The intersection of
sets A and B

Figure 11-1:
Venn
diagrams
showing the
union and
intersection
of sets.

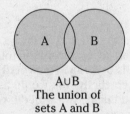

A∪B
The union of
sets A and B

Set B is a subset
of set A
B⊂A

$A = \{0, 1, 2, 3, 4, 5\}$

$B = \{2, 4, 6, 8, 10\}$

$C = A \cap B$

What are the elements of set C?

(A) $\{0, 1, 2, 3, 4, 5, 6, 8, 10\}$

(B) $\{2, 4\}$

(C) $\varnothing$

(D) $\{0, 1, 2, 3, 4, 5\}$

(E) $\{2, 4, 6, 8, 10\}$

This should be fairly simple, as long as you correctly read the symbols that show the relationship between sets. Here, the set C is the intersection of sets A and B. Because the intersection of two sets includes only those elements that show up in both sets, the correct answer is B.

$A = \{5, 10, 15, 20\}$

$B = \{2, 4, 6, 8, 10\}$

$C = A \cup B$

What is the mean average of the elements of set C?

(A) 8.75

(B) 8.89

(C) 10

(D) $\varnothing$

(E) It cannot be determined from the information given.

To solve this one, you need to correctly read the $\cup$ symbol and do some quick math to figure out the mean average. Because you're dealing with the union of two sets, you need to determine all the numbers that appear at one time or another in both sets. The numbers are 2, 4, 6, 8, 10 and 5, 15, 20.

You don't want to count 10 twice, or you'll be double billing that number.

Add 'em up, and you get the grand total of 70. Now divide by the number of elements in the union of the two sets. So, $70 \div 8 = 8.75$.

Your correct answer is A. You would get B if you added in the extra 10, and then took the average of $80 \div 9$. Choice D could be the mean average, but of the members of the intersection, as opposed to the union, of the two sets. The other choices are simply red herrings designed to throw you off when you're weary of test-taking. Stay the course!

$S = \{a, b, c\}$

$T = \{c, e, f\}$

$U = \{d, e, f, g\}$

What is $|S \cup T|$?

(A) 6

(B) $\{a, b, c, d, e, f\}$

(C) $\{c\}$

(D) 5

(E) $\{a, b, c, d, e, f, g\}$

This question is asking for the number (absolute value) of elements in a union of two sets, and not the union of the two sets. The third set, U, is just thrown in as a distracter.

$$S \cup T = \{a, b, c, e, f\}$$

Therefore

$$|S \cup T| = 5 \text{ (the number of elements in } S \text{ and } T)$$

And the correct answer is D.

Choice A is the sum of the number of elements in, or the absolute value of, each of the sets S and T. The equation that results in an answer choice of A is:

$$|S| + |T| = 3 + 3 = 6$$

This is not the correct formula for the number of elements in a union.

Remember that $|S \cup T| = |S| + |T| - |S \cap T|$

c is the only element common to the two sets, and the intersection is spelled out as follows:

$$|S \cap T| = 1$$

Therefore, the absolute value of the union of the two sets is stated as:

$$|S \cup T| = 3 + 3 - 1 = 6 - 1 = 5, \text{ and the correct answer is not 6}$$

Choice C is merely the common element of the two sets, not the number of elements in the union. Choice E is, quite frankly, the union of all three sets.

Possible Arrangements: Permutations

This section deals with certain types of sets that check your ability to do some odd-ball counting exercises that can result in some huge numbers. You can imagine the amount of possible telephone numbers that can result when you have 7 places to fill the phone number with, and 10 possible numbers between 0 and 9 to fill those places. The end result, of course, is 10^7, a pretty huge number.

A permutation is another type of counting exercise. A *permutation* is simply the number of ways you can arrange the elements or objects of a set if you care about what order you arrange them in.

For example, you can arrange the elements of the set {*a, b, c*} in six different ways:

$$a\,b\,c \quad a\,c\,b \quad b\,a\,c \quad b\,c\,a \quad c\,a\,b \quad c\,b\,a$$

We can hear you saying that this isn't really making different combinations; it's simply taking the same three letters and rearranging them and that it should make no difference what order they appear in. But that's exactly what a permutation is. It's arranging the items you have in all the possible ways when you care about the order they show up in. Your doubt is half justified on this one, as you see when you take a look at *combinations* in the next section.

The number of permutations of *n* objects is *n*!. The ! is called a factorial, and the expression is read "*n* factorial." This means you take all the numbers in a series, starting with *n* and counting down by 1 until you get to the number 1, and then multiply all those numbers or factors together.

To use the preceding example of the set {*a, b, c*}, 3 different objects (the letters in the set) are being arranged in as many ways as possible, the number of permutations is 3!, that is, $3 \times 2 \times 1$, which is equal to 6.

Suppose you want to find out how to arrange 5 people in a wedding party in one row so that you have all the possible arrangements of how they'll appear in a photograph. Don't try this out longhand with a wedding party! Trust us, they'd never make it to the altar. The quick answer to your wedding day quandary is 5!. Figuring out the number of permutations for arranging the 5 people in a row for a photograph is a relatively easy venture.

$$5 \times 4 \times 3 \times 2 \times 1 = 120$$

The greater the number of objects you are trying to find the number of permutations for, the more exponential the final result. That's all there is to your basic permutations.

Nothing in life is quite that simple, and the joy of permutations really begins when you have a fixed number of objects *n* to fill a limited number of places *r* and you care about the order the objects are arranged in.

Consider the predicament of the big league baseball coach who has a 20-member baseball team and needs to determine all the different batting orders that these 20 ball players can fill in a 9-slot batting lineup. One way to figure this out longhand is to write out all your factors. Start counting from 20 backwards to get 9 places, because your 20 players can fill only 9 slots in the batting order.

$$20 \times 19 \times 18 \times 17 \times 16 \times 15 \times 14 \times 13 \times 12 = x$$

Just imagine if there were no players on the disabled list. The potential for different permutations would increase astronomically. The coach doesn't have the luxury to figure out all the possibilities in the heat of the game. After all, he'll get nervous just trying to play with the calculator keys that much. There's got to be an easier way.

The number of permutations of *n* things taken *r* at a time is stated as $_nP_r$. (You may even think of a public radio station that has those same call letters to help you remember this mnemonic device.) The *permutation formula* for *n* objects taken *r* at a time is

$$_nP_r = n!/(n - r)!$$

To plug the numbers of players on the baseball team into this formula, you would calculate as follows.

$$n!/(n-r)!$$
$$20!/(20-9)!$$
$$20!/11!$$
$$2.43 \times 10^{18} \div 3.99 \times 10^{7}$$
$$6.09 \times 10^{10}$$

Now, here's your warm-up question.

How many ways can 8 different charms be arranged on a charm bracelet?

(A) 16

(B) 64

(C) 40,320

(D) 16,777,216

(E) 800,000,000

This question is on the easy side to get you started. The hard way to approach this one would be to write this out longhand.

$$8 \times 7 \times 6 \times 5 \times 4 \times 3 \times 2 \times 1$$

That gets a bit unruly, so make use of your calculator with the ! key to make short work of this one.

So, you would have chosen 8! right off the bat.

$$8! = 40,320$$

Your correct answer, of course, is C.

The first two choices, A and B, are way too small. A is simply 8 + 8, while B is 8 × 8. If you know anything about permutations, you know the number gets astronomically large in a hurry. That's why C is not too surprising as the correct answer. But hold on to your horses if you chose D or E, because the numbers shouldn't get that big, unless your calculator is smoking some funny stuff. You would have gotten D if you calculated 8^8, and you would have chosen E if you decided to go out on a limb and plug in 8×10^8. Either way, you have to reign those numbers into something way more probable. You don't want to blow your calculator's gasket.

Now, try something a tad more difficult.

A lawn care company has 5 employees, and there are 10 houses that need care on a given day. How many different ways can the company assign the 5 employees to work at the different houses on that day if each employee works at two houses?

(A) 50

(B) 100

(C) 120

(D) 30,240

(E) 3,628,800

This question looks a little funny in that it seems counterintuitive to the rule just described. In the prior example, you have *n* number of things taken *r* at a time to get the number of permutations. In this problem, it looks like you are taking a smaller number of things *r* (here, the number of employees) and finding out how many times they can be spread around a greater number of places. But that's the purpose of trick questions: to trick you.

This problem may look backward, but it really follows the same formula. Rather than thinking of how to spread 5 workers over 10 houses, think of how many ways you can arrange the 10 houses over the more limited number of workers.

$$_nP_r = n!/(n-r)!$$

$$_{10}P_5 = 10!/(10-5)!$$

$$= 10!/5!$$

$$= 3,628,800 \div 120$$

$$= 30,240$$

Your correct answer is D. If you chose A, you multiplied the number of workers time the number of houses. Anyone can do that, even without a calculator. But it's not quite right. B gets you just a smidgeon closer, but 10×10 is not what you're after, either. Choice C is simply 5!. While this answer is on the right track, the calculation is not complete. Also, if you picked E, you chose the flip side of going way too high with an answer of 10!

Try one more, now that you're on a roll.

A group of 5 bored high school students went trick-or-treating one dark and scary night. One particular house has 8 different snacks to give to trick-or-treaters. How many different ways can the candy be distributed to these kids if they each get one piece?

(A) 336

(B) 5,369

(C) 6,720

(D) 40,200

(E) 40,320

This is a variation on a theme of the previous problem. The smart way to do this is to take the permutation formula and plug in your numbers from the problem.

$$_nP_r = n!/(n-r)!$$

$$_8P_5 = 8!/(8-5)!$$

$$= 8!/3!$$

$$= 40,320 \div 6$$

$$= 6,720$$

If you chose C, then you are really catching on, and you should have no trouble with permutation questions on the SAT II Math test. If you chose A, you did your division a bit wrong and computed 8!/5!. If B was your answer, you did a slightly different thing by dividing 8^8 by 5^5. That won't work for permutations. If you picked D, you did not divide, but you subtracted 8! – 5!. Finally, choice E means you just took the 8! by itself and did not divide it by anything.

Safety in Numbers: Combinations

Combinations are a lot like permutations, only the numbers don't boggle your mind quite as much. The numbers in a combination seem a lot more manageable.

A *combination* is the number of objects you can choose from a total sample of objects and combine them in such a way where the order they are arranged doesn't matter. Because the order doesn't matter with combinations, you get a lot fewer possibilities in your final calculation than you do with permutations. The combination type of question shows up when you want to find out, for example, how many different teams or committees or other types of groups can be formed from a set number of persons.

When you're selecting as many teams as you can from a set number of people, and if you don't care what order the team members are listed in your groupings, you're finding the total number of combinations, as opposed to the permutations, of different teams.

The *formula for combinations* is the number of ways to choose r objects from a group of n objects when the order of the objects doesn't matter. This formula is expressed as follows.

$$_nC_r = n!/r!(n-r)!$$

You can see right away how this is different from the permutation formula. Because you have a larger number in the denominator than you would have with a permutation, the final number is going to be smaller with a combination.

One way to think of this is to ask how many three-member committees can be formed from these three persons: Tom, Dick, and Harry. Because the order of how you list the committee members doesn't matter, you would only have one possible combination of three-member committees from these three persons. A committee composed of Tom, Dick, and Harry is the same as if the committee were composed of Tom, Harry, and Dick, or Dick, Tom, and Harry. You can easily see that a permutation would result in six different possible arrangements, but because order doesn't matter, you only have one possible committee make-up here.

You can prove this by using the combination formula on this example.

$$_nC_r = n!/r!(n-r)!$$
$$= 3!/3!(3-3)!$$
$$(3-3)! = 1$$

It works.

Try an easy one. Suppose a person taking a poll randomly selects a group of 5 people from a group of 10 available persons. How many possible groups of different persons will result? Because the order of the persons in the group doesn't matter, this is a combination problem. Start with the combination formula.

$$_nC_r = n!/r!(n-r)!$$
$$_{10}C_5 = 10!/5!(10-5)!$$
$$= 3,638,800/(120)(120)$$
$$= 3,638,800 \div 14,400$$
$$= 252$$

It's plain to see, then, that the number of different combinations is going to be far less when you are not concerned about the order of persons in the groupings.

A fourth grade class is choosing basketball teams at recess. What is the total possible number of combinations of five-person teams that could be chosen from a classroom of 28 kids, assuming nobody's feelings get hurt?

(A) 120

(B) 98,280

(C) 11,793,600

(D) $2.59 \times 1,022$

(E) $2.54 \times 1,027$

The answer is very systematic. Go ahead and apply the formula for combinations, and see what happens.

$$_nC_r = n!/r!(n - r)!$$
$$_{28}C_5 = 28!/5!(28 - 5)!$$
$$= 28!/(5)! \, (23)!$$
$$= 3.05 \times 10^{29} \div (120) \, (2.59 \times 10^{22})$$
$$= 98,280$$

After you've done the calculations, you find that the correct answer is B.

As you can see, there are all sorts of traps for the unwary, especially for those who have only 20 minutes of recess to try and figure this out. If you chose A, you took the easy route of finding the value of 5!. If you went for choice C, you got a little bit closer by calculating 28!/23!. That looks very peculiarly like a permutation instead of a combination. Choice D means you found the value of 23!, but did not do anything else with it. Finally, choice E is not much better, because it represents 28!/5!. I'm not sure the kids choosing basketball teams could ever count that high, but they would burn up a lot of classroom time trying.

Number of Positions: Factorials

Technically speaking, a *factorial* of any natural number (any integer greater than 0) is the product of all natural numbers in a series from 1 up to and including the number you began with. A factorial is indicated with an exclamation point. The expression $n!$ is stated "n factorial."

An example of a factorial may be 6!. This expression means 6 factorial and is calculated as follows:

$$6 \times 5 \times 4 \times 3 \times 2 \times 1 = 720$$

The factorial of 0 is 0! = 1

A factorial grows exponentially in a very short period of time.

If you got a penny on the first day of a month having 30 days, and each day you got an additional number of pennies equal to the next day of the month multiplied by the number of pennies you had the previous day, how many pennies would you have at the end of the month?

This problem sets up an expression as factorial in reverse.

$$1 \times 2 \times 3 \ldots \times 30$$

This is too hard to figure out in your head for most people who are not the Rainman, so a calculator comes in quite handy.

$$30! = 2.65 \times 10^{32}$$

The concept of the factorial is not that complicated, but it's easy to make a mistake if you're not careful. This question will come up rarely on the Level IIC test. And while you won't run into a question that deals exclusively with finding a factorial on the Level IC test, it does not hurt to get familiar with the concept and the calculator.

Here is an example of why you need to use your calculator instead of working a factorial problem out in your head.

$$(2)! + (3)!/5!$$

At first blush, it may look like you should just add up the two terms in the numerator to get 5!, and then divide by 5!, which would result in an answer of 1. This is woefully inadequate, because the answer is going to be far more than one. As the old saying goes, do the math!

$$((2)! + (3)!) \div 5!$$
$$(2 + 6) \div 120$$
$$8 \div 120$$
$$1 \div 15$$

So while you may think you can beat the SAT II and shave off a few seconds of your time by cutting this kind of corner, it's not worth it. The problem is really easy to get right if you use your calculator rather than risk trying to solve it in your head.

Part V
Practice Makes Perfect

The 5th Wave   By Rich Tennant

"I hear you think you got all the angles figured.
Well, maybe you do and maybe you don't. Maybe the
ratios of the lengths of corresponding sides of an
equiangular right-angled triangle are equal, then
again maybe they're not—let's see your equations."

In this part . . .

The best way to prepare for the SAT II is to practice, and this part provides you with 200 questions to test your knowledge. Each of the four tests (two Level IC and two Level IIC) comes with an answer sheet (so you can even practice filling in the bubbles) and a scoring guide. Give yourself one hour to complete each test. The chapters immediately following each of the four tests provide explanations of the answer choices for each of the questions. You'll find out why the right answers are right and the wrong answers are wrong. We provide a lot of valuable information in each of the explanations, so we suggest that you read through all of them, even those for questions you answered correctly.

Answer Sheet

Begin with Number 1. If the test has fewer than 100 questions, leave the extra spaces blank.

1. Ⓐ Ⓑ Ⓒ Ⓓ Ⓔ	21. Ⓐ Ⓑ Ⓒ Ⓓ Ⓔ	41. Ⓐ Ⓑ Ⓒ Ⓓ Ⓔ	61. Ⓐ Ⓑ Ⓒ Ⓓ Ⓔ	81. Ⓐ Ⓑ Ⓒ Ⓓ Ⓔ
2. Ⓐ Ⓑ Ⓒ Ⓓ Ⓔ	22. Ⓐ Ⓑ Ⓒ Ⓓ Ⓔ	42. Ⓐ Ⓑ Ⓒ Ⓓ Ⓔ	62. Ⓐ Ⓑ Ⓒ Ⓓ Ⓔ	82. Ⓐ Ⓑ Ⓒ Ⓓ Ⓔ
3. Ⓐ Ⓑ Ⓒ Ⓓ Ⓔ	23. Ⓐ Ⓑ Ⓒ Ⓓ Ⓔ	43. Ⓐ Ⓑ Ⓒ Ⓓ Ⓔ	63. Ⓐ Ⓑ Ⓒ Ⓓ Ⓔ	83. Ⓐ Ⓑ Ⓒ Ⓓ Ⓔ
4. Ⓐ Ⓑ Ⓒ Ⓓ Ⓔ	24. Ⓐ Ⓑ Ⓒ Ⓓ Ⓔ	44. Ⓐ Ⓑ Ⓒ Ⓓ Ⓔ	64. Ⓐ Ⓑ Ⓒ Ⓓ Ⓔ	84. Ⓐ Ⓑ Ⓒ Ⓓ Ⓔ
5. Ⓐ Ⓑ Ⓒ Ⓓ Ⓔ	25. Ⓐ Ⓑ Ⓒ Ⓓ Ⓔ	45. Ⓐ Ⓑ Ⓒ Ⓓ Ⓔ	65. Ⓐ Ⓑ Ⓒ Ⓓ Ⓔ	85. Ⓐ Ⓑ Ⓒ Ⓓ Ⓔ
6. Ⓐ Ⓑ Ⓒ Ⓓ Ⓔ	26. Ⓐ Ⓑ Ⓒ Ⓓ Ⓔ	46. Ⓐ Ⓑ Ⓒ Ⓓ Ⓔ	66. Ⓐ Ⓑ Ⓒ Ⓓ Ⓔ	86. Ⓐ Ⓑ Ⓒ Ⓓ Ⓔ
7. Ⓐ Ⓑ Ⓒ Ⓓ Ⓔ	27. Ⓐ Ⓑ Ⓒ Ⓓ Ⓔ	47. Ⓐ Ⓑ Ⓒ Ⓓ Ⓔ	67. Ⓐ Ⓑ Ⓒ Ⓓ Ⓔ	87. Ⓐ Ⓑ Ⓒ Ⓓ Ⓔ
8. Ⓐ Ⓑ Ⓒ Ⓓ Ⓔ	28. Ⓐ Ⓑ Ⓒ Ⓓ Ⓔ	48. Ⓐ Ⓑ Ⓒ Ⓓ Ⓔ	68. Ⓐ Ⓑ Ⓒ Ⓓ Ⓔ	88. Ⓐ Ⓑ Ⓒ Ⓓ Ⓔ
9. Ⓐ Ⓑ Ⓒ Ⓓ Ⓔ	29. Ⓐ Ⓑ Ⓒ Ⓓ Ⓔ	49. Ⓐ Ⓑ Ⓒ Ⓓ Ⓔ	69. Ⓐ Ⓑ Ⓒ Ⓓ Ⓔ	89. Ⓐ Ⓑ Ⓒ Ⓓ Ⓔ
10. Ⓐ Ⓑ Ⓒ Ⓓ Ⓔ	30. Ⓐ Ⓑ Ⓒ Ⓓ Ⓔ	50. Ⓐ Ⓑ Ⓒ Ⓓ Ⓔ	70. Ⓐ Ⓑ Ⓒ Ⓓ Ⓔ	90. Ⓐ Ⓑ Ⓒ Ⓓ Ⓔ
11. Ⓐ Ⓑ Ⓒ Ⓓ Ⓔ	31. Ⓐ Ⓑ Ⓒ Ⓓ Ⓔ	51. Ⓐ Ⓑ Ⓒ Ⓓ Ⓔ	71. Ⓐ Ⓑ Ⓒ Ⓓ Ⓔ	91. Ⓐ Ⓑ Ⓒ Ⓓ Ⓔ
12. Ⓐ Ⓑ Ⓒ Ⓓ Ⓔ	32. Ⓐ Ⓑ Ⓒ Ⓓ Ⓔ	52. Ⓐ Ⓑ Ⓒ Ⓓ Ⓔ	72. Ⓐ Ⓑ Ⓒ Ⓓ Ⓔ	92. Ⓐ Ⓑ Ⓒ Ⓓ Ⓔ
13. Ⓐ Ⓑ Ⓒ Ⓓ Ⓔ	33. Ⓐ Ⓑ Ⓒ Ⓓ Ⓔ	53. Ⓐ Ⓑ Ⓒ Ⓓ Ⓔ	73. Ⓐ Ⓑ Ⓒ Ⓓ Ⓔ	93. Ⓐ Ⓑ Ⓒ Ⓓ Ⓔ
14. Ⓐ Ⓑ Ⓒ Ⓓ Ⓔ	34. Ⓐ Ⓑ Ⓒ Ⓓ Ⓔ	54. Ⓐ Ⓑ Ⓒ Ⓓ Ⓔ	74. Ⓐ Ⓑ Ⓒ Ⓓ Ⓔ	94. Ⓐ Ⓑ Ⓒ Ⓓ Ⓔ
15. Ⓐ Ⓑ Ⓒ Ⓓ Ⓔ	35. Ⓐ Ⓑ Ⓒ Ⓓ Ⓔ	55. Ⓐ Ⓑ Ⓒ Ⓓ Ⓔ	75. Ⓐ Ⓑ Ⓒ Ⓓ Ⓔ	95. Ⓐ Ⓑ Ⓒ Ⓓ Ⓔ
16. Ⓐ Ⓑ Ⓒ Ⓓ Ⓔ	36. Ⓐ Ⓑ Ⓒ Ⓓ Ⓔ	56. Ⓐ Ⓑ Ⓒ Ⓓ Ⓔ	76. Ⓐ Ⓑ Ⓒ Ⓓ Ⓔ	96. Ⓐ Ⓑ Ⓒ Ⓓ Ⓔ
17. Ⓐ Ⓑ Ⓒ Ⓓ Ⓔ	37. Ⓐ Ⓑ Ⓒ Ⓓ Ⓔ	57. Ⓐ Ⓑ Ⓒ Ⓓ Ⓔ	77. Ⓐ Ⓑ Ⓒ Ⓓ Ⓔ	97. Ⓐ Ⓑ Ⓒ Ⓓ Ⓔ
18. Ⓐ Ⓑ Ⓒ Ⓓ Ⓔ	38. Ⓐ Ⓑ Ⓒ Ⓓ Ⓔ	58. Ⓐ Ⓑ Ⓒ Ⓓ Ⓔ	78. Ⓐ Ⓑ Ⓒ Ⓓ Ⓔ	98. Ⓐ Ⓑ Ⓒ Ⓓ Ⓔ
19. Ⓐ Ⓑ Ⓒ Ⓓ Ⓔ	39. Ⓐ Ⓑ Ⓒ Ⓓ Ⓔ	59. Ⓐ Ⓑ Ⓒ Ⓓ Ⓔ	79. Ⓐ Ⓑ Ⓒ Ⓓ Ⓔ	99. Ⓐ Ⓑ Ⓒ Ⓓ Ⓔ
20. Ⓐ Ⓑ Ⓒ Ⓓ Ⓔ	40. Ⓐ Ⓑ Ⓒ Ⓓ Ⓔ	60. Ⓐ Ⓑ Ⓒ Ⓓ Ⓔ	80. Ⓐ Ⓑ Ⓒ Ⓓ Ⓔ	100. Ⓐ Ⓑ Ⓒ Ⓓ Ⓔ

Chapter 12

Practice Test 1, Level IC

● ●

*O*kay, you know your stuff. Now's your chance to shine. The following exam is a 50-question, multiple-choice test. You have 60 minutes to complete it.

To make the most of this practice exam, take the test under similar conditions to the actual test.

1. **Find a place where you won't be distracted (preferably as far from your younger sibling as possible).**

2. **If possible, take the practice test at approximately the same time of day as that of your real SAT II.**

3. **Set an alarm for 60 minutes.**

4. **Mark your answers on the provided answer grid.**

5. **If you finish before time runs out, go back and check your answers.**

6. **When your 60 minutes is over, put your pencil down.**

After you finish, check your answers on the answer key at the end of the test. Use the scoring chart to find out your final score.

Read through all of the explanations in Chapter 13. You learn more by examining the answers to the questions than you do by almost any other method.

Table 12-1	**For Your Reference**
You can use the following information for reference when you are answering the test questions.	
The volume of a right circular cone with radius r and height h: $V = \frac{1}{3}\pi r^2 h$	
The lateral area of a right circular cone with circumference of the base c and slant height l: $S = \frac{1}{2}cl$	
The volume of a sphere with radius r: $V = \frac{4}{3}\pi r^2$	
The surface area of a sphere with radius r: $S = 4\pi r^2$	
The volume of a pyramid with base area B and height h: $V = \frac{1}{3}Bh$	

For each of the following 50 questions, choose the best answer from the choices provided. If no answer choice provides the exact numerical value, choose the answer that is the best approximate value. Fill in the corresponding oval on the answer grid.

<u>Notes:</u>

1. **You must use a calculator to answer some but not all of the questions.**

   You can use programmable calculators and graphing calculators. You should use at least a scientific calculator.

2. **Make sure your calculator is in degree mode because the only angle measure on this test is degree measure.**

3. **Figures accompanying problems are for reference only.**

   Figures are drawn as accurately as possible and lie in a plane unless the text indicates otherwise.

4. **The domain of any function f is assumed to be the set of all real numbers x for which $f(x)$ is a real number unless indicated otherwise.**

5. **You can use the table on the previous page for reference in this test.**

1. If $5p - 2p = 7p + 2p - 24$, then $p =$

   (A) 1 (B) 12 (C) –4 (D) 4 (E) 0

2. In rectangle $ABCD$ in the figure, what are the coordinates of vertex D?

   (A) (6, 3)

   (B) (9, 5)

   (C) (3, 6)

   (D) (8, 3)

   (E) (6, 5)

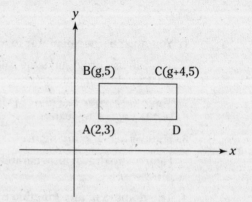

3. For all $y \neq 0$, $\dfrac{3}{\left(\dfrac{4}{y}\right)} =$

   (A) $\dfrac{12}{y}$ (B) $\dfrac{3y}{4}$ (C) $\dfrac{4}{3y}$ (D) $\dfrac{y}{12}$ (E) $\dfrac{3}{4y}$

4. At what point does $3x + 7y = 21$ intersect the x-axis?

   (A) (0, 3)

   (B) (3, 0)

   (C) (7, 0)

   (D) (8, 0)

   (E) (0, 7)

5. If $y = 2$, then $(y - 4)(y + 2) =$

   (A) –24 (B) –8 (C) 2 (D) 8 (E) 24

6. $(x + y + 3)(x + y + 3) =$

   (A) $(x + y)^2 + 9$

   (B) $(x + y)^2 + 6(x + y)$

   (C) $(x + y)^2 + 6(x + y) + 9$

   (D) $x^2 + y^2 + 9$

   (E) $x^2 + y^2 + 6xy$

7. If $3x^3 = 4$, then $4(3x^3)^3 =$

   (A) 16 (B) 32 (C) 64 (D) 256 (E) 1,024

Go on to next page

8. In the figure, if lines *l* and *m* are parallel, then *x* + *y* =

(A) 45° (B) 90° (C) 180° (D) 270° (E) 360°

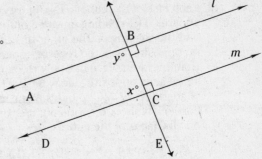

9. If *x* + *y* = 6 and *x* − *y* = 4, then *x* =

(A) 1

(B) 5

(C) 0

(D) −1

(E) −5

10. If the square root of the cube root of a number is 3, what is the number?

(A) 9

(B) 27

(C) 81

(D) 243

(E) 729

11. Amy has half as much money as Hope, who has three-fourths as much as Dakota. If Amy has $60, how much does Dakota have?

(A) $22.50 (B) $90 (C) $150 (D) $160 (E) $360

12. In the figure, ray *OB* is perpendicular to *OA*. If angle *AOC* is 8° greater than angle *AOB*, what is the measure of angle *AOC*?

(A) 82° (B) 90° (C) 98° (D) 172° (E) 188°

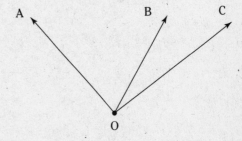

Go on to next page

13. Each face of the cube in the figure consists of 16 small squares. The shading on three of the faces is shown, and the shading on the other three faces is such that on opposite faces the reverse squares are shaded. For example, if one face has only the upper left corner shaded, its opposite face will have the other 15 squares shaded (every square but the one in the upper left corner). What is the total number of shaded squares on all 6 faces of the cube?

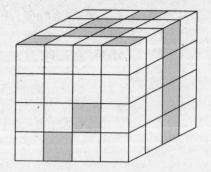

   (A) 32 (B) 48 (C) 64 (D) 80 (E) 96

14. If $x + 2x + 3x = y$, then $2x - y =$

   (A) $-8x$ (B) $-6x$ (C) $-4x$ (D) $-3x$ (E) $4x$

15. If $20^c = 4^3 \cdot 5^3$, what is the value of c?

   (A) 3 (B) 6 (C) 9 (D) 27 (E) 60

16. Imagine if WNBA Champion Detroit Shock conducted a free-throw shooting experiment. Each trial of the experiment consisted of each member of the team taking turns shooting 10 free throws and counting the number of made baskets that resulted. The results for 50 trials are pictured in the figure. In approximately what percent of the trials were there 7 *or more* made baskets?

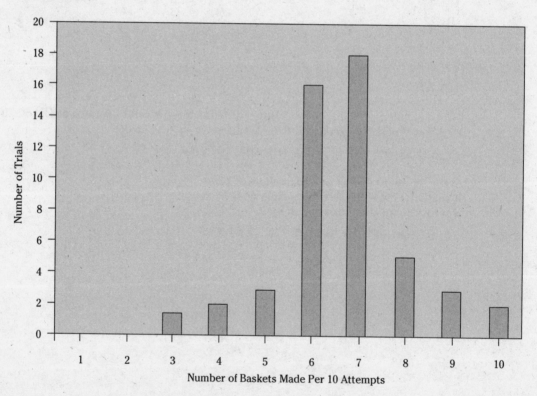

   (A) 20% (B) 28% (C) 50% (D) 56% (E) 88%

17. If $f(x) = \dfrac{1}{x^2}$ for $x > 0$, then $f(0.5) =$

(A) $\dfrac{1}{4}$ (B) $\dfrac{1}{2}$ (C) 1 (D) 2 (E) 4

18. What are all values for z for which $|z - 3| < 4$?

    (A) $z < 7$

    (B) $z > 7$

    (C) $0 < z < 6$

    (D) $-1 < z < 7$

    (E) $0 > z > 6$

19. The circle in the figure has a center C and a radius of 5. What is the length of chord EF?

(A) 5 (B) 7.07 (C) 10 (D) 14.14 (E) 50

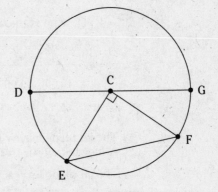

20. License plates in Tinytown consist of three characters. The first is either the letter M or F depending on the gender of the car's owner, the second a single digit between 0–9, and the last is a single letter of the alphabet from A–Z. How many possible license plates can there be?

(A) 38 (B) 468 (C) 520 (D) 780 (E) 6,760

Go on to next page

21. The figure shows the distance of a jet plane from its home airport of Tinytown over a period of time on a given day. The airline services only Tinytown, Bigville, and Medium Lake. Which of the following situations best fits the information?

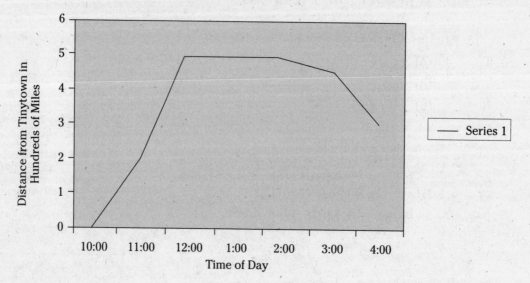

(A) The jet plane leaves Tinytown, flies to Medium Lake, and then flies to Bigville, somewhat closer to Tinytown than Medium Lake.

(B) The jet plane leaves Medium Lake, flies home, and then flies to Bigville.

(C) The jet plane leaves Medium Lake, flies to Bigville, and then flies to Tinytown.

(D) The jet plane leaves Tinytown, flies to Medium Lake, and then flies to Bigville, somewhat farther away from Tinytown than Medium Lake.

(E) The jet plane leaves Bigville, flies to Medium Lake, and then flies back to Bigville.

22. In the figure, triangle *ACE* is equilateral, FB∥EC, and *G* is the midpoint of *FB*. What is the perimeter of trapezoid *FBCE*?

(A) 18 (B) 19 (C) 24 (D) 26 (E) 28

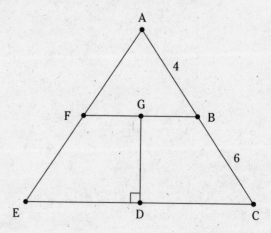

Go on to next page

23. In the figure, two congruent segments are perpendicular to the number line at points 4 and 8, respectively, and end when they intersect rays from points 0 and 6, respectively. The segment at point 8 is to be moved to the right along the number line, and the ray from point 6 is to be rotated until it intersects the top of the segment at its new position and $\cos x° = \cos y°$. How many units to the right must the segment be moved?

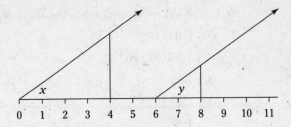

(A) 5 (B) 4 (C) 3 (D) 2 (E) 1

24. If $i^2 = -1$, and if $[(i^2)^5]^x = 1$, the least positive integer value of x is

(A) 0 (B) 1 (C) 2 (D) 3 (E) 4

25. In 1936, Oregon's Cape Blanco Lighthouse was able to provide a boat 4.5 miles away with a flash of light every 20 seconds. How many degrees did the lighthouse rotate in 2 seconds?

(A) 8 (B) 9 (C) 18 (D) 36 (E) 80

26. In the figure, if $\Theta = 36°$, what is the value of x?

(A) 720 (B) 16.18 (C) 14.53 (D) 11.76 (E) 0.04

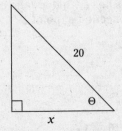

27. Balancing on one foot atop the Empire State Building, towering at a height of 1,454 feet, the height of a baseball after you throw it is a function of the time that has expired from the time you threw it. If t represents the time, in seconds, that has expired because you threw the ball and $h(t)$ represents the height, in feet, of the baseball, then $h(t) = 1,454 - 16t^2 + 8t + 4$. Of the following, which is the closest approximation to the height, in feet, of the baseball at 2 seconds?

(A) 450 (B) 727 (C) 1,410 (D) 1,454 (E) 1,538

28. In the xy-plane, the points O (0, 0), A (2, –2), B (6, 0), and C (4, –6) can be connected to form line segments. Which two segments have the same length?

(A) OA and AB

(B) OC and BC

(C) OB and BC

(D) AC and AB

(E) OA and OB

Go on to next page

29. Of the following, which has the least value?

    (A) $(100 \cdot 10^5)^{10}$

    (B) 1,000,000,000,000,000

    (C) 1000^{100}

    (D) $(10 \cdot 10^{10})^{10}$

    (E) 100^{1000}

30. The figure shows a square prism. Which of the given vertices is located in the plane determined by the vertices S, U, and X?

    (A) W (B) Y (C) Z (D) T (E) V

    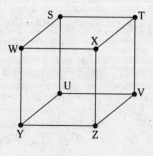

31. In the figure, if square $ABCD$ is reflected across line l, what are the coordinates of the reflection of point D?

    (A) $(1, -2)$

    (B) $(3, -2)$

    (C) $(4, -2)$

    (D) $(7, -2)$

    (E) $(10, -2)$

    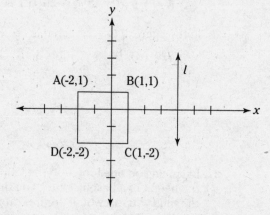

32. In the figure, the cube has an edge of length 4. What is the distance from vertex E to the midpoint G of edge FH?

    (A) 5

    (B) $4\sqrt{2}$

    (C) 6

    (D) 8

    (E) 10

    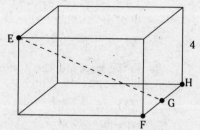

33. A total of 11 students took an exam and their average score (arithmetic mean) was 84. If the average score for 6 of the students was 79, what was the average score of the remaining 5 students?

    (A) 88 (B) 89 (C) 90 (D) 91 (E) 92

Go on to next page

34. Line *l* has a negative slope and a positive *x*-intercept. Line *m* is parallel to *l* and has a negative *x*-intercept. The *y*-intercept of *m* must be

    (A) negative and greater than the *y*-intercept of *l*

    (B) negative and less than the *y*-intercept of *l*

    (C) positive and greater than the *y*-intercept of *l*

    (D) positive and less than the *y*-intercept of *l*

    (E) zero

35. The sum of two roots of a quadratic equation is 3 and their product is 2. Which of the following could be the equation?

    (A) $x^2 + 3x + 2$

    (B) $x^2 - 2x - 3$

    (C) $x^2 + 4x + 3$

    (D) $x^2 + 2x - 3$

    (E) $x^2 - 3x + 2$

36. In the figure, triangles *MNO* and *OPQ* are similar and $c = 3$. What is the value of $\frac{a}{b}$?

    (A) $\frac{1}{3}$ (B) $\frac{3}{4}$ (C) $\frac{1}{4}$ (D) $\frac{4}{3}$ (E) 3

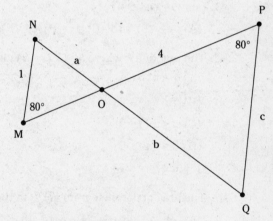

Note: Figure not drawn to scale.

37. The line with equation $x = 6$ is graphed on the same *xy*-coordinate-plane as the circle with center (3, 5) and radius 4. What are the *y*-coordinates of the points of intersection of the line and the circle?

    (A) 8 and 2

    (B) 7 and –1

    (C) 10 and 0

    (D) 7.65 and 2.35

    (E) 6.87 and –0.87

38. In 2000, Esteban invested $1,000 in a bank account that pays him interest at an annual percentage rate (APR) of 2.5% compounded annually (the interest earned every year is added to the principal amount in advance of the next year). Rounded to the nearest dollar, what will his accumulated balance be in 2010?

    (A) $1,025 (B) $1,250 (C) $1,280 (D) $2,000 (E) $9,313

Go on to next page

39. In the figure, if $70 < a + b < 150$, which of the following describes all possible values of $c + d?$

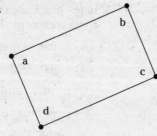

(A) $210 < c + d < 290$

(B) $30 < c + d < 110$

(C) $120 < c + d < 200$

(D) $390 < c + d < 470$

(E) $570 < c + d < 650$

40. If points D, E, and F lie on a circle and if the center of the circle lies on segment DF, angle DEF can measure only (in degrees),

(A) $0 < DEF < 90$

(B) $DEF = 90$

(C) $90 < DEF < 180$

(D) $DEF = 180$

(E) $180 < DEF < 360$

41. $(\sin^2\Theta + \cos^2\Theta - 5)^3 =$

(A) -64 (B) -12 (C) 12 (D) 64 (E) 81

42. The function f, where $f(x) = (x + 2)^2$, is defined for $-3 \leq x \leq 3$. What is the range of f?

(A) $1 \leq f(x) \leq 9$

(B) $1 \leq f(x) \leq 25$

(C) $0 \leq f(x) \leq 9$

(D) $0 \leq f(x) \leq 25$

(E) $0 \leq f(x) \leq 10$

43. The area of parallelogram $EFGH$ in the figure is

(A) 24 (B) 48 (C) $24\sqrt{3}$ (D) 18 (E) $18\sqrt{3}$

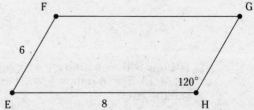

44. The largest pizza ever baked was 122 feet in diameter. If one were to place pepperoni slices on this pizza so that there were 10 slices per square foot, and only one-foot pepperoni sticks that provide 50 pepperoni slices per stick were available, how many one-foot pepperoni sticks are needed?

(A) 77 (B) $2,338$ (C) $4,842$ (D) $7,345$ (E) $9,352$

Go on to next page

45. In the right circular cylinder shown in the figure, C and O are the centers of the bases and segment XY is a diameter of one of the bases. What is the perimeter of triangle XYO if the height of the cylinder is 6 and the radius of the base is 2?

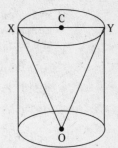

    (A) 6.16

    (B) 6.32

    (C) 10.32

    (D) 16.65

    (E) 20

46. Given the sequence of numbers: 2, 6, 10, 14 . . ., which of the following represents the value of the nth term?

    (A) $2n$

    (B) $2n^2$

    (C) $n(n-1)$

    (D) $n(n+1)$

    (E) $2(2n-1)$

47. If $f(x) = 2x^2 + 3$, and if f^{-1} is the inverse function of f, what is $f^{-1}(9)$?

    (A) 1.41　(B) 1.73　(C) 2.23　(D) 2.45　(E) 165

48. Two positive integers r and s satisfy the relationship $q * r$ if and only if $q = r^2 + 2$. If s, t, and u satisfy the relations $s * t$ and $t * u$, what is the value of s in terms of u?

    (A) $u^2 + 2$

    (B) $u^2 + 4$

    (C) $u^4 + 4u^2 + 4$

    (D) $u^4 + 4u^2 + 6$

    (E) $u^4 + 8u^2 + 16$

49. In the figure, the area of the shaded region bound by the graph of the parabola $y = f(x)$ and the x-axis is 5. What is the area of the region bound by the graph of $y = f(x-3)$ and the x-axis?

    (A) 2　(B) $\frac{5}{3}$　(C) 5　(D) 8　(E) 15

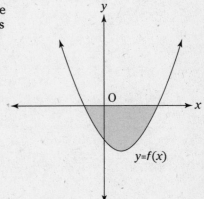

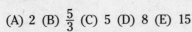

Go on to next page

50. In search of unique and refreshing beverages, many people are mixing iced tea with lemonade. How many liters of lemonade should be mixed with 2 liters of iced tea so that 15% of the combined liquid is lemonade?

(A) 0.3

(B) 0.36

(C) 0.65

(D) 1.65

(E) 0.35

Answer Key for Practice Exam 1

1.	D	26.	B
2.	A	27.	C
3.	B	28.	D
4.	C	29.	B
5.	B	30.	C
6.	C	31.	E
7.	D	32.	C
8.	C	33.	C
9.	B	34.	B
10.	E	35.	E
11.	D	36.	D
12.	C	37.	D
13.	B	38.	C
14.	C	39.	A
15.	A	40.	B
16.	D	41.	A
17.	E	42.	D
18.	D	43.	C
19.	B	44.	B
20.	C	45.	D
21.	A	46.	E
22.	D	47.	B
23.	D	48.	D
24.	C	49.	C
25.	D	50.	E

Scoring Your Exam

First, calculate your "raw score." On this exam, you get one point for each correct answer, and you lose ¼ (0.25) of a point for each wrong answer. Your raw score, then, is determined by the following formula: Raw score = number of correct answers − (0.25 × number of wrong answers). Feel fee to use this Raw Score Wizard to help you determine your raw score:

Raw Score Wizard

1. Use the answer key to count the number of right answers   1) _____

2. Use the answer key to count the number of wrong answers   2) _____

3. Multiply the number of wrong answers by 0.25 (Line 2 × 0.25)   3) _____

4. Subtract this number from the number of right answers
   (Line 1 − Line 3)   4) _____

5. Round this number to the nearest whole number
   (Round Line 4 up or down)   5) _____

   This final number is your raw score.

Use Table 12-2 on the next page to convert your raw score to one of the College Board Scaled Scores, which range from 200–800. Ultimately, when you take the test for real, this scaled score will be the one the College Board reports to you and the colleges/universities you request.

Note: The College Board explains that it converts students' raw scores to scaled scores in order to make sure that a score earned on one particular subject test is comparable to the same scaled score on all other versions of the same subject test. Scaled scores are adjusted so that they indicate the same level of performance regardless of the individual test taken and the ability of the specific group of students that takes it. In other words, a score of 500 on one version of SAT II Math Level IC administered at one time and place indicates the same level of achievement as a score of 500 on a different version of the test administered at a different time and place.

Table 12-2		Scaled Score Conversion Table, Mathematics Level IC Test			
Raw	*Scaled*	*Raw*	*Scaled*	*Raw*	*Scaled*
50	800	29	600	8	410
49	790	28	590	7	400
48	780	27	580	6	390
47	780	26	570	5	380
46	770	25	560	4	380
45	750	24	550	3	370
44	740	23	540	2	360
43	740	22	530	1	350
42	730	21	520	0	340
41	720	20	510	−1	340
40	710	19	500	−2	330
39	710	18	490	−3	320
38	700	17	480	−4	310
37	690	16	470	−5	300
36	680	15	460	−6	300
35	670	14	460	−7	280
34	660	13	450	−8	270
33	650	12	440	−9	260
32	640	11	430	−10	260
31	630	10	420	−11	250
30	620	9	420	−12	240

Chapter 13

Explaining the Answers to Practice Test 1, Level IC

● ●

*T*he following explanations are for Practice Test 1, Level IC, in Chapter 12. Make sure you read through all the explanations. You may pick up some valuable information.

1. **D.** First, combine your like terms. On the left side of the equation, $5p - 2p$ combines to $3p$, and on the right side, $7p + 2p$ combines to $9p$. You are left with $3p = 9p - 24$. Next, get all of your p's on one side of the equation and your numbers on the other. Subtract $3p$ from both sides of the equation: $0 = 6p - 24$. Now, add 24 to both sides of the equation: $24 = 6p$. Finally, to get the variable by itself, divide both sides of the equation by 6 to get $4 = p$.

   A is wrong because you incorrectly combined like terms by multiplying the coefficients (the numbers in front of the variable) together instead of adding/subtracting them. B is also incorrect because you likely saw what you thought was the same term "$2p$" on both sides of the equation, and tried to have them cancel each other out. However, they have different signs. Thus they are different and cannot cancel each other out. Neither is C correct; you probably confused your signs. On the right side of the equation, 24 is being subtracted. Therefore, to move it to the right side of the equation, you have to add it. Keep track of your signs! Finally, E serves as an incorrect decoy.

   If there is any confusion, give the variable the name of a common object; for example, change the letter p to "pillow." On the left side, 5 pillows minus 2 pillows leaves you with three pillows. On the right, 7 pillows plus 2 pillows equals 9 pillows.

2. **A.** After you identify that coordinates B and C both have the same y-coordinate of 5, you can take for granted that the sides of the rectangle are parallel to the axes and you're good to go. Vertex B and vertex A must have the same x-coordinate. Therefore, $g = 2$. Logically, then, vertex C has the coordinates $(6, 5)$ because $g + 4 = 2 + 4 = 6$. Finally, D must have the same x-coordinate of C and the same y-coordinate of A; thus D is $(6, 3)$.

   B is incorrect because you would have to mistakenly give g a value of 3 instead of 2 by using the y-coordinate instead of the x. C uses correct math but switches the x- and y-coordinates at the end. D is incorrect as a result of adding $g + 4$. E is wrong because it incorrectly uses the 5 from vertex C. In problems dealing with the coordinate plane, be mindful of which axis you are using.

3. **B.** When you divide by a fraction, you multiply by the reciprocal. In this case, change the big division bar to a multiplication symbol, and then flip the fraction underneath from $\frac{4}{y}$ to $\frac{y}{4}$. Then, multiply across: you get $\frac{3y}{4}$.

   Dividing by a fraction involves three steps:

   1. Change the division bar to a multiplication symbol.

   2. Flip the fraction upside down.

   3. Multiply across.

Hints for solving equations

Use these tips for solving equations:

✔ Combine like terms.

✔ Get the variable by itself on one side of the equation.

To do this, "do the opposite." If a term is being added, subtract it; if a term is being multiplied, divide it. Similarly, if a term is being subtracted, add it, and if a term is being multiplied, divide it.

Remember: What you do to one side of the equation, you must do to the other! Think of the equation like one of those gold Lady Justice scales that always must be kept in balance. If you add something to one side only, it will tip the scale out of balance.

✔ Be conscious of your plus and minus signs.

A is incorrect because it neglects to flip $\frac{4}{y}$ before multiplying across. C is wrong because it flips the top term instead of the bottom. D results from flipping $\frac{4}{y}$, and then leaving the 3 on the bottom instead of the top (in the denominator instead of the numerator). Finally, E is wrong because it breaks up the term in the parenthesis and uses the wrong division bar for the problem. The parenthesis clearly indicates that you should avoid dividing $\frac{3}{4}$ by y.

4. **C.** The equation $3x + 7y = 21$ intersects the x-axis at the point whose y-coordinate is 0, or $y = 0$. Therefore, simply substitute 0 for y in the given equation and you have $3x + 7(0) = 21$. This simplifies to $3x = 21$. Divide both sides by 3 and $x = 7$. The point's coordinates are $(7, 0)$.

   A is incorrect because it substitutes 0 for x instead of y. B substitutes 0 for x instead of y, and then switches the coordinates around. D is wrong due to faulty math, and E is incorrect because it reverses the coordinates.

5. **B.** Simply substitute and then follow the order of operations (that whole "Aunt Sally" thing found in Chapter 3). Substituting 2 for y, you get $(2 - 4)(2 + 2)$. Solve inside the parenthesis first, and you get $(-2)(4)$, which equals -8. If you are at all confused, see the sidebar about the rules combining signs or flip to Chapter 3.

   If you chose A or E, you probably subtracted the terms too quickly. You were thinking that the 2 was negative instead of positive and reached a total of -6 instead of the correct total of -2. With D and C, the problems lies in multiplying the signs, and E has a problem with subtracting and multiplying signs.

6. **C.** Use the FOIL method (First, Outer, Inner, Last). With three terms in each parenthesis, just distribute (multiply) one term at a time from one set of parenthesis to the other three terms in the other set of parenthesis.

   After you complete that step in this problem, the expression will look like this: $x^2 + xy + 3x + yx + y^2 + 3y + 3x + 3y + 9$.

   Then, combine your like terms. Remember that xy and yx are the same, so you get $x^2 + 2xy + y^2 + 6x + 6y + 9$.

   At this point, you should recognize the telltale expression: $x^2 + 2xy + y^2$, which simplifies to $(x + y)^2$. You are then left with $(x + y)^2 + 6x + 6y + 9$.

   The only other thing you can do at this point is to pull the common factor 6 out of the mini-expression $6x + 6y$, which leaves you with $(x + y)^2 + 6(x + y) + 9$. (You could also pull the common factor 6 out of the 9, but you would be left with $(x + y)^2 + 6(x + y + \frac{3}{2})$ and that is not one of the possible answers.)

   A, B, D, and E are all decoy answers waiting for you to make careless errors when you multiply out the two factors and combine like terms. Take your time and be careful.

Rules for combining signs for addition, subtraction, multiplication, and division

Addition:

positive + positive =   add the numbers and keep the positive sign

negative + negative =   add the numbers and keep the negative sign

positive + negative =   subtract the numbers and take the sign from the larger number

negative + positive =   subtract the numbers and take the sign from the larger number

Subtraction:

If the subtraction problem isn't simple (for example: $7 - 5 = 2$), just, "add the opposite." Change the subtraction to addition and flip the sign of the second number ($6 - 8$ becomes $6 + -8$). Then follow the rules for addition.

Multiplication/division:

Two numbers with the same sign have a positive product or quotient.

$$+ \times + = +$$

$$+/+ = +$$

$$- \times - = +$$

$$-/- = +$$

Two numbers with different signs have a negative product or quotient.

$$+ \times - = -$$

$$- \times + = -$$

$$+/- = -$$

$$-/+ = -$$

7. **D.** All those 3's and 4's? And . . . what?! Raised what to the 3rd power?! Aaaaahhhhh! Okay, settle down, don't freak out. The SAT II uses similar numbers a few times in different ways to try to confuse you. But if you take your time, and follow the rules, you are good to go. Admittedly, "$4(3x^3)^3 =$" looks intimidating.

But wait! Look what they give you in the first part of the question: "If $3x^3 = 4$." If you can recognize that the $3x^3$ in the second part is the same as the $3x^3$ in the first, you can substitute and solve this problem lickety-split. $4(3x^3)^3$ becomes $4(4)^3$.

Now, follow the order of operations. Raise 4 to the 3rd power before you multiply, and the expression simplifies to $4(64)$, which equals 256. (Another way to solve it from this point is to recognize that first 4 can be thought of as 4^1; and $4^1 \times 4^3 = 4^4$, which equals 256.)

A is wrong because it ignores the exponent in the expression. B is incorrect because it adds some of the 4s together instead of multiplying them together. C is wrong because it ignores the first 4 and only deals with the exponent. E is just a decoy representing 4^5.

8. **C.** If two parallel lines are cut by a transversal (a line cutting across them), then corresponding angles are congruent (equal). Additionally, supplementary angles (the ones next to each other along a straight line) add up to $180°$.

You can do this problem one of two ways. The first is to recognize that angle y, the bottom of the four angles formed when line g intersects line l (angle ABC), corresponds to the bottom angle formed when line g intersects line m (angle DCE), which means that the two angles have equal measures, which is $y°$.

Recognize that angle x (angle BCD) shares a vertex C with angle DCE (which you now know has the same measure as angle y), and the two are back-to-back (you can also think of them as being pasted together to form a straight line). In any event, angle x and angle y are supplementary and add up to $180°$.

A, B, D, and E are all incorrect decoys. To avoid them, turn to Chapter 6 and look over the rules on parallel lines and angles.

9. **B.** There are two ways to solve these simultaneous equations. The easiest way is by elimination — to stack them one on top of the other and add them. The $+y$ and $-y$ cancel out and you are left with $2x = 10$. From there, divide both sides by 2, and $x = 5$.

   The other way is to solve the first equation for y "in terms of x," and then substitute what you get for y into the second equation. In the first equation, when you solve for y, you get $y = 6 - x$. You now have a value for y "in terms of x." Now substitute $6 - x$ for y in the second equation.

   $x - (6 - x) = 4$

   $x - 6 + x = 4$

   Combine like terms.

   $2x - 6 = 4$.

   To solve the equation, add six to both sides to get the variable by itself, which leaves you with $2x = 10$. Finally, divide both sides by 2 and you get $x = 5$.

   A is wrong because it is derived by subtracting the numbers on the right side of the equations instead of adding them. C is an incorrect decoy. D comes from giving the wrong sign to the 6 while employing the second method just given. E is another decoy that takes the right answer and gives it the wrong sign.

10. **E.** The problem states that it takes the square root of the cube root of a number and gets 3. So, start with the number 3 and work backward. Replace the phrase "the cube root of a number" with "something" (or x).

   The problem then becomes: the square root of "something" is 3.

   The question then becomes 3 is the square root of what? Well that's easy, $3 \times 3 = 9$. "Something" equals 9.

   Now do the second part of the problem. The "something" is 9. But, remember "something" (9) is "the cube root of a number." Therefore, if "the cube root of a number" is 9, multiply $9 \times 9 \times 9$ and you get 729.

   A, B, C, and D are all incorrect as they are the number 3 raised to the 2nd, 3rd, 4th, and 5th powers, respectively.

11. **D.** If Amy has half as much money as Hope, then Hope has twice as much as Amy. Make an equation. Let A = Amy and H = Hope. If $A = \frac{1}{2}H$, multiply both sides of the equation by 2 or divide both sides by $\frac{1}{2}$. You get $2A = H$. Because Amy has $60, Hope has $120.

   Next, if Hope has $\frac{3}{4}$ as much as Dakota, then Dakota has $\frac{4}{3}$ as much as Hope. If $H = \frac{3}{4}D$, multiply both sides by $\frac{4}{3}$ and you get $\frac{4}{3}H = D$. $\frac{4}{3} \times 120 = \$160$. Dakota has $160.

   A is wrong because the student solved the numbers going in the wrong direction. B is wrong because it multiplies $120 by $\frac{3}{4}$ instead of $\frac{4}{3}$. If you chose C, you correctly doubled $60 to get $120, but you figured that if Hope's $120 is $\frac{3}{4}$ of Dakota, you would just need to add $\frac{1}{4}$ of $120 onto it. Finally, E is just a high-valued incorrect decoy.

12. **C.** Despite that the band is passé, 98° is the correct answer. If ray *OB* and ray *OA* are perpendicular, they form a right angle measuring 90°. If angle *AOC* is 8° greater than angle *AOB,* it must measure 98°.

   A is incorrect because it subtracts 8° from 90° instead of adding it. B is incorrect because it is the measure of right angle AOB. D is incorrect twice over because it starts with 180° (the measure of a straight angle or a straight line), and then subtracts 8°. E is incorrect because it starts with 180° and adds 8°.

13. **B.** Basically, if each side and its opposite have the reverse squares shaded, then for every two sides of the cube, the equivalent to one side of the cube is completely shaded. With three sets of opposite sides, that is the equivalent of three sides completely shaded. With 16 squares per side, $3 \times 16 = 48$ cubes shaded.

A, C, D, and E are all incorrect, representing two, four, five, and six sides of the cube, respectively.

14. **C.** As all of the possible answers are in terms of *x*, your best bet is to solve for *y* in terms of *x* in the first equation and substitute what you get into the second equation. When you add $x + 2x + 3x$ together, the first equation simplifies to $6x = y$. Now, given that you know $y = 6x$, substitute $6x$ for *y* in the second equation.

$2x - y$ becomes $2x - 6x$, which $= -4x$.

A is wrong because it violates the rules for combining signs using addition during the last step. B is incorrect as it ignores the $2x$ in the second equation and just subtracts out the $-6x$. You get D if you forget that *x* is equivalent to $1x$, so you end up with $5x = y$ instead of $6x = y$. E is an incorrect decoy, the exact opposite of the correct answer.

15. **A.** The fastest way to solve this problem is to recognize when you multiply exponents the exponent stays the same. $20^c = (4 \cdot 5)^c = 4^c \cdot 5^c$. If you keep the exponents the same on either side of the equation, *c* must be 3.

The other (and longer) way to figure this out is to simplify the right side of the equation and then play with your calculator to find out what power of 20 that is. The right side of the equation simplifies as follows: $4^3 \cdot 5^3 = 64 \cdot 125 = 8,000$. So, $20^c = 8,000$. Turns out 8,000 is $20 \times 20 \times 20$, which is 20^3; $c = 3$

B adds 3 and 3 together, C multiplies them together, and D raises one to the power of the other, all of which are mathematically incorrect. Check out the rules for bases and exponents in Chapter 3. E is just a false decoy representing 20×3.

16. **D.** To determine in what percent of the trials there were 7 or more made baskets, you have to divide the number of trials with 7 or more made baskets by the total number of trials, which is 50.

Starting with the 5th bar from the left, the number of trials where the players made 7 or more baskets is $18 + 5 + 3 + 2 = 28$ trials.

Finally, $28 \div 50 = 0.56$ or 56%.

Remember to convert from a decimal to a percentage, move the decimal two digits to the right and add a percentage sign. Reverse these directions to go from a percentage to a decimal.

A is not the correct answer because it represents the percentage of trials where *more than 7* made baskets occurred (as opposed to *7 or more*.) B is incorrect because the student divided the 28 trials by 100 total trials instead of 50. C is an incorrect decoy, and E is wrong because it represents the percentage of *6 or more* made baskets.

17. **E.** The function says that its value is ½ for any value of *x* you plug in that is greater than zero. When you plug in 0.5, you end up with or $1/(0.5)^2$ or $1/(½)^2$, depending on your preference. At this point, you can use your calculator (be sure to solve the items in parenthesis first) or solve it longhand: $(½)^2 = ¼$. $1/(¼) = 4$. (See our explanation for #3.)

A is wrong because it fails to multiply by the reciprocal. B is wrong because in squaring ½, it adds across instead of multiplying across. C is wrong because in squaring ½, it adds instead of multiplying. D is wrong because it ignores the exponent altogether.

18. **D.** Remember, the absolute value of a number is always positive. It simply refers to a number's absolute distance from 0. Therefore, the absolute value bars will make the expression $z - 3$ positive in all cases. Just be careful.

Start with what values make $|z - 3| = 4$. They are 7 and –1. Those two numbers become your limits. 7 and higher or –1 and lower makes the absolute value equal to or greater than 4. You need to find the values that make it less than 4. Any value that is less than 7 but greater than –1 will work. Therefore, the answer is $-1 < z < 7$.

A is wrong because it completely ignores the absolute value bars. Don't fall for that trick. B is similar but uses a greater-than symbol as a decoy. C is tricky as it moves one unit in from each side of the correct range. But you don't need to do that because the < and > symbols take care of that. Finally, E is C flipped around as another decoy.

19. **B.** The little half-square symbol in angle *ECF* indicates that it is a right angle measuring 90°. Therefore, triangle *ECF* is a right triangle.

The first thing you should notice is that segments CE and CF are both radii of the circle and, therefore, have the same length, 5. What we have here is a 45:45:90 triangle for which the ratio of the sides is $s : s : s\sqrt{2}$.

Chord *EF* is the other side of the triangle. To find its length, use your calculator to find the length for EF represented in the ratio: $5\sqrt{2}$. The answer is 7.07.

If you forget the ratio, you can apply the Pythagorean theorem $a^2 + b^2 = c^2$ to solve for chord *EF*. Here's how you do it, but to save time on the test memorize the ratios given to you in Chapter 6.

$$CE^2 + CF^2 = EF^2$$

Segments *CE* and *CF* are both radii of the circle and, therefore, have a length of 5. Substituting, you get $5^2 + 5^2 = EF^2$. Solving, you get $25 + 25 = EF^2$; simplifying, you get $50 = EF^2$. Now take the square root of both sides, and you get that $7.07 = EF$.

A is wrong because *EF* is not a radius of the circle. Also, the hypotenuse *EF* must be greater than the two legs of the right triangle. C is wrong because it adds the legs together, which is incorrect. D is wrong because it uses the value of the circle's diameter, 10, for the length of the two equal sides instead of the radius 5. Finally, E is wrong in that it fails to perform the final step of the second method, that of taking the square root of both sides of the equation.

20. **C.** In this arrangement with repetition, you have 2 choices for the first character (M or F), 10 choices for the second character (0–9), and 26 choices for the third character. Using the multiplication principle, you get $2 \times 10 \times 26 = 520$.

A is incorrect because it adds 2, 10, and 26 instead of multiplying them. B is incorrect because it neglects to account for the 0 in 0–9, using 9 as the second value instead of 10. D incorrectly adds the squares of 2, 10, and 26, and E mistakenly treats the first character as if it can be from any letter of the alphabet (26) instead of just M or F(2).

21. **A.** The graph shows that at the beginning of the day the jet plane is 0 miles from home — meaning it is home. Employ a little process of elimination. That makes B, C, and E incorrect in all cases. Finally, D is incorrect because it states that after the plane leaves home and flies to Medium Lake, it then flies to Bigville, "somewhat farther away from Tinytown than Medium Lake." That situation would have to be represented by a third segment that continues to increase in its distance from Tinytown, which is not the case.

22. **D.** To determine the perimeter of trapezoid FBCE, you need to solve for the lengths of segments *FB, BC, EC,* and *FE*.

If triangle ACE is equilateral, all angles are equal and measure 60°. Then this problem harps back to the wise adage, "If two parallel lines are cut by a transversal, the corresponding angles are congruent."

If angle FED measures 60°, then angle AFG must also measure 60°. You already know that angle FAB measures 60°; therefore, angle ABF must also measure 60°. Therefore, triangle ABF must be equilateral, and segment *FB* = 4.

Furthermore, segment *BC* = 6 (supplied by the problem). Given that FB∥EC, you know that if *BC* = 6, then *FE* must also equal 6. You know that because the two lines are parallel and angle FED = angle BCD = 60°.

Next, if *BC* = 6, then *EC,* the bottom side of the big triangle, must equal the other two sides of the big triangle (*AE* and *AC*), and you know the left side of the triangle is equal to 4 + 6, which equals 10. Finally, you add 4 + 6 + 10 + 6 and you get 26.

By the way, the fact that G is the midpoint of *FB* is irrelevant to the problem and simply serves to try to trick you.

A and B are just incorrect decoys. C is wrong because it either uses 8 (2×4) as the value of *EC* instead of 10 or 2 as the value of *FB* (trapping you with the irrelevant G). E is wrong because it uses 12 (2×6) as the value of EC instead of 10.

23. **D.** This problem, while long-winded, is fairly easy. Basically, for the cosines of *x* and *y* to be equal, the angles have to be equal. The angles will be equal when the vertical and horizontal legs of the right triangle are equal. Now, the vertical legs are already equal according to the problem, so all you have to do is make the horizontal legs equal. That simply involves moving the segment in question two units to the right on the number line — from 8 to 10 — thus increasing the distance of the horizontal segment from 2 to 4. A, B, C, and E are incorrect decoys just lying in wait for you to make silly mistakes.

24. **C.** This is a substitution problem. Substitute –1 for i^2. (Don't get caught thinking $i^2 = -1$.)

    The equation simplifies to $[(-1)^5]^x = 1$

    $(-1)^5 = -1$.

    To make –1 a positive number, its exponent has to be even. The value of *x* has to be an even number; the least positive integer that is also an even number is 2.

A is incorrect because zero is not positive. (Remember that zero is neither positive nor negative.) Zero is a tempting answer, however, because by definition, any number *n* raised to the zero power = 1 ($n^0 = 1$), but don't let this fool you. B is incorrect because anything raised to the 1st power is itself ($n^1 = n$), so the expression would still equal –1. D is just an incorrect decoy. E makes the equation true, but it is not the *least* positive integer value of *x*.

25. **D.** An entire circle measures 360°. If it takes 20 seconds to complete one full rotation, the lighthouse rotates 360°/20 = 18° per second. Multiply this by 2 seconds, and you get $18° \times 2 = 36°$.

    A is incorrect because it takes the correct answer of 36° and further divides it by the 4.5-mile distance, a piece of information that has no bearing on the solution to the problem. B is incorrect because it is the result of dividing 180° by 20 (instead of 360° by 20). C is incorrect because 18° is only the distance the lighthouse rotates in one second, not two. E is incorrect because it divides 360° by the 4.5-mile distance, a piece of information that has no bearing on the solution to the problem.

26. **B.** Recognizing that the triangle is right, you should use the equation: cosΘ = adjacent ÷ hypotenuse.

    Plugging in the correct values you get: cos 36° = *x* ÷ 20.

    Solving the equation, you get: 0.8090 . . . = *x* ÷ 20.

    Multiply both sides of the equation by 20 and you get: *x* = 16.18.

    A is incorrect because it multiplies 36° by 20, which is wholly incorrect. C is incorrect because it uses the equation for tangent instead of cosine. D is incorrect because it uses the equation for sine instead of cosine. E is incorrect because it divides cos 36° by 20 instead of multiplying by 20.

27. **C.** To solve this function, substitute 2 for *t* on the right side of the equation, and then solve for h(*t*): h(*t*) = 1,454 – 16(2)² + 8(2) + 4; h(*t*) = 1,410. Remember to follow the order of operations (parentheses, exponents, multiplications, and divisions from left to right, additions and subtractions from left to right).

    A is wrong because it uses the wrong order of operations, multiplying 16×2 first instead of solving 2^2. B is wrong because it divides the height of the Empire State Building by 2 instead of using the quadratic function. D, the height of the skyscraper, is an incorrect decoy. E is incorrect because it switches the sign of the whole term $-16t^2$ when the 2 is squared, thus adding the term instead of subtracting it.

28. **D.** To solve this problem, use the distance formula derived from the properties of the right triangle in the coordinate plane. To find out more about the distance formula, read Chapter 7.

$$d = \sqrt{(y_2 - y_1)^2 + (x_2 - x_1)^2}$$

Then solve for d.

$$d_{AC} = \sqrt{[-6 - (-2)]^2 + (4 - 2)^2}$$

$$d_{AC} = \sqrt{(-4)^2 + 2^2}$$

$$d_{AC} = \sqrt{16 + 4} = \sqrt{20}$$

$$d_{AC} = d_{AB} = \sqrt{20}$$

A is incorrect as the distance of OA is $\sqrt{8}$ and the distance of AB is $\sqrt{20}$. B is incorrect as the distance of OC is $\sqrt{52}$ and the distance of BC is $\sqrt{40}$. C is incorrect as the distance of OB is 6 and the distance of BC is $\sqrt{40}$. E is incorrect because the distance of OA is $\sqrt{8}$ and the distance of OB is 6.

29. **B.** This problem focuses on the rules of bases and exponents. The easiest way to solve this problem is to reduce each expression to a single power of 10 (10^x). Then you can compare the quantities. 1,000,000,000,000,000 is actually 10^{15} and ends up being the lowest value.

Remember, when you multiply something by 10 just add a zero. In this case, you can just count the number of zeros to find the exponent. Fifteen zeros means 15 is the exponent. You can learn more about bases and exponents in Chapter 3.

A is incorrect because it simplifies to 10^{70} [$100 = 10^2$; $10^2 \cdot 10^5 = 10^{2+5} = 10^7$; $(10^7)^{10} = 10^{70}$]. C is incorrect because it simplifies to 10^{300} [$1000 = 10^3$; $(10^3)^{100} = 10^{300}$]. D is incorrect because it simplifies to 10^{110} [$10 = 10^1$; $10^1 \cdot 10^{10} = 10^{1+10} = 10^{11}$; $(10^{11})^{10} = 10^{110}$]. E is wrong as it simplifies to $10^{2,000}$ [$100 = 10^2$; $(10^2)^{1,000} = 10^{2,000}$].

30. **C.** The plane that contains vertices S, U, and X cuts diagonally through the square prism from back-left to front-right. That plane will also contain vertex Z in the lower right corner. A, B, D, and E are all incorrect because none of the other vertices falls in this plane.

31. **E.** In a reflection problem, think of a mirror. The near points stay near and the far points stay far, and the distances from the reflection line remain the same. Therefore, think of the shape flipping and moving to the other side just as close or far away to the reflection line as it was on the original side.

As the shape is reflected on line l, which runs parallel to the y-axis, the y-coordinates will remain exactly the same. So, you are concerned only with the x-coordinates of point D. Point D is the lower left point of the square. When reflected, it will become the lower right point of the new square. In the original square, point D coordinates are $(-2, -2)$, which is 3 units to the left of the beginning of the square, which itself begins three units to the left of the reflection line, $x = 4$ (it runs vertically and it intercepts the x-axis at 4, 0).

Don't let the y-axis confuse you; line l is the critical line in this problem. Therefore, when reflected, point D will end up 3 points to the right of the beginning of a square, which itself begins 3 points to the right of the reflection line, $x = 4$. Finally, just add $4 + 3 + 3 = 10$. Point D would be at $(10, -2)$.

A is wrong because it reflects the square directly on itself and, further, it fails to flip the square. B is incorrect because it reflects the square using the y-axis as the new starting point. C is incorrect for two reasons: First, it incorrectly reflects the square, putting the left side of it dead on line l (instead of 3 units to the right of line l); second, it fails to flip the square. Finally, D is wrong because it correctly reflects the square beginning 3 units to the right of line l, but fails to flip the square.

32. **C.** To solve this problem, create a right triangle with segment EG as the hypotenuse; Leg 1 will be the segment created by point E and the unnamed point directly below it, and Leg 2 will be the segment created by that same unnamed point below E and midpoint G. (Mark those two lines in with dashes so you can see them!)

The distance of Leg 1 of the right triangle is 4 units, because it is an edge of the given cube with sides measuring 4 units. The distance of Leg 2 is trickier; you actually need to set up a *second* right triangle to solve for that. No problem. Quoting Rob Schneider, "We can do it!"

Leg 1*b* of this second right triangle is the segment created by the unnamed point below *E* and point *F*, and Leg 2*b* will be segment *FG*. These two values we know: Leg 1*b* is an edge of the cube and measures 4 units; Leg 2*b* is segment *FG*, which measures 2 units because *G* is the midpoint of the edge of the given cube, which measures 4 units.

A right triangle with sides of 2 and 4 does not obviously meet the specifications of one of those special ratio triangles, so use the Pythagorean Theorem ($c^2 = a^2 + b^2$) to solve for the length of leg 2 using what you know about Leg 1*b* and Leg 2*b*.

$c^2 = 4^2 + 2^2$

$c^2 = 16 + 4$

$c^2 = 20$

Take the square root of both sides and you get $c = \sqrt{20}$.

With your marked calculations, your figure now looks like Figure 13-1.

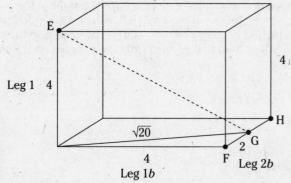

Figure 13-1: Sample drawing.

Segment *EG* is the hypotenuse of a right triangle with sides measuring 4 and $\sqrt{20}$. Again, this triangle does not easily fit into the special triangle category, so apply Pythagorean again.

$c^2 = 4^2 + \left(\sqrt{20}\right)^2$

$c^2 = 16 + 20$

$c^2 = 36$

Take the square root of both sides and you get $c = 6$.

Whew! Solving this problem eats up some minutes, but if you apply the concepts you know about right triangles, you can systematically solve these types of problems in relatively little time.

A is an incorrect decoy. B is wrong because it solves for the diagonal of one of the faces of the cube and not segment *EG*. D just mistakenly adds 4 + 4 as a decoy, and E mistakenly adds 4 + 4 + 2 as a decoy.

33. **C.** By definition, an average (or arithmetic mean) equals the sum of all the scores divided by the number of scores. For a more detailed discussion of averages and means, see Chapter 10. Set up an equation and plug in what you know.

$$average = \frac{sum}{number\ of\ terms}$$

On the left of the equation is the arithmetic mean of the scores, which the problem tells you is 84. On the right is the sum of all eleven scores (we suggest putting them in parenthesis) divided by the total number of scores which is 11.

The problem tells you that the average of these first six scores is 79; therefore, in your equation, the sum of the first six scores can be represented by 6×79. Similarly, the sum of the last 5 scores can be represented by $5x$ (meaning 5 times x), with x representing the average score of the five remaining students. The equation looks like this.

$$84 = \frac{(6 \times 79 + 5x)}{11}$$

Now, multiply both sides of the equation by 11 and simplify 6×79 within the parenthesis.

$924 = 474 + 5x$

Subtract 474 from both sides.

$450 = 5x,\ 90 = x.$

A, B, D, and E are decoy answers waiting for you to make careless arithmetical errors. Be careful!

34. **B.** Drawing a diagram is your best strategy for this problem. As you can see in Figure 13-2, if line *l* has a negative slope and a positive *x*-intercept (to the right of the *y*-axis), its *y*-intercept has to be positive as well (above the *x*-axis). If line *m* is parallel to line *l*, but it has a negative *x*-intercept (to the left of the *y*-axis), its *y*-intercept must also be negative (below the *x*-axis) and is clearly less than that of line *l* (it is lower down on the *y*-axis).

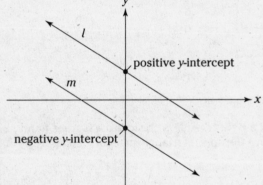

Figure 13-2:
Drawing a
diagram.

A, C, and D do not match the answer revealed by the diagram, and E is there to distract you.

35. **E.** The expression factors into $(x - 2)(x - 1)$, which yields roots of 2 and 1, the sum of which is 3 and product of which is 2.

Quick and dirty factoring of all five options will work here, but it is a bit tedious. If you can, try to recognize that factoring A and C, which have all positive terms, will yield only negative roots. In that case the product will be positive, but the sum will be negative. Knowing this tidbit of information at least eliminates A and C, so you only have to try B, D, and E. B and D, when factored, do not satisfy the requirements of the problem.

36. **D.** If two triangles are similar, one is an enlargement of the other with the same angles and respective sides in the same proportion to one another. Therefore, set up the proportion you need to solve the problem.

The problem asks you to find the length of *MO*. Therefore, set up a proportion using two similar sides (one from each triangle, just like *a* and *b*) to figure out the ratio of the two triangles. Between the diagram and information from the problem, you know the lengths of *NM* (1) and *PQ* (3). These are corresponding sides of both triangles. Be sure and stay consistent in lining up your proportion. Keep the sides from one triangle on the top and the sides from the other on the bottom. The ratio of the small triangle to the large triangle, therefore, is ⅓.

Now apply the ratio of ⅓ to the length of *MO*. You know that *PO* equals 4. Its corresponding side on the smaller triangle is *MO*, the unknown length (*x*).

$$\frac{1}{3} = \frac{x}{4}$$

Cross multiply: $3x = 4$

$x = ⅘$

A designates the ratio of the two triangles in proportion to one another, but it does not go the step further to apply that proportion. B uses the correct method for determining the length, but does not relate corresponding sides of the two triangles. C provides an incorrect ratio based on comparing incongruent sides of the two triangles. E is put there to trap you into guessing based on looking at the figure.

37. **D.** You can rough in the graph of this problem, and it will help you visualize the problem, but if you can't solve it visually, graphing eats up precious time.

This problem involves two equations and substitution. The first equation is that of the line.

$x = 6$

The second equation is that of the circle. The general equation for a circle is $(x-h)^2 + (y-k)^2 = r^2$, with (h, k) being the *x*- and *y*-coordinates of the center of the circle and *r* being the radius. Thus, the equation for this specific circle is as follows:

$$(x-3)^2 + (y-5)^2 = 4^2$$

Now, using the first equation $x = 6$, substitute 6 for *x* in the equation of the circle. You can now work out the equation.

$$(6-3)^2 + (y-5)^2 = 4^2$$
$$3^2 + (y-5)^2 = 16$$
$$9 + (y-5)^2 = 16$$

Subtract 9 from both sides.

$$(y-5)^2 = 7$$

Take the square root of both sides.

$$y - 5 = \pm\sqrt{7}$$

Add 5 to both sides.

$y = \pm\sqrt{7} + 5$ or $y = 5 \pm \sqrt{7}$, meaning $5 + \sqrt{7}$ and $5 - \sqrt{7}$, which equals 7.65 and 2.35.

A is wrong and is the result of trying to eyeball a rough graph. B is wrong and is the result of trying to eyeball the graph after mistakenly drawing the line $y = 6$ instead of $x = 6$. C is an incorrect decoy adding and subtracting the radius 5 to the *y*-coordinates. Last, E is wrong because it substitutes 6 for *y* instead of 6 for *x*.

38. **C.** For this problem you have to use the typical formula for exponential growth NV = IV × (1 + P)^Y, where NV = new value, IV = initial value, P = percent increase, and Y = number of years. You may also see this formula associated with exponential population growth. Note that when you see a percentage growth from year to year, you have to use a formula for exponential growth. If you see an increase by an absolute amount from year to year (that is $200 or 1,000 people), the problem is talking about linear growth and you can simply multiply and add.

Plug the numbers into the formula.

$NV = 1,000 \times (1 + 0.025)^{10}$

$NV = 1,000 \times (1.025)^{10}$

$NV = 1,000 \times 1.280084544$

$NV = 1,280.08$, which rounds to $1,280.

B is incorrect because it simply multiplies $0.025 \times \$1000 \times 10$ years and adds that product to the $1,000 principal. Again, that would be the case if it were linear growth, increasing by an absolute amount each year (in this case called *simple interest*). A is incorrect for the same reason, plus it uses 0.0025 for the percentage rate instead of 0.025. (Remember the rules for converting percentages to decimals. See Chapter 3 if you have questions.) D is an incorrect decoy. E is wrong because it uses 0.25 as the percentage increase instead of 0.025.

39. **A.** The sum of the interior angles of a polygon $= 180° \times (n - 2)$, where n is the number of sides. If you can't remember that formula, simply divide the given shape into triangles. Each triangle measures 180°, so for each triangle add 180°, and you get the sum of all the angles in the polygon.

In this problem, with 4 sides, you get a sum of 360°. Now look at the first half of the given statement. If $a + b$ is more than 70°, the lowest it can be is 70° and the highest $c + d$ can be is $360 - 70 = 290$; the higher $a + b$ gets, the lower $c + d$ must get; so the first part of the answer is $c + d < 290$. The second half of the equation says that $a + b$ has to be less than 150°. If $a + b$ is less than 150°, then the highest it can be is 150°, and the lowest $c + d$ can be is $360 - 150 = 210$; the lower $a + b$ gets then the higher $c + d$ must get; therefore, the second half of the answer is $c + d > 210$. Combine those two deductions, and you have your answer.

B, C, D, and E are all incorrect because they incorrectly use 180°, 270°, 540°, and 720°, respectively, as the total for the interior angles of the polygon instead of 360°.

40. **B.** Without getting into the math behind Thales' theorem, what you need to know for this problem is that an inscribed angle (an angle with all three points on the circumference of a circle) in a semicircle is a right angle. For more details, be sure to read Chapter 6. Thus, A, C, D, and E are all incorrect.

41. **A.** By definition, $\sin^2\Theta + \cos^2\Theta = 1$, so the expression simplifies to $(1 - 5)^3$, which simplifies to $(-4)^3$ which equals -64. B and C are incorrect because they multiply two numbers together instead of treating the 3 like an exponent. D is incorrect because when you raise -4 to the third power $(-4 \times -4 \times -4)$, the product is negative, not positive. E is incorrect because it reverses the two numbers before solving.

42. **D.** -2 from the domain produces the lowest value of the range, which is 0 $(-2 + 2 = 0$ and $0^2 = 0)$. On the high end, 3 from the domain produces the highest value of the range, which is 25 $(3 + 2 = 5$ and $5^2 = 25)$. Despite that -3 is the lowest value in the domain, it does not produce the lowest value in the range; therefore A and B are incorrect. C is incorrect because it just squares 3 and forgets to add it to the 2. E is incorrect because it multiplies 5×2 instead of raising 5 to the second power.

43. **C.** The area of a parallelogram equals $b \times h$. The problem gives us the base, which is 8. To find the height, you have to somehow create a right triangle.

Angle *EHG*, measuring 120°, is the starting point. In a parallelogram, opposite angles are congruent (equal) and all four angles add up to 360°. Therefore, angle EFG also measures 120°, which means angle *HEF* (and angle *FGH*) measure 60°. Next, drop a straight line down from the upper left corner to the bottom of the parallelogram in order to create a 30-60-90 right triangle on the left side of the parallelogram. Remember our special right triangles? If not, go to Chapter 6 to find out all about them.

The special properties of this right triangle will help you determine the height. The smaller side (opposite the 30° angle) is equal to half the hypotenuse and the longer side (opposite the 60°angle) is equal to the smaller side multiplied by $\sqrt{3}$. So the side ratio for a 30:60:90 triangle is $s: s\sqrt{3}: 2s$. Therefore, the smaller side is ⁶⁄₂ = 3 and the longer side, which is the height, is $3 \times \sqrt{3} = 3\sqrt{3}$.

Now go back to the original formula for area.

$A = 8 \times 3\sqrt{3} = 24\sqrt{3}$.

A is not the answer because it multiplies the base by the smaller side of the right triangle, which is not the height. B is wrong because it multiplies the two sides of the parallelogram together and ignores the height. D is incorrect because it uses two sides that are not functional in determining area the way the solution was set up. E uses the wrong base according to the way the solution was set up.

44. **B.** First, solve for the area of the pizza using the formula for the area of a circle: $A = \pi \times r^2$. You have to divide the diameter of 122 feet by 2 to get the correct radius of 61 feet.

Now plug in: $A = \pi \times (61)^2 = 11{,}689.86626$ square feet. You want 10 slices per square foot, so multiply that number by 10, which equals 116,898.6626 total slices of pepperoni on the pizza. Finally, divide that large number by 50 to figure out how many pepperoni sticks are needed; the answer is 2,337.97. Don't forget to round up to 2,338 so you don't run out of pepperoni!

A is incorrect because it uses the formula for circumference instead of area. C is just an incorrect middle value put in as a decoy. D is incorrect because it uses the wrong order of operations when solving for the area, multiplying $61 \times \pi$ first instead of squaring 61 first. E is incorrect because it uses the 122 foot value of the diameter in the area formula instead of dividing it by 2 to get the radius.

45. **D.** To solve this problem, draw in a right triangle to solve for the length of the diagonal sides *XO* and *YO*. (They will be equal.) Leg 1 will be the segment that starts with *Y* at the top and ends at the point where the segment heading down the cylinder meets the lower base; Leg 2 is a radius that starts from that point and ends at center *O* of the lower base.

Now use the Pythagorean theorem: $c^2 = a^2 + b^2$. Side *a*, which is Leg 1, has a value of the height of the cone, which is 6. Side *b*, Leg 2, has a value of the radius of the circle, which is 2. Now solve.

$c^2 = 6^2 + 2^2$

$c^2 = 36 + 4$

$c^2 = 40$

Take the square root of both sides and $c = \sqrt{40}$.

You now know that $YO = XO = \sqrt{40}$. Now add up the three sides, the top side being the diameter of the circle, which is 4. You know that because the diameter is twice the radius.

$4 + \sqrt{40} + \sqrt{40} = 16.65$

A is incorrect because it assigned the value 6 to the hypotenuse of the right triangle instead of to one of the legs. B is incorrect because it solves only for one of the three sides of the triangle and forgets to add it to the other two. C is an incorrect decoy. E is wrong because it mistakenly adds 6 + 2 to get 8 for the value of the diagonal sides of the triangle instead of using the Pythagorean theorem.

46. **E.** Just using a quick number substitution for each possible answer is the fastest way to solve this problem. Stop when plugging in 1, 2, and 3 fails to give you 2, 6, and 10. A, B, C, and D all fail to work for all the terms in the sequence.

47. **B.** To solve for an inverse function, you replace $f(x)$ with y, and then solve for x in terms of y, in effect reversing the function.

$y = 2x^2 + 3; y - 3 = 2x^2$

$(y - 3) \div 2 = x^2$

$\sqrt{(y - 3) \div 2} = x$

Now plug in 9 for y and solve for x.

$x = \sqrt{(9 - 3) \div 2} = \sqrt{6 \div 2} = \sqrt{3} = 1.73$

A, C, and D are all incorrect because they are the square roots of 2, 5, and 6, respectively. E is incorrect because it fails to reverse the function and plugs 9 into the function as is.

48. **D.** You have two equations: $s \cdot t$ means $s = t^2 + 2$ and $t \cdot u$ means $t = u^2 + 2$. Because you want to solve for s in terms of u, you have to make the t's disappear. The second equation already has solved for t in terms of u, so use that and substitute $u^2 + 2$ for t in the first equation and you will have your answer of s in terms of u.

$s = t^2 + 2$

$s = (u^2 + 2)^2 + 2$

Now multiply it out and combine like terms.

$s = u^4 + 4u^2 + 4 + 2$

$s = u^4 + 4u^2 + 6$

A is wrong as it is just the right side of the initial equations without anything further done. B, C, and E all involve errors which break major mathematical principles.

49. **C.** The area of the shaded region bound by the graph of the parabola $y = f(x)$ and the x-axis is still 5. Replacing $f(x)$ with $f(x - 2)$ does nothing other than shift the parabola over 3 units to the right; thus, the area remains the same. Don't let the presence of the y-axis fool you; it does not affect the area described in this problem. A, B, C, and D are all incorrect decoys that subtract, divide, add, and multiply, respectively, 5 and 3.

50. **E.** You want lemonade to end up to be 15% of the total beverage, so the best way to solve the problem is to set up a proportion that will have lemonade on top and the total liquid on the bottom — on both sides of the equal sign.

On the left side, put $\frac{15}{100}$, because you want lemonade to be 15% of the total or 15 parts per 100. On the right side, on top, you need to put the total amount of lemonade that will end up in the beverage; we'll call that x because that is what the problem asks us to solve. On the bottom is the total amount of combined liquid. We represent that with $2 + x$, which is the two liters of iced tea plus x amount of lemonade.

From there, cross multiply.

$15(2 + x) = 100x$

Distribute the 15.

$30 + 15x = 100x$

Combine your x's on one side of the equation.

$30 = 85x.$

Divide both sides by 85.

$x = 0.35$

A is incorrect because it multiplies the 2 liters of iced tea by 15%, not accounting for the addition of liquid in making the combination. B is wrong due to faulty rounding of decimals. C and D are incorrect decoys.

Chapter 14

Practice Test 2, Level IIC

. . . .

*O*kay, you know your stuff. Now's your chance to shine. The following exam is a 50-question, multiple-choice test. You have 60 minutes to complete it.

To make the most of this practice exam, take the test under similar conditions to the actual test.

1. **Find a place where you won't be distracted (preferably as far from your younger sibling as possible).**

2. **If possible, take the practice test at approximately the same time of day as that of your real SAT II.**

3. **Set an alarm for 60 minutes.**

4. **Mark your answers on the provided answer grid.**

5. **If you finish before time runs out, go back and check your answers.**

6. **When your 60 minutes is over, put your pencil down.**

After you finish, check your answers on the answer key at the end of the test. Use the scoring chart to find out your final score.

Read through all of the explanations in Chapter 15. You learn more by examining the answers to the questions than you do by almost any other method.

Table 14-1	For Your Reference

You can use the following information for reference when you are answering the test questions.

The volume of a right circular cone with radius r and height h: $V = \frac{1}{3}\pi r^2 h$

The lateral area of a right circular cone with circumference of the base c and slant height l: $S = \frac{1}{2}cl$

The volume of a sphere with radius r: $V = \frac{4}{3}\pi r^2$

The surface area of a sphere with radius r: $S = 4\pi r^2$

The volume of a pyramid with base area B and height h: $V = \frac{1}{3}Bh$

For each of the following 50 questions, choose the best answer from the choices provided. If no answer choice provides the exact numerical value, choose the answer that is the best approximate value. Fill in the corresponding oval on the answer grid.

<u>Notes:</u>

1. **You must use a calculator to answer some but not all of the questions.**

   You can use programmable calculators and graphing calculators. You should use at least a scientific calculator.

2. **Make sure your calculator is in degree mode because the only angle measure on this test is degree measure.**

3. **Figures accompanying problems are for reference only.**

   Figures are drawn as accurately as possible and lie in a plane unless the text indicates otherwise.

4. **The domain of any function f is assumed to be the set of all real numbers x for which $f(x)$ is a real number unless indicated otherwise.**

5. **You can use the table on the previous page for reference in this test.**

1. If $1 - \frac{2}{a} = 3 - \frac{4}{a}$, then $1 - \frac{2}{a} =$

   (A) –1 (B) 0 (C) 1 (D) ⅔ (E) ¾

   $1 - 2 = -1$

2. $2x\left(\frac{3}{y} + \frac{4}{z}\right) =$

   $a - 2 = 3a - 4$
   $2a = 2$
   $a = 1$

   (A) $\frac{24x}{yz}$

   (B) $\frac{24x}{y+z}$

   (C) $\frac{14x}{y+z}$

   $\frac{6x}{y} + \frac{8x}{z}$

   (D) $\frac{8xy + 6xz}{yz}$

   (E) $\frac{14x}{y+z}$

3. The figure shows one cycle of the graph of the function $y = \sin x$ for $0 \le x \le 2\pi$. If the maximum value of the function occurs at point A, the coordinates of A are

   (A) $\left(\frac{\pi}{4}, 1\right)$

   (B) $\left(\frac{\pi}{4}, \pi\right)$

   (C) $\left(\frac{\pi}{2}, 0\right)$

   (D) $\left(\frac{\pi}{2}, 1\right)$

   (E) $\left(\frac{\pi}{2}, \pi\right)$

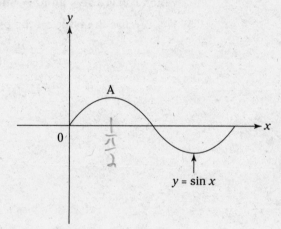

Note: Figure not drawn to scale.

4. If $\sqrt{5p} = 3.67$, then $p =$

   (A) 0.73 (B) 1.64 (C) 2.69 (D) 8.21 (E) 18.35

5. In the figure, $z \sin\Theta =$

   (A) x

   (B) y

   $\sin\theta = \frac{y}{z}$

   (C) z

   (D) $\frac{y}{z}$

   $z \cdot \frac{y}{z}$

   (E) $y + z$

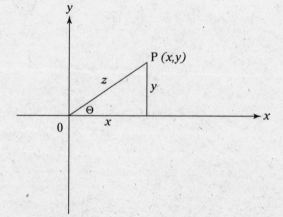

6. If a, b, and c are nonzero real numbers and if $a^3 b^5 c^6 = \frac{a^2 c^6}{3b^{-5}}$, then $a =$

   $a^1 = \frac{1}{3}$ a

Go on to next page

(A) $\frac{b^{10}}{3}$ (B) $3b^{10}$ (C) $\frac{c^{12}}{3}$ (D) ⅓ (E) 3

7. If 4 and –3 are both zeros of the polynomial $p(x)$, then a factor of $p(x)$ is

    (A) $x^2 - 12$

    (B) $x^2 + x + 12$

    (C) $x^2 + x - 12$

    (D) $x^2 - x + 12$

    (E) $x^2 - x - 12$

$p(x) = (x-4)(x+3)$
$x^2 - x - 12$ ✓

8. Parasailing is somewhat like waterskiing, except that it is connected to a parachute instead of waterskiis and trails the boat high up in the air instead of being directly behind it. If the rope onto which a parasailer is holding is 50 meters in length and the rope makes an angle of 53° with the surface of the water, what is the distance in meters from the parasailer to the surface of the water? (Assume that the rope is taut and the surface of the water is level.)

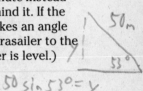

    (A) 66 (B) 50 (C) 40 (D) 30 (E) 20

$50 \sin 53° = y$

9. Picture a large square in a single plane. P and Q are two different points in that square. The set of all points inside this square that is closer to P than to Q is

    (A) the interior of a smaller square inside the square

    (B) the interior of a smaller circle inside the square

    (C) the region of the plane on one side of a line

    (D) a pie slice region of the plane

    (E) the region of the plane bound by a parabola

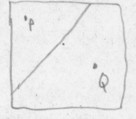

10. If $f(x) = \sqrt{0.4x^2 - x}$ and $g(x) = \frac{2x+2}{x-6}$, then $f(g(20)) =$

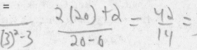

    (A) 0.77 (B) 1.06 (C) 1.34 (D) 2.87 (E) 4.40

$\sqrt{.4(3)^2 - 3}$ $\frac{2(20)+2}{20-6} = \frac{42}{14} = 3$

11. Where defined, $\sec(3\Theta)\cos(3\Theta) =$

    (A) 0

    (B) $\sec^2(3\Theta)$

    (C) $\cos^2(3\Theta)$

    (D) 1

    (E) $2\sin(3\Theta)$

12. If $f(x) = 2x + 4$ and $f(g(2)) = 24$, which of the following could be $g(x)$?

    (A) $-4x + 2$ $-4 \cdot 2 + 2 = -6$

    (B) $4x + 2$ $4 \cdot 2 + 2 = 10$

    (C) $2x + 4$

    (D) $-6x + 4$

    (E) $6x - 4$

13. The figure shows a cube with an edge of length 5 centimeters. If points W and Z are midpoints

Go on to next page ⟶

of the edges of the cube, what is the area of
the region *WXYZ*?

(A) 5.59 cm

(B) 11.18 cm

(C) 22.36 cm

(D) 31.25 cm

(E) 976.56 cm

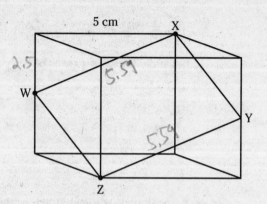

14. In the figure, the equation of line *l* is

(A) $x + y = 4$

(B) $x = 0$

(C) $x = 4$

(D) $y = 4$

(E) $y = x + 4$

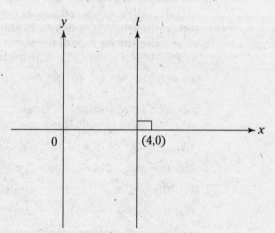

15. The mean salary of the 14 employees of Simon's Crougar Industrial Smoothing, Inc., is
$51,000. When George, a new employee, is hired, the mean increases to $51,200. What was
the salary, in dollars, at which George was hired?

(A) 54,000

(B) 53,000

(C) 52,000

(D) 51,700

(E) 51,400

16. If $-2\pi < x < 2\pi$ and $\sin x = 0.866$, what is the value of $\tan\left(\frac{x}{1.5}\right)$?

(A) 0.0002

(B) 0.4196

(C) 0.8390

(D) 0.9968

(E) 1.1546

Go on to next page

17. Recently, on a day when many in Congress were sick or back in their home states, 410 members of the House of Representatives and 90 members of the Senate voted on a tax increase. A total of 350 members from the two governmental bodies voted no. If the same percentage of the members in each governmental body voted no, how many of the members of the House of Representatives voted no?

    (A) 349

    (B) 287

    (C) 206

    (D) 70

    (E) 63

18. If $f(x,y) \Vdash (x, y + 3x)$ for every (x,y) in the plane, for what points (x,y) is it true that $(x,y) \Vdash (x,y)$?

    (A) $(0,0)$ only

    (B) $(1,1)$ only

    (C) the set of points (x,y) such that $x = \frac{1}{3}$

    (D) the set of points (x,y) such that $y = 0$

    (E) the set of points (x,y) such that $x = 0$

19. What number should be added to each of the three numbers 1, 9, and 33 so that the resulting three numbers form a geometric progression?

    (A) 1 (B) 2 (C) 3 (D) 4 (E) 5

20. If $f(x) = ax^2 + bx + c$ for all real numbers x and if $f(0) = 2$ and $f(1) = 5$, then $a + b =$

    (A) 0 (B) 1 (C) 2 (D) 3 (E) 4

21. What is the degree measure of the smallest angle of a triangle that has sides of length 2, 5, and 5?

    (A) 11.31

    (B) 11.54

    (C) 23.07

    (D) 47.16

    (E) 78.46

22. The figure shows a portion of the graph of $y = 2^x$. What is the sum of the areas of the three inscribed triangles shown?

    (A) 3.5 (B) 6 (C) 7 (D) 14 (E) 21

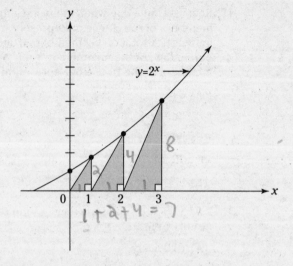

23. Which of the following lines are vertical asymptotes of the graph of $y = \dfrac{x^2 + 2x - 3}{x^2 - 5x - 6}$?

    I. $x = 6$

    II. $x = -1$

    III. $x = 3$

    (A) I only

    (B) II only

    (C) III only

    (D) I and II only

    (E) I, II, and III

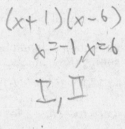

24. If, in the triangle in the figure, OA = AB = BO, what is the slope of segment OA?

    (A) $\sqrt{3}$

    (B) $\dfrac{\sqrt{3}}{2}$

    (C) $-\sqrt{3}$

    (D) $-\dfrac{\sqrt{3}}{2}$

    (E) It cannot be determined from the information given.

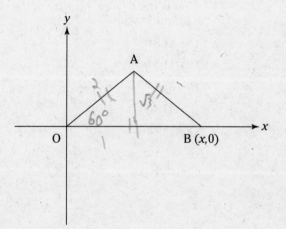

Go on to next page

25. The graph $y = f(x)$ is shown in the figure. Which of the following could be a graph of $y = |f(x)|$?

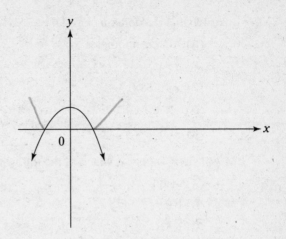

(A)

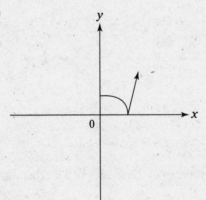

(B)

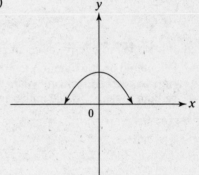

(C)

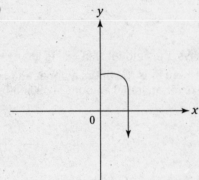

(D)

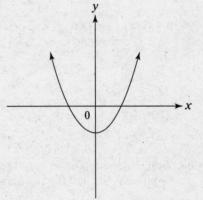

(E)

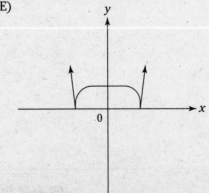

Go on to next page

26. What is the domain of $f(x) = \sqrt[3]{5 - 2x^2}$?

(A) All real numbers

(B) $x > 0$

(C) $x < 1.58$

(D) $-1.36 < x < 1.36$

(E) $-1.58 < x < 1.58$

27. If $\sin x = \cot an\ x$, which of the following is a possible radian value of x?

(A) -1.00

(B) 0.62

(C) 0.67

(D) 0.90

(E) 1.00

$\sin x = \cot x$

28. When the radioactive element radium decays, the amount that exists at any time t can be calculated by the function $R(t) = R_0\, e^{\frac{-4t}{1000}}$, where R_0 is the initial amount and t is the elapsed time in years. How many years would it take for an initial amount of 500 milligrams of radium to decay to 200 milligrams?

(A) 0.4

(B) 99

(C) 229

(D) 598

(E) 916

$t = -250$

$200 = 500\, e^{\frac{-4t}{1000}}$

$e^{\frac{-4t}{1000}} = \frac{2}{5}$

$\frac{-4t}{1000} = \ln\left(\frac{2}{5}\right)$

29. If $f(x - 1) = x + 1$ for all real numbers x, then $f(x) =$

(A) $-x - 1$

(B) $x + 2$

(C) $x + 1$

(D) $x - 2$

(E) $2x + 1$

$f(x-1) = x+1$

$f(x) = x+2$

30. Which of the following could be the coordinates of the center of a circle tangent to the negative y-axis and the positive x-axis?

(A) $(-3, 3)$

(B) $(0, -3)$

(C) $(3, -2)$

(D) $(3, -3)$

(E) $(3, 0)$

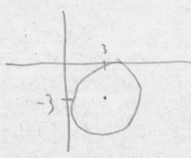

Go on to next page

31. If $2x - 6y + 3 = 0$ and $3y - x^2 = 0$ for $x \le 0$, then $x =$

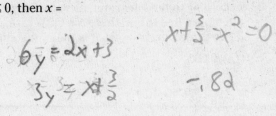

(A) −0.62

(B) −0.82

(C) −1.52

(D) −2.22

(E) 3.61

32. What is the range of the function defined by $f(x) \begin{cases} x^{\frac{1}{2}}, x \ge 4 \\ 3x + 1, x < 4 \end{cases}$?

(A) $y \ge 0$

(B) $y \ge 2$

(C) $y < 13$

(D) $2 \le y < 13$

(E) all real numbers

33. If $f(x) = \log_3 x$ for $x > 0$, then $f^{-1}(x) =$

(A) 3^x

(B) x^3

(C) $\frac{x}{3}$

(D) $\frac{3}{x}$

(E) $\log_x 3$

34. If $x_0 = 2$ and $x_{n+1} = \sqrt{5 + x_n}$, then $x_3 =$

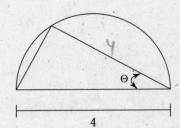

(A) 2.236

(B) 2.646

(C) 2.765

(D) 2.787

(E) 7.118

35. The figure shows a triangle inscribed in a semicircle. What is the perimeter of the triangle in terms of Θ?

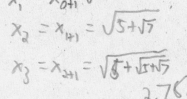

(A) 8

(B) $8 \sin\Theta \cos\Theta$

(C) $4(\sin\Theta + \cos\Theta + 1)$

(D) $4 \tan\Theta + 4$

(E) 16

Go on to next page

36. According to the latest statistics, each time a particular professional basketball player attempts a free-throw, there is a 0.537 probability that he will make the shot. If he shoots 3 free-throws in a row, what is the probability that he makes them all?

(A) 0.083

(B) 0.155

(C) 0.288

(D) 0.537

(E) 1.611

37. If the magnitudes of vectors x and y are 13 and 4, respectively, then the magnitude of $(x - y)$ could NOT be

(A) 17

(B) 13

(C) 11

(D) 9

(E) 4

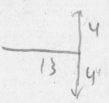

38. If $(3.45)^a = (6.78)^{2b}$, what is the value of $\frac{a}{b}$?

(A) 0.647

(B) 0.773

(C) 1.55

(D) 1.29

(E) 3.09

39. $\dfrac{[n!]^2}{\left[(n-2)!\right]^2} =$

(A) $[n][n-1]$

(B) $[n]^2$

(C) $[n]^2[n-1]^2$

(D) $\dfrac{1}{n^2}$

(E) $\dfrac{[n!]^2}{\left[(n-1)!\right]^2}$

40. If the 40th term of an arithmetic sequence is 90 and the 65th term of the sequence is 160, what is the first term of the sequence?

(A) 22

(B) 19.8

(C) 2.8

(D) –19.8

(E) –22

Go on to next page

41. If $\arcsin(\sin x) = \pi$, and $\pi < x < 2\pi$, then x could equal

    (A) π

    (B) $\frac{5\pi}{4}$

    (C) $\frac{4\pi}{3}$

    (D) $\frac{3\pi}{2}$

    (E) 2π

42. For all Θ, $\sin\Theta - \sin(-\Theta) + \cos\Theta - \cos(-\Theta) + \tan\Theta - \tan(-\Theta)$

    (A) 0

    (B) $2(\sin\Theta + \cos\Theta + \tan\Theta)$

    (C) $2\cos\Theta$

    (D) $2\sin\Theta + 2\tan\Theta$

    (E) 3

43. If n distinct lines in a plane intersect at a point, and another line ℓ intersects one of these lines in a single point, what is the *least* number of these n lines that ℓ could intersect?

    (A) $\frac{n-1}{2}$

    (B) $\frac{n}{2}$

    (C) $n-1$

    (D) n

    (E) $n+1$

44. The radius of the base of a right circular cone is 10 and the radius of a parallel cross section is 4. If the distance between the cross section and the point of the cone is 6, what is the volume of the cone?

    (A) 80π

    (B) 240π

    (C) 333π

    (D) 500π

    (E) 700π

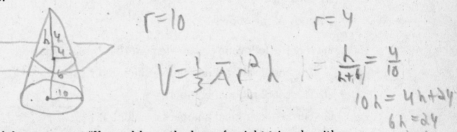

45. An indirect proof of the statement "If a and b are the legs of a right triangle with a hypotenuse c, then $a^2 + b^2 = c^2$" could begin with the assumption that

    (A) $a^2 + b^2 = c^2$

    (B) $a^2 + b^2 \neq c^2$

    (C) a and b and c are all > 0

    (D) a and b form a right angle

    (E) $a + b = c$

Go on to next page

46. Suppose the graph $f(x) = -x^2 + 1$ is translated 1 unit right and 2 units down. If the resulting graph represents $g(x)$, what is the value of $g(1.4)$?

    (A) –6.76

    (B) –2.76

    (C) –1.16

    (D) 2.76

    (E) 2.84

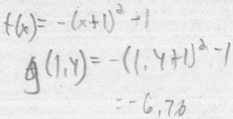

47. In how many ways can 8 people be divided into two groups, one with 5 people and the other with 3 people?

    (A) 40 (B) 56 (C) 120 (D) 336 (E) 6720

48. What is the equation of a vertical ellipse centered on $(3, -2)$ and whose two axes measure 10 and 20?

    (A) $(x-3)^2 + (y+2)^2 = 10$

    (B) $\dfrac{(x+3)^2}{100} + \dfrac{(y-2)^2}{25} = 1$

    (C) $\dfrac{(x+3)^2}{25} + \dfrac{(y+2)^2}{100} = 1$

    (D) $\dfrac{(x-3)^2}{100} + \dfrac{(y+2)^2}{25} = 1$

    (E) $\dfrac{(x-3)^2}{25} + \dfrac{(y+2)^2}{100} = 1$

49. Under which of the following conditions is $\dfrac{yx}{y-x}$ negative?

    (A) $0 < y < x$

    (B) $0 < x < y$

    (C) $x < y < 0$

    (D) $y < 0 < x$

    (E) none of the above

50. Which of the following has an element that is more than any other element in the set?

    I. The set of rational numbers r such that $r \le \pi$

    II. The set of rational numbers

    III. The set of rational numbers r such that $r < 2$

    (A) I only

    (B) II only

    (C) III only

    (D) I and III

    (E) none

Answer Key for Practice Exam 2

1.	A	26.	A
2.	D	27.	D
3.	D	28.	C
4.	C	29.	B
5.	B	30.	D
6.	D	31.	B
7.	E	32.	E
8.	C	33.	A
9.	C	34.	D
10.	A	35.	C
11.	D	36.	B
12.	B	37.	E
13.	D	38.	E
14.	C	39.	C
15.	A	40.	E
16.	C	41.	A
17.	B	42.	D
18.	E	43.	C
19.	C	44.	D
20.	D	45.	B
21.	C	46.	C
22.	C	47.	B
23.	D	48.	E
24.	A	49.	A
25.	E	50.	E

Scoring Your Exam

First, calculate your "raw score." On this exam, you get one point for each correct answer, and you lose ¼ (0.25) of a point for each wrong answer. Your raw score, then, is determined by the following formula: Raw score = number of correct answers − (0.25 × number of wrong answers). Feel free to use this Raw Score Wizard to help you determine your raw score:

Raw Score Wizard

1. Use the answer key to count the number of right answers   1) _44_

2. Use the answer key to count the number of wrong answers   2) _5_

3. Multiply the number of wrong answers by 0.25 (Line 2 × 0.25)   3) _1.25_

4. Subtract this number from the number of right answers
   (Line 1 − Line 3)   4) _42.75_

5. Round this number to the nearest whole number
   (Round Line 4 up or down)   5) _43_

   This final number is your raw score.

Use Table 14-2 on the next page to convert your raw score to one of the College Board Scaled Scores, which range from 200–800. Ultimately, when you take the test for real, this scaled score will be the one the College Board reports to you and the colleges/universities you request.

Note: The College Board explains that it converts students' raw scores to scaled scores in order to make sure that a score earned on one particular subject test is comparable to the same scaled score on all other versions of the same subject test. Scaled scores are adjusted so that they indicate the same level of performance regardless of the individual test taken and the ability of the specific group of students that takes it. In other words, a score of 500 on one version of SAT II Math Level IIC administered at one time and place indicates the same level of achievement as a score of 500 on a different version of the test administered at a different time and place.

Table 14-2   Scaled Score Conversion Table, Mathematics Level IIC Test

Raw	Scaled	Raw	Scaled	Raw	Scaled
50	800	29	600	8	410
49	790	28	590	7	400
48	780	27	580	6	390
47	780	26	570	5	380
46	770	25	560	4	380
45	750	24	550	3	370
44	740	23	540	2	360
43	740	22	530	1	350
42	730	21	520	0	340
41	720	20	510	−1	340
40	710	19	500	−2	330
39	710	18	490	−3	320
38	700	17	480	−4	310
37	690	16	470	−5	300
36	680	15	460	−6	300
35	670	14	460	−7	280
34	660	13	450	−8	270
33	650	12	440	−9	260
32	640	11	430	−10	260
31	630	10	420	−11	250
30	620	9	420	−12	240

Chapter 15

Explaining the Answers to Practice Test 2, Level IIC

● ●

*T*he following explanations are for Practice Test 2, Level IIC, which is in Chapter 14. Make sure you read through all of the explanations. You may pick up some valuable information.

1. **A.** Solve the equation for a, then evaluate the desired expression. Add 4/*a* to both sides and subtract 1 from both sides, which leaves you with 2*a*/2. Now, cross multiply and you get 2*a* = 2; divide both sides by 2 to get *a* = 1.

    Wait, you're not done yet! Despite that a = 1, that is not what they ask you. They ask you the value of 1 − 2/*a*. So, substitute 1 for *a* and you get $1 - \frac{2}{1} = 1 - 2 = -1$. Be careful to always answer the question they ask.

2. **D.** To evaluate the expression, distribute (multiply) the 2*x* to both terms inside the parenthesis. You get $\frac{6x}{y} + \frac{8x}{z}$. Now find the lowest common denominator (also known as LCD, or, the term below the fraction bar) which is *yz*. Now, convert each individual term so that it has the LCD of *yz*. For the first term, multiply top and bottom by *z* (effectively multiplying the first term by *z/z*, which is the same as multiplying by 1). You get $\frac{6x}{y} \left(\frac{z}{z} \right) = \frac{6xz}{yz}$. For the second term, multiply top and bottom by *y* and you get $\frac{8xy}{yz}$. Finally, add the two together and you get $\frac{8xy + 6xz}{yz}$.

    A, B, C, and E all follow various incorrect procedures. A just multiplies all of the numerators and all the denominators together. B multiplies the numerators together but adds the denominators. C adds the numerators and denominators together. E adds the numerators together and multiplies the denominators together. Don't do any of these things! If need be, review your rules for combining fractions with different denominators (see Chapter 3).

3. **D.** The information to help you find the answer is hidden in the problem. You have to remember that sin π = 0 (or use your calculator to figure it out). Read through the problem and add information to the figure they provide.

    Following the *x*-axis from left to right, you have three *x*-intercepts (where the graph of the function crosses the *x*-axis and *y* = 0). The first *x*-intercept is the origin (0, 0). The last *x*-intercept, the problem tells you, has coordinates of (2π, 0) (2π meaning two times the value of π). Given that sin π = 0, you can deduce that the middle *x*-intercept has coordinates of (π, 0). Point *A* then must have *x*-coordinates between 0 and π and a *y*-coordinate greater than 0. Now, recall that the range of values for sine lies only between −1 and 1 (−1 ≤ sin *x* ≤ 1). Because *A* represents the maximum *y*-value of sin *x*, the *y*-coordinate will be 1.

    The *x*-coordinate appears to be halfway between 0 and π, so you are probably thinking that it will be $\left(\frac{\pi}{2}, 1 \right)$, but you can't trust that, because the figure notes that it is not drawn to scale.

To confirm your guess, use your calculator to determine the sine of $\frac{\pi}{2}$ and, sure enough, it equals 1. If you guessed A or B, you probably thought the graph represented $0 \le x \le \pi$ as opposed to $0 \le x \le 2\pi$. C can't be right because clearly $y \ne 0$, and E can't be right because $\sin x \ne \pi$ — π is approximately 3.14 and the sine of a number cannot exceed 1.

4. **C.** Don't let the radical throw you! Remember your rules for solving equations: Get the variable by itself on one side of the equation by performing the opposite function to both sides of the equation. In this case, you have to "break out" the variable p from the radical "jail" in which it is trapped. What, then, is the opposite of taking the square root of something? That's right, squaring it (raising it to the 2nd power). So, square everything on both sides of the equation. You end up with $\sqrt{5p} \times \sqrt{5p} = 3.67 \times 3.67$, which simplifies to $5p = 13.4689$. Finally, divide both sides of the equation by 5 and you get 2.69.

If you guessed B, you probably thought that p was being multiplied by $\sqrt{5}$, but that is not the case because the p is also under the radical symbol.

5. **B.** $\sin \theta = \dfrac{opposite}{hypotenuse} = \dfrac{y}{z}$. Therefore, $z \sin \theta = z \times \dfrac{y}{z} = y$. A would be the answer if the equation used cosine. The rest are decoys.

6. **D.** For this problem, start with a reminder of how to multiply and divide like numbers (bases) with different exponents (powers):

$n^x \times n^y = n^{x+y}$: If you are multiplying the same base, add the exponents.

$n^x \div n^y = n^{x-y}$: If you are dividing the same base, subtract the exponents.

Now add to that a reminder of dealing with negative exponents: $n^{-x} = \dfrac{1}{n^x}$. If you have a base raised to a negative exponent, it equals one over the base raised to the positive version of the exponent.

Now solve the problem. To get a by itself, first divide both sides of the equation by c^6 and you're left with $a^3 b^5 = \dfrac{a^2}{3b^{-5}}$. Now convert b^{-5} into $\dfrac{1}{b^5}$. Then, because you are now dividing by a fraction, multiply by the reciprocal, and you are left with $a^3 b^5 = \dfrac{a^2 b^5}{3}$. Now divide both sides by b^5, and they will cancel out. Finally, divide both sides by a^2, and you will be left with a by itself on the left and ⅓ on the right.

E incorrectly flips the coefficient 3 on the bottom along with the negative exponent; A incorrectly reverses the process of working with the negative exponent; B combines both of those errors.

7. **E.** If 4 and –3 are both zeros of the polynomial $p(x)$, the factor of $p(x)$ must be $(x - 4)(x + 3)$, which, when multiplied out using the FOIL method, equals $x^2 - x - 12$.

Remember that when the factors of a polynomial are $(x - a)(x - b)$, a and b are the zeros of the polynomial.

8. **C.** Arts and crafts time! Draw a picture! Sketch the little boat . . . add the water . . . draw the rope . . . draw you in the air . . . and behold: You end up with a right triangle, the vertices of which are the boat, you, and the point in the water directly beneath you.

As the problem explains, the angle that the rope makes with the surface of the water is 53°. Therefore, to determine the distance between you and the water, use the equation $\sin = \dfrac{opposite}{hypotenuse}$. Plug in $\sin 53° = \dfrac{x}{50}$. Cross multiply and you get $\sin 53° \times 50 = x$. So $x = 39.93 . . .$, which rounds to 40.

A uses the formula for tangent, and D uses the formula for cosine. E is a mean decoy that results from having your calculator in Radian mode by mistake.

Remember, in almost all cases on the SAT II Math test, make sure that your calculator is in Degree mode and not Radian mode. Be familiar with the functions of your calculator!

Finally, if you ever get confused about sine, cosine, and tangent, remember: "sohcahtoa." The word may sound like a Native American name to you. It's not, but it is still a helpful mnemonic device for remembering the formulas for sine, cosine, and tangent. If you break the word up into syllables it's soh-cah-toa:

$$\sin = \frac{opposite}{hypotenuse} \quad \cos = \frac{adjacent}{hypotenuse} \quad \tan = \frac{opposite}{adjacent}$$

9. **C.** Draw a picture again. Draw your big square and points P and Q inside the square (draw P on the left and Q on the right for now, though it does not really matter). Next, draw in the line segment that connects them. Now draw a long line perpendicular (forming a right angle) to this connecting segment and have it intersect the segment halfway between the two points. Now examine what you have drawn, and note that everything on the long perpendicular line is the same distance from P and Q. Everything to the left of the line — the half plane that contains P — is closer to P. Your drawing should look something like Figure 15-1.

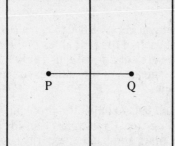

Figure 15-1: Sample drawing.

The same cannot be said for the other geometric figures in the answer choices: the interior of the smaller circle or square, the pie slice, or the parabola.

10. **A.** The best way to solve the type of problem that combines functions is to work from inside out (that's easy because you are used to working this way given the order of operations). The problem asks you to find f of g of *20*. In other words, solve $g(x)$ for $g(20)$, and then plug what you get in for x in the other equation. Plugging in 20 for x in $g(x)$, you get $g(x) = \frac{2(20)+2}{20-6} = \frac{42}{14} = 3$. Now substitute 3 for x in the equation $f(x)$, and you get

$f(x) = \sqrt{0.4(3)^2 - (3)} = \sqrt{0.4(9) - 3} = \sqrt{3.6 - 3} = \sqrt{0.6} = 0.77$

If the problem had asked for $g[f(20)]$ (or g of f of *20*), you would have switched the order in which you used the equations. In fact, E performs exactly this switch of equations as a decoy. C uses the incorrect order of operations when solving the first equation.

11. **D.** Ah, more trigonometric functions, isn't that wonderful? Okay, we heard that! Do you kiss your mother with that mouth?!

In addition to sine, cosine, and tangent, there are also cosecant (csc), secant (sec), and cotangent (cot). These three are essentially defined as the reciprocals ($1/x$) of the sine, cosine, and tangent functions, respectively — and that is the best way to remember them. Their formulas are listed below:

$$\csc\Theta = \frac{1}{\sin\Theta} = \frac{hypotenuse}{opposite}$$

$$\sec\Theta = \frac{1}{\cos\Theta} = \frac{hypotenuse}{adjacent}$$

$$\cot\Theta = \frac{1}{\tan\Theta} = \frac{adjacent}{opposite}$$

After you know these equations, a problem like #11 becomes a piece of cake (mmm . . . cake). Substituting for $\sec(3\Theta)$, you get: $\frac{1}{\cos 3\Theta} \times \cos 3\Theta$. The two $\cos 3\Theta$s cancel out, and you're left with 1 (remember that anything divided by itself equals 1, except for 0). A serves as a decoy for those who make the mistake of thinking that anything divided by itself is 0 instead of 1. B and C involve big time boo-boo's in the reciprocal process.

12. **B.** When combining functions, work from the inside out. Find $g(2)$ first using the potential answers for $g(x)$, and then plug in the value you get each time into $f(x)$ and see whether you get 24. When you plug 2 into $4x + 2$, you get $4(2) + 2 = 10$. Then substitute this 10 in for x in $f(x)$ and you get $f(x) = 2(10) + 4 = 24$. None of the other options adds up to 24 when you follow this proper procedure.

13. **D.** To find the area of WXYZ, first find the length of it sides. As it turns out, all four sides are the hypotenuse of a right triangle — in fact, the hypotenuse of congruent (equal) right triangles. Each triangle has one leg that measures 5 cm (the length of one edge of the cube) and a second leg that measures 2.5 cm (half the length of one side of the cube). Using the Pythagorean theorem, you get:

$$c^2 = a^2 + b^2 = (5)^2 + (2.5)^2 = 25 + 6.25 = 31.25$$

If $c^2 = 31.25$, take the square root of both sides, and you get $c = 5.5901 \ldots$ cm (leave that number in your calculator), which is the length of each of the four sides of the region. If all 4 sides are equal, the region is a square. For the area of a square, use either formula $A = s^2$ or $A = L \times W$. Either way, when you calculate the area of the square, you multiply the side by itself, and you end up right back at 31.25 cm.

A represents the length of one side of the region and fails to finish the problem. B multiplies that side by 2 and not by itself when calculating area. C multiplies the side by 4 and not by itself (the formula for perimeter instead of area), and E unnecessarily squares 31.25 a second time.

14. **C.** In the case of a vertical line, the x-coordinate always stays the same; it is only the y-coordinate that changes. Line l will always have the x-coordinate 4; thus, you can represent the line with the equation $x = 4$. D represents the horizontal line, where 4 is always the y-coordinate. B is the equation of the vertical y-axis. A and E are the equations of a line with a y-intercept of 4 and slopes of -1 and 1, respectively.

15. **A.** By definition, an arithmetic mean equals the sum of all the values divided by the actual number of values. For a more detailed discussion of averages and means, read Chapter 10. Set up an equation and plug in what you know.

$$average = \frac{sum}{number\ of\ terms}$$

On the left of the equation is the arithmetic mean of the salaries, which the problem tells you is $51,200. On the right is the sum of all 15 salaries (we suggest putting them in parenthesis) divided by the total number of salaries, which is 15.

The problem tells you that the average of these first fourteen salaries is $51,000; therefore, in your equation, the sum of the first 14 salaries can be represented by $14 \times 51,000$. Now just add George's salary (which you can call x) to the sum. The equation looks like this.

$$51,200 = \frac{(6 \times 51,000 + x)}{15}$$

Now, multiply both sides of the equation by 15 and simplify $14 \times 51,000$ within the parenthesis. You get:

$768,000 = 714,000 + x$

Subtract 714,000 from both sides.

$54,000 = x$

If you chose E, you probably made the mistake of only adding George's salary (x) to one 51,000 salary instead of 14 of them. Be careful!

16. **C.** How do you solve for x if it is embedded in a trigonometric function? You take the inverse — that's how!

To free the x from the bonds of its sine slavery, take the "inverse sine" ($\sin^{-1}$) of both sides (the same thing could be done if it were cosine or tangent).

On the left side, taking the inverse sine of sin x leaves you with just x. On the right side, you take the inverse sin of 0.866. Type 0.866 on your calculator and press the sin-1 function (or reverse the order depending on the calculator), and your calculator will read 59.9970 . . . , which, by the way, represents the angle whose sine is 0.866. The problem then asks you to find the $\tan\left(\dfrac{x}{1.5}\right)$, so you divide the 59.9970 . . . number by 1.5 (parenthesis first), and then take the tangent of that number.

A mistakenly takes the sine of 0.866 instead of the inverse sine. E starts off well taking the inverse sign, but it then messes up the order of operations taking the tangent of 59.9970 . . . and dividing by 1.5 instead of the opposite.

17. **B.** You first want to know the percentage of no-voters that the two governmental bodies have in common. Then you multiply that percentage by the number of members in the House of Representatives to get your answer. Set up an equation that represents the problem. Call x the percentage of members that voted no in each body. The equation will look something like this: $x(410) + x(90) = 350$.

In other words, if you multiply the percentage of members who voted no by the total number of members in the House and Senate, respectively, and you add them together, it equals 350. Solving the equation you get $410x + 90x = 350$. Combine your like terms, and you get $500x = 350$. Divide both sides by 500 and you arrive at $x = 0.7$ (which is 70%). Finally, multiply the number of members in the House by 0.7, and you get: $410 \times 0.7 = 287$.

E multiplies the percentage by the number of members of the Senate, and D is the percentage of members who voted no — not the actual number which the question requests.

18. **E.** For $(x, y) \not\vdash (x, y)$ to be true under the established condition $f(x, y) \not\vdash (x, y + 3x)$, then $(x, y + 3x)$ has to equal (x, y), which means that $y + 3x = y$. Now solve for x. Subtract y from both sides of the equation and you get $3x = 0$; divide both sides by 3, and you get $x = 0$.

A satisfies the condition ($x = 0$) but is not the "only" option.

19. **C.** A geometric progression is a sequence of numbers, where each term after the first is found by _multiplying_ the previous number by a fixed term called a _common ratio_. Examples are 2, 4, 8, 16, 32 (common ration is 2) . . . or 3, 9, 27, 81 (common ratio is three).

When you add 3 to 1, 9, and 33, you get 4, 12, and 36, a geometric sequence with the common ratio 3. None of the other answers satisfies the conditions for the geometric sequence.

While we're on the subject, another major type of progression is the arithmetic progression. An arithmetic progression is a sequence of numbers where each new term after the first is formed by _adding_ a fixed amount called the _common difference_ to the previous term in the sequence. For example, the sequence 3, 5, 7, 9, 11 . . . is an arithmetic progression. Each new term is found by adding 2 to the previous term, so the common difference is 2. _Note:_ In the geometric and arithmetic progressions, the common ratio and common difference can be negative.

20. **D.** The problem solving process for this question is strictly algebraic, with two steps. Basically, the problem gives you x and $f(x)$ and wants to know what $a + b$ is; so, all you have to figure out is what c equals and you're in business. Start with the equation the problem gives you, plug in 0 and 2 [for x and $f(x)$ respectively], and solve for c:

$$2 = a(0)^2 + b(0) + c.$$

Now simplify:

$2 = c$. Now do the same thing using 1 and 5[for x and $f(x)$ respectively], but this time instead of c, plug in 2:

$5 = a(1)^2 + b(1) + 2$.

Now simplify:

$5 = a + b + 2$.

Subtract 3 from both sides and you are left with

$3 = a + b$.

21. **C.** The key to this problem is to recognize that the triangle is isosceles, which means it has two equal sides as well as two equal angles (the two angles opposite the two equal sides). Further, when you drop an altitude down (a straight line down from the vertex where the two equal sides meet to the side opposite that vertex, creating a right angle with that side) it bisects the third side at the midpoint, creating equal halves on either side. You end up with the picture in Figure 15-2. To find the measure of the largest angle, solve for the measure of the two equal θ angles and subtract their sum from 180°.

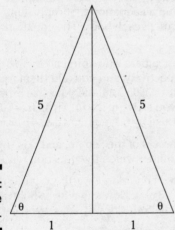

Figure 15-2:
Sample
drawing.

The same cannot be said for the other geometric figures in the answer choices: the interior of the smaller circle or square, or the pie slice, or the parabola.

Note: The three angles of a triangle always add up to 180°.

Because you now know the values of the adjacent side and the hypotenuse, solve for Θ using the cosine equation: $\cos\Theta = \dfrac{adjacent}{hypotenuse} = \dfrac{1}{5} = 0.2$. Now take the inverse cosine $(\cos^{-1})$ of both sides, and you get: $\Theta = 78.4630\ldots$(leave the number in your calculator). Multiply this angle by two (because there are two of them in the triangle), subtract that sum from 180, and it equals 23.07.

A and B incorrectly uses the tangent and sine function, respectively, instead of the cosine. D divides 2 by 5 instead of 1 by 5, and E represents the largest angle in the triangle instead of the smallest.

22. **C.** Although this problem may look daunting, it is actually straightforward. Ultimately, to add up the area of the triangles shown, just use the formula A = ½*bh* for each and add them up. The bases of the three triangles are all the same, measuring 1 unit each (0 to 1, 1 to 2, and 2 to 3 on the graph). To calculate the height of each of the triangles, all you need to do

is recognize the height of the triangles is the *y*-coordinate of the graph $y = 2^x$. To determine the *y*-coordinate at each point, plug the appropriate *x*-coordinate into the function $y = 2^x$. The coordinates (heights) end up being 2, 4, and 8, respectively. Thus:

$$A_1 + A_2 + A_3 = \tfrac{1}{2}b_1h_1 + \tfrac{1}{2}b_2h_2 + \tfrac{1}{2}b_3h_3 = \tfrac{1}{2}(1)(2) + \tfrac{1}{2}(1)(4) + \tfrac{1}{2}(1)(8) = 7$$

23. **D.** Vertical asymptotes is a fancy way of saying the "zeros" of the denominator of a rational function. The idea is that you cannot have zero in the denominator of a rational function (because when you divide by zero the result is undefined). So, if you set the denominator of the fraction equal to zero and solve, this will inform you as to the values that *x* cannot be. In this case, set $x^2 - 5x - 6$ equal to zero and solve:

$$x^2 - 5x - 6 = 0$$

Factoring the expression, you get:

$$(x - 6)(x + 1) = 0$$

So $x = 6$ and $x = -1$

24. **A.** Don't be fooled! Just because you do not know the coordinates of the vertices does not mean that you cannot determine the slope of the segment. You biggest clue is that all the segments of the triangle are equal, meaning it is an equilateral triangle. Therefore, you can determine the lengths of the sides in terms of *x*, either using the properties of an equilateral triangle or using the Pythagorean theorem, and, in turn, when you calculate the slope of the segment, the *x*'s will cancel, and you will get a number.

$slope = \dfrac{\Delta y}{\Delta x}$, or the change in the *y*-coordinates divided by the change in the *x*-coordinates.

Okay, the first thing you need to do is to drop an altitude down from *A* to the midpoint of *OB* (draw in the point *M* halfway in between *O* and *B*). Label *OA* as measuring *x*; label *OB* as measuring *x*, and label each half of *OB* (*OM* and *MB*) as measuring ½*x*. See Figure 15-3 as a reference.

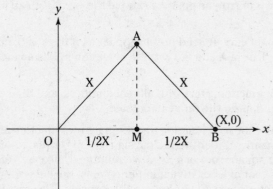

Figure 15-3:
Sample
drawing.

As you may now recognize, the slope of *AB* will equal *AM* divided by *OM*. Because you know *OM* = ½*x*, all you have to do is solve for *AM*. Triangle *OAB* is equilateral, so you may recognize that the new (half) triangle *AOM*, created by the altitude, is a 30-60-90 right triangle (angle *AOM* is 60° as it is part of the original equilateral right, angle *AMO* is of course 90°, and this makes angle *OAM* equal to 30°). You also may know that in this special 30-60-90 right triangle, the longer leg is equal to the measure of the shorter leg multiplied by $\sqrt{3}$.

In this case: $\dfrac{1}{2}x \cdot \sqrt{3} = \dfrac{\sqrt{3}}{2}x$.

If not, you can use the Pythagorean theorem to figure it out:

$$\left(\frac{1}{2}x\right)^2 + AM^2 = x^2$$

The next step is to multiply out the parenthesis:

$$\frac{1}{4}x^2 + AM^2 = x^2$$

Now subtract $\frac{1}{4}x^2$ from both sides of the equation and you are left with $AM^2 = \frac{3}{4}x^2$

Take the square root of both sides and you get $AM = \frac{\sqrt{3}}{2}x$

Finally, plug this into your slope formula from before:

$$slope = \frac{\Delta y}{\Delta x} = \frac{AM}{OM} = \frac{\frac{\sqrt{3}}{2}x}{\frac{1}{2}x}$$

To solve, cancel out your x's, change "dividing by a fraction" to "multiplying by the reciprocal," and you get:

$$\frac{\frac{\sqrt{3}}{2}x}{\frac{1}{2}x} = \frac{\sqrt{3}}{2} \cdot \frac{2}{1} = \sqrt{3}$$

25. **E.** The absolute value of a quantity is always positive. You can think of it as the "positive version" of that number. For example, the absolute value of both –3 and positive 3 is equal to 3.

In this problem, then, they ask you the difference between $y = f(x)$ and $y = |f(x)|$. You can determine the answer as follows: Wherever the y in $y = f(x)$ is negative, it will become positive for $y = |f(x)|$. Therefore, look at the graph of the original figure. Wherever y is negative (below the y-axis), reflect it, like a mirror, as positive (above the y-axis). E satisfies that perfectly.

Where y is negative in the original figure, it need not disappear as it does in B, nor is that which is *left* of the x-axis affected, as in A and C. Last, the inversion in D is an incorrect decoy.

26. **A.** To determine the domain of a function, start with all real numbers, and then exclude any members of this group that would make the range undefined.

You may be tempted to treat this cube root function like a square root function and exclude all values for x that make the quantity under the radical sign $(5 - 2x^2)$ equal to or less than zero because you can't take the square root of a negative number. Unlike a square root, however, you can take the cube root of a negative number (that is, the cube root of –8 is –2 because $-2 \times -2 \times -2 = -8$). Therefore, you do not imitate this process. The correct answer is all real numbers.

E is the result of the aforementioned process for the square root, and C is similar to E but incomplete.

27. **D.** Ah! Now this is one of the rare occasions when you can switch your calculator to Radian mode (the word "radian" in the problem is your clue!).

To solve the problem, not surprisingly, you have to solve for x. Admittedly, it gets a little tricky. First, recall the definitions of your trigonometric functions. Cotangent is the inverse tangent function: $\cot x = \frac{1}{\tan x} = \frac{\cos x}{\sin x}$

Next, substitute that into the original equation:

$$\sin x = \frac{\cos x}{\sin x}$$

Now solve. Start by cross-multiplying, and you get:

$$\sin^2 x = \cos x$$

Okay, at this point you have two ways to answer the original question:

- ✔ Perhaps the easiest thing to do is to plug each of the answers into this equation you just came up with ($\sin^2 x = \cos x$) and see which answer choice makes the equation true. Again, make sure your calculator is in Radian mode, because the problem specifically requires it to be.

- ✔ The other method is more time-consuming, but we include it just so you know how to solve the problem. The idea is to continue to solve the equation ($\sin^2 x = \cos x$) by getting one of the trigonometric functions in terms of the other. There exists an equation that explains the relationship between sine and cosine: $\sin^2 x + \cos^2 x = 1$. Solve this equation for $\sin^2 x$ by subtracting $\cos^2 x$ from both sides of the equation. You get $\sin^2 x = 1 - \cos^2 x$.

Now substitute $1 - \cos^2 x$ for $\sin^2 x$ in the equation you are solving:

$$\sin^2 x = \cos x \text{ becomes } 1 - \cos^2 x = \cos x$$

Now combine everything on the right side of the equation by subtracting $1 - \cos^2 x$ from both sides of the equation:

$$0 = \cos^2 x + \cos x - 1$$

This should look like a quadratic equation to you (except that you're not solving immediately for x but for $\cos x$), but hold on to that weird idea for just a second. Use the quadratic formula to solve.

The quadratic equation is $x = \dfrac{-b \pm \sqrt{b^2 - 4ac}}{2a}$ (based on $ax^2 + bx + c = 0$)

Now plug in your coefficients (a, b, and c are 1, 1, and –1, respectively) and solve:

$$x = \frac{-1 \pm \sqrt{1^2 - 4(1)(-1)}}{2(1)}$$

Simplifying, you get $x = \dfrac{-1 \pm \sqrt{5}}{2}$, or $x = 0.11803$ and $x = -2.11803$.

This is where you can add back the fact that you had been solving for $\cos x$ (and not x itself), so you have $\cos x = 0.61803$ and $\cos x = -1.61803$. Take the inverse cosine of both sides of both these equations (with your calculator in Radian mode), and you realize that x is only defined for the first one and $x = 0.90$.

Whew! You can see that plugging in the answer choices may turn out to be faster than solving the problem. If you find yourself spending way too much time trying to solve a problem, try plugging in answer choices to see which one works. If you still struggle, skip the question and go on to another less time-eating one.

A and E are decoys. B solves for $\cos x$ but doesn't finish out the problem, and C incorrectly solves for the inverse sine instead of the inverse cosine.

Remember to switch your calculator back to Degree mode!

28. **C.** Plug in and solve.

Oh, and remember that the use of the real number constant e requires that you use the natural log function ("ln" button) on your calculator instead of the usual base 10 log function ("log" button).

Starting with $R(t) = R_0 e^{\frac{-4t}{1000}}$ and plugging in you get $200 = 500 e^{\frac{-4t}{1000}}$.

Divide both sides by 500 and you get:

$0.4 = e^{\frac{-4t}{1,000}}$

Take the natural log of both sides and you get

$\ln(0.4) = \frac{-4t}{1,000}$ or $-0.91629 = \frac{-4t}{1,000}$

Cross multiply and divide by –4 and you get:

$t = 229$.

A fails to finish the problem; so does E. B uses the log base 10 instead of the natural log.

29. **B.** Simply substitute $x - 1$ for x in each of the possible answers for $f(x)$ and see which one spits out the answer $x + 1$. When you take B, $x + 2$ and substitute in $x - 1$ for x, you get $(x - 1) + 2$ or $x - 1 + 2 = x + 1$. None of the others yields this result.

30. **D.** To the tune of Handel's "Halleluiah Chorus" from his *Messiah:* "Draw-a-Pic-ture!"

The circle in question is "tangent to the negative *y*-axis and the positive *x*-axis." Thus, it sits in the lower right quadrant of the coordinate plane, and two of its points (its uppermost and leftmost points) are touching the axes. See Figure 15-4. You can further deduce that because the center is equidistant (the same distance) from all points on the circle, the center must be equidistant from the axes themselves. The only answer that satisfies both of these requirements is D.

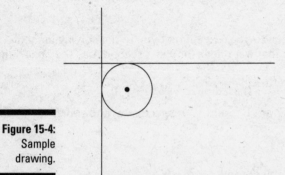

Figure 15-4:
Sample
drawing.

A would place the circle in the upper left quadrant; B and E would have the circle split in half by each of the axes. C makes it impossible for the circle to be tangent to BOTH axes at the same time.

31. **B.** To solve these simultaneous equations, solve for one variable in terms of the other, substitute into the other equation, and then use the quadratic formula (see the explanation for question #27). Pay attention to the specific requirement "for $x \le 0$" explicitly stated by the problem. Solving the second equation for y, you get:

$3y = x^2$

$y = \frac{x^2}{3}$

Substituting that into the first equation, you get:

$2x - 6\left(\frac{x^2}{3}\right) + 3 = 0$

Simplifying, you get:

$2x - 2x^2 + 3 = 0$

Reordered to match the $ax^2 + bx + c$ form required by the quadratic formula, you get:

$-2x^2 + 2x + 3 = 0$

Finally, you may find it easier to divide the equation by –1 so that the first term is positive:

$2x^2 - 2x - 3 = 0$ (To be sure; the answer will be the same either way).

Anyhow, plugging in the coefficients to the quadratic equation, you get:

$$x = \frac{2 \pm \sqrt{(-2)^2 - 4(2)(-3)}}{2(2)}$$

Simplified, you get:

$$x = \frac{2 \pm \sqrt{28}}{4}$$

$x = 1.82$ and $x = -0.82$.

Because the problem says that $x \le 0$, you discount the first one.

The other answers involve a myriad of careless errors involving the application of the quadratic formula. As always, be careful!

32. **E.** The easiest way to solve this problem is to use your graphing calculator. Have it graph both functions at the same time (don't worry about the domain restrictions yet). After you have the graph in front of you, see what each is doing for its designated x's (now observe the domain restrictions). You will quickly see that all possible real numbers are accounted for; none is left out.

33. **A.** To review, $f^{-1}(x)$ represents the inverse function, or what happens when you take your function and invert it, solving x for y.

$f(x) = \log_3 x$ becomes $y = \log_3 x$.

Now, take the inverse log of both sides. As you are dealing with base 3, you get:

$3^y = 3^{\log 3x}$

The right side simplifies (see chapter 5), and you get:

$3^y = x$

Finally, it is customary to switch x and y, so that the equation takes the form of the usual function. The answer is $y = 3^x$.

34. **D.** The problem tells you the value of the first term (x_0) in this sequence and an equation that represents the value of one term (x_{n+1}) in terms of the previous term (x_n). To solve this sequence problem, then, start with the second term (x_1) and go from there:

$x_1 = \sqrt{5 + x_0}$

$x_1 = \sqrt{5 + 2} = \sqrt{7}$

Now that you know the value of x_1, you can repeat the process for x_2.

$x_2 = \sqrt{5 + x_1}$

$x_2 = \sqrt{5 + \sqrt{7}}$

And, now that you know the value of x_2, you can repeat the process for x_3.

$x_3 = \sqrt{5 + x_2}$

$x_3 = \sqrt{5 + \sqrt{5 + \sqrt{7}}} = 2.787$

Just be careful of those radicals inside the other radicals when you use your calculator. E is the result of this type of mistake.

A is the square root of 5, and B and C are the values of x_1 and x_2, respectively.

35. **C.** As the explanation of question #40 in Chapter 13 reveals, Thales' Theorem states that an inscribed angle (an angle with all three points on the circumference of a circle) in a semicircle is a right angle.

If you were unaware of this theorem, you may have guessed that one of the angles of the given triangle was a right angle because some of the possible answers involve sine and cosine. Knowing that the triangle is a right triangle and that its hypotenuse is the diameter of the circle and equal to 4, you can easily calculate the perimeter (P). Because the adjacent side equals $4 \cos \theta$ and the opposite side is $4 \sin \theta$, you can add that to the hypotenuse and get that $P = 4 \cos \theta + 4 \sin \theta + 4$. When you pull out the common factor (4), it equals: $4(\sin \theta + \cos \theta + 1)$.

B is just the figure you get when you determine the area of the triangle (½bh).

36. **B.** When you combine AND probabilities (x must happen AND y must happen), you multiply the probabilities together. In this case, the basketball player must hit all three free throws, which means he must hit the first AND the second AND the third.

$0.537 \times 0.537 \times 0.537 = 0.155$.

E is the result of adding the probabilities together instead of multiplying them. D represents only the probability of hitting one free-throw; C represents only the probability of hitting two in a row; and A represents 4 in a row.

37. **E.** When you combine two vectors, the maximum amount they can add up to (if they are headed in the same direction) is one plus the other, and the least they can add up to is one minus the other (if they are headed in opposite directions).

In this case, the most these vectors could add up to is $13 + 4 = 17$ and the least is $13 - 4 = 9$.

38. **E.** First and foremost, recognize that you are solving for %, not just a or b.

To solve, take the log of both sides:

$\log(3.45)^a = \log(6.78)^{2b}$

which becomes

$a \log(3.45) = 2b \log(6.78)$. (Refer to Chapter 4.)

As you want to solve for a/b, divide both sides by b and you get:

$\frac{a}{b} \log(3.45) = 2 \log(6.78)$

Finally, divide both sides by $\log(3.45)$ and whip out the old scientific calculator:

$\frac{a}{b} = \frac{2 \log(6.78)}{\log(3.45)} = 3.09$

A is the result of having the right side of the equation upside down at the end. B is the result of having the 2 on the wrong side of the bar, and C ignores the 2 altogether.

39. **C.** The factorial of the natural number n (indicated by "$n!$"), is defined as the product of n and all the positive integers less than n ($3! = 3 \times 2 \times 1$; $4! = 4 \times 3 \times 2 \times 1$; and so on).

The key to this problem is to expand the expression only as much as you need to, and then let equal items on the top and bottom cancel each other out. Remember that the factorial symbol "!" indicates that you will keep multiplying the expression by 1 less than the previous number (in other words, $n - 1$). Behold:

$$\frac{[n!]^2}{[(n-2)!]^2}$$

can be expanded to

$$\frac{[n!]^2 [n-1]^2 [n-2!]^2}{[(n-2)!]^2}$$

At this point, the two $[n-2!]^2$ cancel each other out and you are left with:

$[n]^2[n-1]^2$

If this confuses you, further expansion of the exponents may help clear things up:

$$\frac{[n!]^2}{\left[(n-2)!\right]^2}$$

can be expanded to

$$\frac{[n][n][n-1][n-1][n-2!][n-2!]}{\left[(n-2)!\right][n-2!]}$$

The two sets of $[n-2!]$ on top and bottom cancel each other out.

40. **E.** An *arithmetic sequence* (or *arithmetic progression*) is a sequence of real numbers for which each term is equal to the previous term plus a constant (called the *common difference*). For example, starting with 1 and using a common difference of 4, you get the arithmetic sequence: 1, 5, 9, 13, and so on. In general, then, you can create a formula for an arithmetic sequence $a_n = a_0 + nd$ where n is the number of the current term, a_0 is the value of the first term, and d is the common difference.

With this equation and the information given in the problem, you can set up simultaneous equations and solve for the unknowns.

Knowing that the 40th term of an arithmetic sequence is 90, you can generate the equations $90 = a_0 + 40d$. Knowing that the 65th term of the sequence is 160, you can generate the equation $160 = a_0 + 65d$. Solve both equations for a_0 and set them equal to each other ($a_0 = a_0$):

For the 40th term

$90 = a_0 + 40d$

$a_0 = 90 - 40d$

For the 65th term

$160 = a_0 + 65d$

$a_0 = 160 - 65d$

Therefore,

$90 - 40d = 160 - 65d$

Combine your like terms:

$25d = 70$ and

$d = 2.8$

So, the common difference, or numerical difference, from one term to the next is 2.8.

The last step of the problem is to plug this answer for d back into one of the two initial equations to solve for a_0:

$90 = a_0 + 40d$

$90 = a_0 + 40(2.8)$

$90 = a_0 + 112$

$a_0 = -22$

41. **A.** The arcsine is another way of saying the inverse sine or $\sin^{-1}$ (the same goes for arcos and arctan). Given that, the arcsin (sin x) is going to equal (simply) x, because you're taking the inverse sine of the sin of x.

Here it is mathematically:

arcsine (sin x) = π

x = π

The last step of the problem is to make sure that x can have no other values. "What?" you ask, "Other values? If x = π, haven't I finished?"

Keep in mind that the trigonometric functions and their inverses are all periodic functions with a period of 2π (the function repeats itself every 2π on the x-axis). Therefore, you have to make sure that the domain of the function (values for x) has been limited in some way, or else x will have multiple values (in this case 3π, 5π, 7π, and so on).

Sure enough, this problem dictates that x is limited according to π < x < 2π; so, x can only equal π.

42. **D.** This problem harps back to your trigonometric identities; specifically, the reduction formulas. This problem seeks to determine how the values of sine, cosine, and tangent are changed when you apply the function to the opposite of θ (or negative θ). The three formulas that apply in this case are:

sin (–θ) = –sin θ

cos (–θ) = –cos θ

tan (–θ) = –tan θ

In other words, the sine of –θ equals the opposite of sin θ (the same goes for tangent), and the cosine of –θ just equals cos θ (the value does not change).

Using the preceding three formulas, substitute these values in for the original equation as follows:

sin θ – sin (–θ) + cos θ – cos (–θ) + tan θ – tan (–θ)

becomes

sin θ – (–sin θ) + cos θ – (cos θ) + tan θ – (–tan θ)

When you subtract a negative, you are really adding a positive; therefore, the two cosines cancel each other out, and you are left with 2 sinθ + 2 tan θ.

43. **C.** To picture the geometric figure in this problem, think of an asterisk with an infinite number of lines running through the center (and these lines head off to infinity on both ends). Now ask yourself: If you draw another line intersecting one of these lines, how many of the other lines will it intersect?

Given that lines run infinitely in both directions (away from the center), you may think that this new line will intersect all of the lines in the asterisk. You are almost right. This new line that you have drawn will end up being parallel to one of the lines in the existing asterisk; therefore, it will not intersect that one line to which it is parallel. Thus, the answer is n – 1.

44. **D.** Yep, you guessed it, draw a picture. To find the volume of the cone ($V = \frac{1}{3} \pi r^2 h$ as per the instructions at the beginning of the exam), you need to determine the height h of the cone. When you draw your picture, you see that this problem involves using similar triangles, so you will be able to solve for h using a proportion. Your picture, according to the problem, should look like Figure 15-5. (The letters are added for reference.)

Because the two cross-sections are parallel, triangles ABE and ACD must be similar. Therefore, you can set up a proportion as follows:

$$\frac{BE}{CD} = \frac{AB}{AC}$$

In other words, the respective sides of the two triangles are in the same direct proportion to one another.

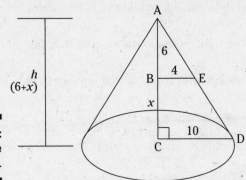

Figure 15-5:
Sample
drawing.

Filling in the values, and adding x for unknown segment BC (because you have to add AB and BC to get AC), you get

$$\frac{4}{10} = \frac{6}{6 + x}$$

Cross multiplying, you get:

$4(6 + x) = 10(6)$

Simplifying, you get:

$24 + 4x = 60$

$4x = 36$

$x = 9$

Now, if $x = 9$, the height of the cone is $6 + 9 = 15$.

Returning to the original formula for the volume, you can plug in and solve:

$$V = \frac{1}{3} \pi r^2 h$$

$$V = \frac{1}{3} \pi (10)^2 15$$

$$V = \frac{1}{3} \pi (100)(15)$$

$$V = 5\pi (100)$$

$$V = 500\pi$$

The other possible answers involve a myriad of careless errors, including labeling the segments in your picture incorrectly, setting up your proportion wrong, and making mathematical errors.

45. **B.** By definition, to perform an indirect proof, you assume that what you intend to prove is false. You then show that something contradictory (absurd) happens as a result.

Therefore, the first step of an indirect proof is to assume what you need to prove is false, which in the case of proving the Pythagorean theorem would be this inequality: $a^2 + b^2 \neq c^2$.

46. **C.** A parabola is the graph of a quadratic function. When this quadratic function looks like $f(x) = a(x - h)^2 + k$ (and is not equal to zero), it is said to be in standard form. If a is positive, the graph opens upward, and if a is negative, it opens downward. The line of symmetry is the vertical line $x = h$, and the vertex is the point (h, k).

Therefore, the following are true about vertical (up and down) shifts:

✔ When you shift the parabola up, you *add* that amount to k.

✔ When you shift the parabola down, you *subtract* that amount from k.

In dealing with the horizontal (moving right and left), you have to be careful because it essentially reverses:

✔ When you shift the parabola to the right, you add the amount to h, but you really further *subtract* it from x.

✔ When you shift the parabola to the left, you subtract that amount from h, but you really *add* that amount to x.

In this problem, you start with $f(x) = -x^2 + 1$. When you translate (or shift) 1 unit to the right, the equation changes to $-(x-1)^2 + 1$ (h increases but you are really subtracting it from x). And, when you translate it two units down, the equation changes to $-(x-1)^2 - 1$ (as $1 - 2 = -1$).

Finally, the problem tells you that this new graph represents $g(x)$, and then asks you to find the value of $g(1.4)$. So, what you need to do is substitute 1.4 for x in the new equation and you get:

$$-(1.4 - 1)^2 - 1 = -(0.4)^2 - 1 = -0.16 - 1 = -1.16$$

The other possible answers involve changing the equation of the parabola in the wrong directions, adding when you should be subtracting, and vice-versa.

47. **B.** This is a straightforward combination problem. If you are dividing people into groups, the order in which they are grouped does not matter, so it is not a permutation. Furthermore, you can't use any member more than once, so it is not an arrangement with repetition.

The formula for combination is $_nC_r = \dfrac{n!}{(n-r)! \times r!}$ where n is the number of total items from which you are choosing and r is the number of items you choose. Another way of reading it would be "the number of combinations of n items taken r at a time."

Now just plug in and solve:

$$_nC_r = \frac{n!}{(n-r)! \times r!}$$

$$_8C_5 = \frac{8!}{(8-5)! \times 5!}$$

Simplify the parenthesis:

$$_8C_5 = \frac{8!}{3! \times 5!}$$

Expand your greatest factorial (in this case 8!) so you can cancel and simplify. (And don't you dare try to make $3! \times 5! = 8!$ Because it doesn't work that way.)

$$_8C_5 = \frac{8 \times 7 \times 6 \times 5!!}{3! \times 5!}$$

Cancel your 5!'s

$$_8C_5 = \frac{8 \times 7 \times 6}{3!}$$

Expand your last factorial and solve

$$_8C_5 = \frac{8 \times 7 \times 6}{3 \times 2 \times 1} = 56$$

Don't let the fact that there are two groups throw you. When you choose five people to be in that first group, each combination you come up with naturally leaves another combination of three in the other group.

48. **E.** The standard equation for an ellipse is $\dfrac{(x-h)^2}{a^2} + \dfrac{(y-k)^2}{b^2} = 1$ where (h, k) is the center (or intersection point of the major and minor axis), a is the length of the semi-major axis (half of the major axis), and b is the length of the semi-minor axis (half of the minor axis). As the two words indicate, the major axis is the longer of the two and the minor is the smaller. It follows, then, that if the ellipse is horizontal, then a will end up under the term with the x-coordinate (the first one), and if the ellipse is vertical, a will end up under the term with the y-coordinate (the second one). Figure 15-6 illustrates the ellipse according to the terms of the problem. The center is $(3, -2)$; the semi-major axis (a) is 10 (half of 20); the semi-minor axis (b) is 5 (half of 10). The equation of this vertical ellipse is then $\dfrac{(x-3)^2}{25} + \dfrac{(y-2)^2}{100} = 1$.

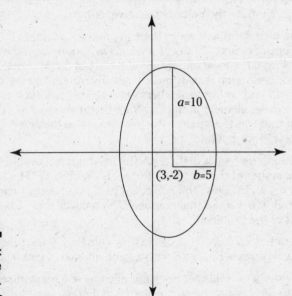

Figure 15-6:
Sample
drawing.

49. **A.** This question isn't hard; it is just time-consuming. If you are running out of time at the end of the test, you may want to jump ahead to #50 to see whether you can answer that question more quickly. Then go back to #49 if you have time.

This problem refers to the rules for combining signs for the various mathematical operations. (Read Chapter 3 if you need more information about combining signs.) The bottom line is that for the quotient $\dfrac{yx}{y-x}$ to be negative, either the numerator has to be positive and the denominator negative or the numerator has to be negative and the denominator positive.

Now you have to break it down further:

- For the numerator (the product yx) to be positive, both y and x have to be positive or both have to be negative.

- For the numerator to be negative, either y or x has to be negative.

- For the denominator (the difference $y - x$) to be positive, x has to be less than y if x and y are both positive or both negative, and y cannot be negative if x is positive.

- For $y - x$ to be negative, y has to be less than x if they are both positive or both negative, and y cannot be positive if x is negative.

Now if it hurts your head to hang on to all of that information, it may help to plug numbers into the answer choices to see how they work in the equation, a little number substitution.

A gives you $0 < y < x$. Substitute 1 for y and 2 for x in the equation. This would make the numerator positive (2) and the denominator negative (1 – 2) and the result would be a negative number. This seems to work. At this point, you would be best to mark A as your answer and go on to #50 before it's too late, but for instruction purposes, we continue with the other answer choices.

For B, substitute 1 for x and 2 for y. This would make the numerator positive (2). The denominator would also be positive (2 – 1). B would not make the number negative.

For C, substitute –2 for x and –1 for y. The numerator is positive (2); the denominator is positive [–1 – (–2) or 1]. C doesn't make the number negative.

For D, substitute 2 for x and –1 for y. The numerator is negative (–2); the denominator is negative (–1, – 2, or –3). D doesn't make the number negative, either.

50. **E.** To determine if a set has an element that is more than any other element in the set, you have to determine the "least upper bound." Think of "bound" as "boundary," the least upper boundary that exists. Now, if the least upper bound is part of the set, you're in business (meaning that the least upper bound is truly more than any other element in the set). Conversely, if the "least upper bound" is not a member of the set, the set does *not* have an element that is more than any other element in the set. Note that if you had to determine whether a set had an element that was less than any other element in the set, you would have to determine the "greatest lower bound."

I states, "The set of rational numbers r such that $r \leq \pi$" The least upper bound here is π. The question then, is: Is π a member of the set? Well, the set is defined as "The set of rational numbers r such that $r \leq \pi$" The "less than or equal to sign certainly includes π," so you are okay on that front. But, is π a rational number? The answer is no. Therefore, I does not meet the conditions of the problem.

At this point eliminate answer choices A and D because they contain I. If the proctor calls time at this point, you can quickly guess B, C, or E with a good chance of getting it right.

II states, "The set of rational numbers." This set does not even have a least upper bound as it goes off to infinity in both directions; thus II does not satisfy the conditions of the problem. Eliminate B. You are now down to III only or "none," a 50 percent guess if time were called.

Finally, III states, "The set of rational numbers r such that $r < 2$." The least upper bound is 2. Now: Is 2 a member of the set? Well, the set is defined as "The set of rational numbers r such that $r < 2$." According to the "less than" sign (as opposed to the "less than or equal to" sign), 2 is *not* part of the set. Thus, III does not satisfy the condition of the problem. C is wrong and the answer must be "none."

Answer Sheet

Begin with Number 1. If the test has fewer than 100 questions, leave the extra spaces blank.

1. Ⓐ Ⓑ Ⓒ Ⓓ Ⓔ	21. Ⓐ Ⓑ Ⓒ Ⓓ Ⓔ	41. Ⓐ Ⓑ Ⓒ Ⓓ Ⓔ	61. Ⓐ Ⓑ Ⓒ Ⓓ Ⓔ	81. Ⓐ Ⓑ Ⓒ Ⓓ Ⓔ
2. Ⓐ Ⓑ Ⓒ Ⓓ Ⓔ	22. Ⓐ Ⓑ Ⓒ Ⓓ Ⓔ	42. Ⓐ Ⓑ Ⓒ Ⓓ Ⓔ	62. Ⓐ Ⓑ Ⓒ Ⓓ Ⓔ	82. Ⓐ Ⓑ Ⓒ Ⓓ Ⓔ
3. Ⓐ Ⓑ Ⓒ Ⓓ Ⓔ	23. Ⓐ Ⓑ Ⓒ Ⓓ Ⓔ	43. Ⓐ Ⓑ Ⓒ Ⓓ Ⓔ	63. Ⓐ Ⓑ Ⓒ Ⓓ Ⓔ	83. Ⓐ Ⓑ Ⓒ Ⓓ Ⓔ
4. Ⓐ Ⓑ Ⓒ Ⓓ Ⓔ	24. Ⓐ Ⓑ Ⓒ Ⓓ Ⓔ	44. Ⓐ Ⓑ Ⓒ Ⓓ Ⓔ	64. Ⓐ Ⓑ Ⓒ Ⓓ Ⓔ	84. Ⓐ Ⓑ Ⓒ Ⓓ Ⓔ
5. Ⓐ Ⓑ Ⓒ Ⓓ Ⓔ	25. Ⓐ Ⓑ Ⓒ Ⓓ Ⓔ	45. Ⓐ Ⓑ Ⓒ Ⓓ Ⓔ	65. Ⓐ Ⓑ Ⓒ Ⓓ Ⓔ	85. Ⓐ Ⓑ Ⓒ Ⓓ Ⓔ
6. Ⓐ Ⓑ Ⓒ Ⓓ Ⓔ	26. Ⓐ Ⓑ Ⓒ Ⓓ Ⓔ	46. Ⓐ Ⓑ Ⓒ Ⓓ Ⓔ	66. Ⓐ Ⓑ Ⓒ Ⓓ Ⓔ	86. Ⓐ Ⓑ Ⓒ Ⓓ Ⓔ
7. Ⓐ Ⓑ Ⓒ Ⓓ Ⓔ	27. Ⓐ Ⓑ Ⓒ Ⓓ Ⓔ	47. Ⓐ Ⓑ Ⓒ Ⓓ Ⓔ	67. Ⓐ Ⓑ Ⓒ Ⓓ Ⓔ	87. Ⓐ Ⓑ Ⓒ Ⓓ Ⓔ
8. Ⓐ Ⓑ Ⓒ Ⓓ Ⓔ	28. Ⓐ Ⓑ Ⓒ Ⓓ Ⓔ	48. Ⓐ Ⓑ Ⓒ Ⓓ Ⓔ	68. Ⓐ Ⓑ Ⓒ Ⓓ Ⓔ	88. Ⓐ Ⓑ Ⓒ Ⓓ Ⓔ
9. Ⓐ Ⓑ Ⓒ Ⓓ Ⓔ	29. Ⓐ Ⓑ Ⓒ Ⓓ Ⓔ	49. Ⓐ Ⓑ Ⓒ Ⓓ Ⓔ	69. Ⓐ Ⓑ Ⓒ Ⓓ Ⓔ	89. Ⓐ Ⓑ Ⓒ Ⓓ Ⓔ
10. Ⓐ Ⓑ Ⓒ Ⓓ Ⓔ	30. Ⓐ Ⓑ Ⓒ Ⓓ Ⓔ	50. Ⓐ Ⓑ Ⓒ Ⓓ Ⓔ	70. Ⓐ Ⓑ Ⓒ Ⓓ Ⓔ	90. Ⓐ Ⓑ Ⓒ Ⓓ Ⓔ
11. Ⓐ Ⓑ Ⓒ Ⓓ Ⓔ	31. Ⓐ Ⓑ Ⓒ Ⓓ Ⓔ	51. Ⓐ Ⓑ Ⓒ Ⓓ Ⓔ	71. Ⓐ Ⓑ Ⓒ Ⓓ Ⓔ	91. Ⓐ Ⓑ Ⓒ Ⓓ Ⓔ
12. Ⓐ Ⓑ Ⓒ Ⓓ Ⓔ	32. Ⓐ Ⓑ Ⓒ Ⓓ Ⓔ	52. Ⓐ Ⓑ Ⓒ Ⓓ Ⓔ	72. Ⓐ Ⓑ Ⓒ Ⓓ Ⓔ	92. Ⓐ Ⓑ Ⓒ Ⓓ Ⓔ
13. Ⓐ Ⓑ Ⓒ Ⓓ Ⓔ	33. Ⓐ Ⓑ Ⓒ Ⓓ Ⓔ	53. Ⓐ Ⓑ Ⓒ Ⓓ Ⓔ	73. Ⓐ Ⓑ Ⓒ Ⓓ Ⓔ	93. Ⓐ Ⓑ Ⓒ Ⓓ Ⓔ
14. Ⓐ Ⓑ Ⓒ Ⓓ Ⓔ	34. Ⓐ Ⓑ Ⓒ Ⓓ Ⓔ	54. Ⓐ Ⓑ Ⓒ Ⓓ Ⓔ	74. Ⓐ Ⓑ Ⓒ Ⓓ Ⓔ	94. Ⓐ Ⓑ Ⓒ Ⓓ Ⓔ
15. Ⓐ Ⓑ Ⓒ Ⓓ Ⓔ	35. Ⓐ Ⓑ Ⓒ Ⓓ Ⓔ	55. Ⓐ Ⓑ Ⓒ Ⓓ Ⓔ	75. Ⓐ Ⓑ Ⓒ Ⓓ Ⓔ	95. Ⓐ Ⓑ Ⓒ Ⓓ Ⓔ
16. Ⓐ Ⓑ Ⓒ Ⓓ Ⓔ	36. Ⓐ Ⓑ Ⓒ Ⓓ Ⓔ	56. Ⓐ Ⓑ Ⓒ Ⓓ Ⓔ	76. Ⓐ Ⓑ Ⓒ Ⓓ Ⓔ	96. Ⓐ Ⓑ Ⓒ Ⓓ Ⓔ
17. Ⓐ Ⓑ Ⓒ Ⓓ Ⓔ	37. Ⓐ Ⓑ Ⓒ Ⓓ Ⓔ	57. Ⓐ Ⓑ Ⓒ Ⓓ Ⓔ	77. Ⓐ Ⓑ Ⓒ Ⓓ Ⓔ	97. Ⓐ Ⓑ Ⓒ Ⓓ Ⓔ
18. Ⓐ Ⓑ Ⓒ Ⓓ Ⓔ	38. Ⓐ Ⓑ Ⓒ Ⓓ Ⓔ	58. Ⓐ Ⓑ Ⓒ Ⓓ Ⓔ	78. Ⓐ Ⓑ Ⓒ Ⓓ Ⓔ	98. Ⓐ Ⓑ Ⓒ Ⓓ Ⓔ
19. Ⓐ Ⓑ Ⓒ Ⓓ Ⓔ	39. Ⓐ Ⓑ Ⓒ Ⓓ Ⓔ	59. Ⓐ Ⓑ Ⓒ Ⓓ Ⓔ	79. Ⓐ Ⓑ Ⓒ Ⓓ Ⓔ	99. Ⓐ Ⓑ Ⓒ Ⓓ Ⓔ
20. Ⓐ Ⓑ Ⓒ Ⓓ Ⓔ	40. Ⓐ Ⓑ Ⓒ Ⓓ Ⓔ	60. Ⓐ Ⓑ Ⓒ Ⓓ Ⓔ	80. Ⓐ Ⓑ Ⓒ Ⓓ Ⓔ	100. Ⓐ Ⓑ Ⓒ Ⓓ Ⓔ

Chapter 16

Practice Test 3, Level IC

• •

Okay, you know your stuff. Now is your chance to shine. The following exam is a 50-question, multiple-choice test. You have one hour to complete it.

To make the most of this practice exam, take the test under similar conditions to the actual test.

1. **Find a place where you won't be distracted (preferably as far from your younger sibling as possible).**

2. **If possible, take the practice test at approximately the same time of day as that of your real SAT II.**

3. **Set an alarm for 60 minutes.**

4. **Mark your answers on the provided answer grid.**

5. **If you finish before time runs out, go back and check your answers.**

6. **When your 60 minutes is over, put your pencil down.**

After you finish, check your answers on the answer key at the end of the test. Use the scoring chart to find out your final score.

Read through all of the explanations in Chapter 17. You learn more by examining the answers to the questions than you do by almost any other method.

Table 16-1	For Your Reference

You can use the following information for reference when you are answering the test questions.

The volume of a right circular cone with radius r and height h: $V = \frac{1}{3}\pi r^2 h$

The lateral area of a right circular cone with circumference of the base c and slant height l: $S = \frac{1}{2}cl$

The volume of a sphere with radius r: $V = \frac{4}{3}\pi r^2$

The surface area of a sphere with radius r: $S = 4\pi r^2$

The volume of a pyramid with base area B and height h: $V = \frac{1}{3}Bh$

For each of the following 50 questions, choose the best answer from the choices provided. If no answer choice provides the exact numerical value, choose the answer that is the best approximate value. Fill in the corresponding oval on the answer grid.

Notes:

1. **You must use a calculator to answer some but not all of the questions.**

   You can use programmable calculators and graphing calculators. You should use at least a scientific calculator.

2. **Make sure your calculator is in degree mode because the only angle measure on this test is degree measure.**

3. **Figures accompanying problems are for reference only.**

   Figures are drawn as accurately as possible and lie in a plane unless the text indicates otherwise.

4. **The domain of any function *f* is assumed to be the set of all real numbers *x* for which *f(x)* is a real number unless indicated otherwise.**

5. **You can use the table on the previous page for reference in this test.**

1. If $6c - 9c = 12 - 2c + c$, then $c =$

   (A) -9

   (B) -6

   (C) -3

   (D) 0

   (E) 3

2. If $z = 3$, then $z(z + 5)(z - 4) =$

   (A) -24

   (B) -8

   (C) 0

   (D) 8

   (E) 24

3. For all $r \neq 0$, $\dfrac{2}{\left(\dfrac{4r}{6}\right)} =$

   (A) $\dfrac{1}{12r}$ (B) $\dfrac{1}{3r}$ (C) $\dfrac{4r}{3}$ (D) ¾   (E) $6r$

4. In circle C in the figure, what are the coordinates of point C?

   (A) $(7, 4)$

   (B) $(6, 5)$

   (C) $(7, 5)$

   (D) $(8, 5)$

   (E) $(9, 4)$

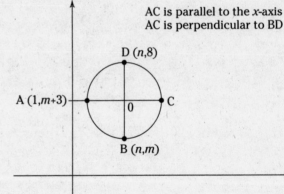

AC is parallel to the x-axis
AC is perpendicular to BD

D $(n,8)$

A $(1,m+3)$

B (n,m)

5. At what point does $3x - 5y = 15$ intersect the y-axis?

   (A) $(-3, 0)$

   (B) $(0, -3)$

   (C) $(0, 3)$

   (D) $(5, 0)$

   (E) $(0, 5)$

Go on to next page

6. $(x + y + 4)(x + y - 4) =$

   (A) $x^2 + y^2 - 16$

   (B) $x^2 + y^2 + 16$

   (C) $(x + y)^2 + 8(x + y) - 16$

   (D) $(x + y)^2 + 8(x - y) - 16$

   (E) $(x + y)^2 - 16$

7. If $10x^4 = 5$, then $5(10x^4)^2 =$

   (A) 5   (B) 25   (C) 125   (D) 625   (E) 3,125

8. If lines l and m are parallel and are intersected by line t, what is the sum of the measures of the exterior angles on the same side of line t?

   (A) 360°   (B) 270°   (C) 180°   (D) 135°   (E) 90°

9. If $a + b = -3$ and $a - b = 1$, then $b =$

   (A) 2

   (B) 1

   (C) 0

   (D) −1

   (E) −2

10. If the cube root of the cube root of a number is −2, what is the number?

   (A) −512

   (B) −64

   (C) 64

   (D) 512

   (E) The number is irrational.

11. Each face of the cube in the figure consists of 16 small squares. The shading on three of the faces is shown, and the shading on the other three faces is such that on the opposite faces, only half of the number of squares are shaded. What is the total number of shaded squares on all six faces of the cube?

   (A) 72   (B) 60   (C) 48   (D) 36   (E) 24

Go on to next page

12. Larry's Labradors has two-thirds as many dogs as Greg's Goldens, which has three-fourths as many as Ralph's Retrievers. If Larry has 50 dogs, how many does Ralph have?

    (A) 25 (B) 75 (C) 100 (D) 125 (E) 150

13. In the figure, when ray *OA* is rotated counterclockwise 29° about point *O,* rays *OA* and *OB* will form a straight line. What is the measure of angle AOB before this rotation?

    (A) 90° (B) 119° (C) 151° (D) 180° (E) 209°

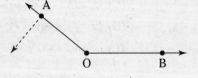

14. If $35^c = 5^2 \cdot 7^2$, what is the value of *c*?

    (A) 32 (B) 16 (C) 8 (D) 4 (E) 2

15. If $2y + 4y - y = x$, then $2y - x =$

    (A) $-4y$ (B) $-3y$ (C) $-2y$ (D) $-y$ (E) y

16. A probability and statistics class conducted a die-rolling experiment. Each trial of the experiment consisted of rolling a single, fair, six-sided die six times and counting the number of 3s that resulted. The results for 100 trials are pictured in the figure. In approximately what percent of trials were more than two 3s rolled?

    (A) 3% (B) 5% (C) 10%

    (D) 15% (E) 65%

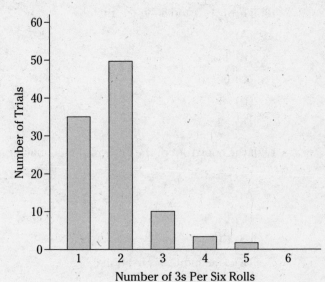

17. If $f(x) = \dfrac{1}{1.5x}$ for $x > 0$, then $f(0.25) =$

    (A) ¼ (B) ⅜ (C) ⅝ (D) ⅚ (E) ⅔

18. The circle in the figure has a center *C* and a chord *EF* of length 15. What is the length of radius *r*?

    (A) 5.30 (B) 7.5 (C) 10.61 (D) 30 (E) 225

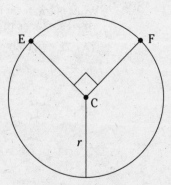

Go on to next page

19. What are all values for z for which $|z - 4| < 5$?

(A) $0 < z < 8$

(B) $-1 < z < 9$

(C) $0 \leq z \leq 8$

(D) $z < 9$

(E) $z > 9$

20. At Make-Your-Own-Burrito Restaurant, the stations of the assembly line provide you with your choice of either flour or corn tortillas; either steak, chicken, pork, or guacamole; either black beans or pinto beans; and any one of their salsas. You can only pick one item from each station. If there are 48 different burrito possibilities, from how many different salsas can you choose?

(A) 6   (B) 5   (C) 4   (D) 3   (E) 2

21. Chris works in regional sales. The graph in the figure shows the distance of Chris' car from his home in Omaha, Nebraska, over a period of time in a given week. Which of the following situations best fits the information?

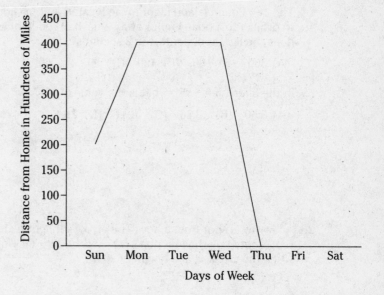

(A) Chris leaves one of his clients, drives to another client, stays there a few days, and then returns home.

(B) Chris leaves one of his clients, drives to another client, and then drives back home.

(C) Chris leaves one of his clients, drives back home, stays for a few days, and then drives to another client.

(D) Chris leaves home, drives to a client, stays there a few days, and then drives to another client.

(E) Chris leaves home, drives to a client, stays there a few days, and then drives home.

Go on to next page ⇒

22. In the figure, triangle CDE is equilateral, and CE∥BF∥AG. What is the perimeter of triangle ADG?

AD and DG are straight lines.

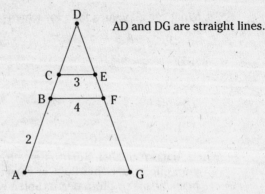

   (A) 9   (B) 12   (C) 15   (D) 18   (E) 21

23. If $i^2 = -1$, then $[(i^2)^3]^2 =$

   (A) 2   (B) 1   (C) 0   (D) –1   (E) –2

24. The Ten Pound Island Lighthouse located within Gloucester Harbor, Massachusetts, is able to project to a boat 4 miles away a flash of red light every 6 seconds. How many degrees does the lighthouse rotate in 2 seconds?

   (A) 360   (B) 180   (C) 120   (D) 90   (E) 60

25. In the figure, if θ = 35°, what is the value of x?

   (A) 8.60   (B) 13.46   (C) 18.31   (D) 21.42   (E) 26.15

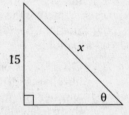

26. An arrow is shot from a bow. The arrow's height h at time t seconds after the arrow is shot is represented by the function $h(t) = -5t^2 + 20t + 60$. How much time, in seconds, has gone by when the arrow is at a height of 20 ft.?

   (A) 5.46

   (B) 4

   (C) 3.10

   (D) 1

   (E) 0.45

27. In the xy-plane, the points O (2, –1), A (1, –1), and B (–1, 0), can be connected to form line segments. Which of the following represents the lengths of the segments from least to greatest?

   (A) AB, OB, OA

   (B) AB, OA, OB

   (C) OA, OB, AB

   (D) OA, AB, OB

   (E) OB, OA, AB

Go on to next page

28. On a baseball team, the statistician calculated that the average (arithmetic mean) batting average for the 9 starters is 0.252, and the average batting average for the 11 reserves is 0.225. What is the average batting average for the whole team?

    (A) 0.236

    (B) 0.237

    (C) 0.238

    (D) 0.230

    (E) 0.240

29. Line l has a positive slope and a positive x-intercept. Line m is parallel to l and has a positive y-intercept. The x-intercept of m must be

    (A) positive and less than the x-intercept of l

    (B) positive and greater than the x-intercept of l

    (C) negative and less than the x-intercept of l

    (D) negative and greater than the x-intercept of l

    (E) zero

30. Of the following, which has the least value?

    (A) $3,333^3$

    (B) 3^{33}

    (C) $3^{(3^3)}$

    (D) $3,333,333,333$

    (E) $\left(33 \cdot 3^3\right)^3$

31. The figure shown is a rectangular prism. Which of the following vertices are all located in the same plane?

    (A) W, X, Z, and V

    (B) S, T, U, and Z

    (C) W, S, V, and X

    (D) T, X, Y, and U

    (E) X, S, T, and Z

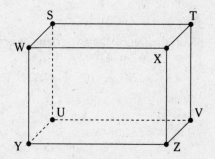

32. Given the quadratic equation $x^2 + 2x - 15 = 0$, which of the following are the sum and product of its roots?

    (A) Sum: –2   Product: –15

    (B) Sum: –2   Product: 15

    (C) Sum: 2   Product: –15

    (D) Sum: 2   Product: 15

    (E) Sum: –8   Product: 15

Go on to next page

33. In the figure, if the circle, with center C and radius $r = 2$, is reflected across line l, what would be the coordinates of the reflection of point A?

(A) $(-4, 3)$

(B) $(-3, 4)$

(C) $(-6, 3)$

(D) $(-5, 4)$

(E) $(-2, 2)$

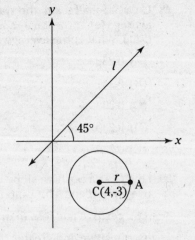

34. In the figure, the cube has an edge of length 3. What is the distance from vertex E to the point G on FH, such that G is one-third of the way from F to H on segment FH?

(A) 2.65

(B) 3.16

(C) 3.76

(D) 4.24

(E) 4.36

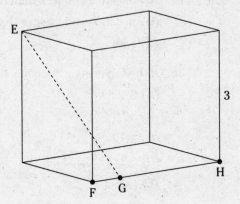

35. In the figure, the two triangles are similar, and $w = 6$. What is the value of z/x?

(A) ⅔ (B) 5/7 (C) 6/7 (D) ½ (E) ¼

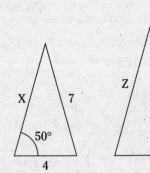

36. In the figure, in order for the equation sin x = sin y to be true, what must be done to triangle DEF?

    (A) increase DF by 3

    (B) increase DF by 2

    (C) increase DF by 1

    (D) decrease DF by 1

    (E) decrease DF by 2

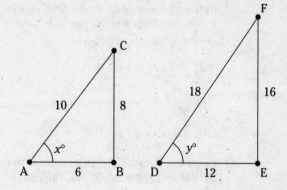

37. $\left[\cos^2\Theta + 6 - \left(-\sin^2\Theta + 2\right)\right]^2 =$

    (A) $\cos^2\Theta - \sin^2\Theta + 4$

    (B) $\cos^2\Theta - \sin^2\Theta + 8$

    (C) 0

    (D) 25

    (E) 81

38. The line with the equation $y = 4$ is graphed on the same xy-plane as the circle with center (2, 3). If the circle and line intersect at the point where the x-coordinate is 5, what is the radius of the circle?

    (A) 3

    (B) 3.09

    (C) 3.16

    (D) 3.57

    (E) 4

39. In 1987, four pairs of red wolves, an endangered species, were reintroduced to the wild on the Alligator River National Wildlife Refuge in North Carolina. If the population of red wolves increases at the rate of 12% each year, what will the population be at the end of 2004?

    (A) 14

    (B) 24

    (C) 34

    (D) 44

    (E) 54

Go on to next page

40. In the figure, if $160 < r + s < 240$, which of the following describes all possible values of $t + u + v$?

    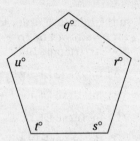

    (A) $120 < t + u + v < 300$

    (B) $300 < t + u + v < 380$

    (C) $480 < t + u + v < 560$

    (D) $660 < t + u + v < 740$

    (E) $840 < t + u + v < 920$

41. Points A, B, and C lie on a circle, and the center of the circle lies on segment AC. If segment $AB = 3$ and segment $BC = 4$, the radius of the circle equals

    (A) 2.5

    (B) 3.0

    (C) 4.0

    (D) 4.5

    (E) 5

42. The function f, where $f(x) = (x - 1)^4$, is defined for $-2 \leq x \leq 2$. What is the range of f?

    (A) $1 \leq f(x) \leq 27$

    (B) $1 \leq f(x) \leq 81$

    (C) $0 \leq f(x) \leq 16$

    (D) $0 \leq f(x) \leq 27$

    (E) $0 \leq f(x) \leq 81$

43. The area of parallelogram EFGH in the figure is

    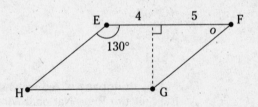

    (A) 28.93

    (B) 34.47

    (C) 53.63

    (D) 58.74

    (E) 70

44. Sunflowers are planted in a garden so that there are 5 sunflowers per square foot. If sunflowers are available only in packs of 10, and 7 packs are purchased and all are planted except for 1 sunflower, what is the circumference, in feet, of the garden?

    (A) 13.08

    (B) 13.17

    (C) 13.48

    (D) 13.8

    (E) 27.6

Go on to next page

45. In the right circular cylinder shown in the figure, C and O are the centers of the bases and segment xy is a diameter of one of the bases. If triangle XYO is equilateral, and the radius of the base is 4, what is the area of the triangle?

    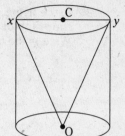

    (A) 8

    (B) 9.23

    (C) 13.86

    (D) 20.79

    (E) 27.71

46. The figure shows a sequential arrangement of squares that are formed according to a pattern. Each arrangement after the first is generated by adding one to both the number of rows and columns of the previous arrange-ment. If this pattern continues, which of the following gives the number of squares in the nth arrangement?

    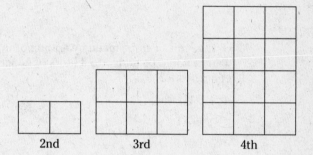

    2nd   3rd   4th

    (A) $2n$

    (B) $2n^2$

    (C) $n(n-1)$

    (D) $n(n+1)$

    (E) $2(2n-1)$

47. If $f(x) = 2x^4 + 2$, and if f^{-1} is the inverse function of f, what is $f^{-1}(5)$?

    (A) 0   (B) 1.11   (C) 1.32   (D) 1.50   (E) 627

48. Two positive integers r and s satisfy the relationship $q \cdot r$ if and only if $q = r^3$. If s, t, and u satisfy the relations $s \cdot t$ and $t \cdot u$, what is the value of s in terms of u?

    (A) u^0

    (B) u

    (C) u^6

    (D) u^9

    (E) u^{27}

Go on to next page

49. In the figure, the area of the shaded region bound by the graph of the parabola $x = f(y)$ and the y-axis is 6. What is the area of the region bound by the graph of $x = f(y) - 3$ and the y-axis?

(A) 3 (B) ⅔ (C) 6 (D) ¹⁵⁄₂ (E) 9

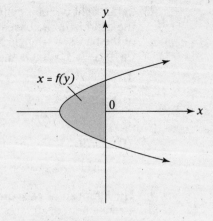

50. Orange juice can be made by adding orange juice concentrate to water. How many liters of water should be added to 3 liters of concentrate, so that 27% of the resulting orange juice is concentrate?

(A) 8.11

(B) 11.11

(C) 14.11

(D) 17.11

(E) 20.11

Answer Key for Practice Exam 3

1.	B	26.	A
2.	A	27.	D
3.	D	28.	B
4.	C	29.	C
5.	B	30.	E
6.	E	31.	D
7.	C	32.	A
8.	C	33.	C
9.	E	34.	E
10.	A	35.	D
11.	D	36.	B
12.	C	37.	D
13.	C	38.	C
14.	E	39.	E
15.	B	40.	B
16.	D	41.	A
17.	E	42.	E
18.	C	43.	C
19.	B	44.	B
20.	D	45.	E
21.	A	46.	C
22.	D	47.	B
23.	B	48.	D
24.	C	49.	C
25.	E	50.	A

Scoring Your Exam

First, calculate your "raw score." On this exam, you get one point for each correct answer, and you lose ¼ (0.25) of a point for each wrong answer. Your raw score, then, is determined by the following formula: Raw score = number of correct answers – (0.25 × number of wrong answers). Feel free to use this Raw Score Wizard to help you determine your raw score:

Raw Score Wizard

1. Use the answer key to count the number of right answers   1) _____

2. Use the answer key to count the number of wrong answers   2) _____

3. Multiply the number of wrong answers by 0.25 (Line 2 × 0.25)   3) _____

4. Subtract this number from the number of right answers   4) _____
   (Line 1 – Line 3)

5. Round this number to the nearest whole number   5) _____
   (Round Line 4 up or down)

This final number is your raw score.

Use Table 16-2 on the next page to convert your raw score to one of the College Board Scaled Scores, which range from 200–800. Ultimately, when you take the test for real, this scaled score will be the one the College Board reports to you and the colleges/universities you request.

Note: The College Board explains that it converts students' raw scores to scaled scores in order to make sure that a score earned on one particular subject test is comparable to the same scaled score on all other versions of the same subject test. Scaled scores are adjusted so that they indicate the same level of performance regardless of the individual test taken and the ability of the specific group of students that takes it. In other words, a score of 500 on one version of SAT II Math Level IC administered at one time and place indicates the same level of achievement as a score of 500 on a different version of the test administered at a different time and place.

Table 16-2		Scaled Score Conversion Table, Mathematics Level IC Test			
Raw	Scaled	Raw	Scaled	Raw	Scaled
50	800	29	600	8	410
49	790	28	590	7	400
48	780	27	580	6	390
47	780	26	570	5	380
46	770	25	560	4	380
45	750	24	550	3	370
44	740	23	540	2	360
43	740	22	530	1	350
42	730	21	520	0	340
41	720	20	510	−1	340
40	710	19	500	−2	330
39	710	18	490	−3	320
38	700	17	480	−4	310
37	690	16	470	−5	300
36	680	15	460	−6	300
35	670	14	460	−7	280
34	660	13	450	−8	270
33	650	12	440	−9	260
32	640	11	430	−10	260
31	630	10	420	−11	250
30	620	9	420	−12	240

Chapter 17

Explaining the Answers to Practice Test 3, Level IC

The following explanations are for Practice Test 3, Level IC, which is in Chapter 16. Make sure you read through all of the explanations. You may pick up some valuable information.

1. **B.** First, combine like terms. On the left side of the equation, $6c - 9c$ combines to $-3c$, and on the right side, $-2c + c$ combines to $-c$. You are left with $-3c = -c + 12$.

    Next, get all of your c's on one side of the equation and all your numbers on the other. Add c to both sides of the equation: $-2c = 12$.

    Finally, divide both sides of the equation by -2: $c = -6$.

2. **A.** Simply substitute and then follow the order of operations (that whole "Aunt Sally" thing found in Chapter 3).

    Substituting 3 for z, you get $3(3 + 5)(3 - 4)$.

    Solve inside the parenthesis first, and you get $3(8)(-1)$, which equals -24. (If you are at all confused, flip to Chapter 3.)

3. **D.** When you divide by a fraction, you need to "multiply by the reciprocal."

    In this case, change the big division bar to a multiplication symbol and flip the fraction underneath from $\frac{4r}{6}$ to $\frac{6}{4r}$. So, now you multiply 2 and $\frac{6}{4r}$.

    Multiply the numerators: 2×6, which is 12.

    Multiply the denominators: $1 \times 4r$, which is $4r$.

    $12 \times 4r$ gives you $\frac{3}{r}$.

4. **C.** The first thing is to determine either the radius or diameter of the circle. (After you know one, you know the other, because $d = 2r$.) You immediately realize this, because C's x-coordinate is the value of A's x-coordinate plus the diameter of the circle.

    From there, it's easy. You know that C's x-coordinate will be the x-coordinate of A plus the value of two of the circle's radii (plural of radius) or its diameter. And you can also tell that C's y-coordinate is the same as the y-coordinate of O. Because O is one radius length below D, you can find O's y-coordinate by subtracting the value of one radius from the y-coordinate of $D(8)$.

    To find the radius of the circle, look at the coordinates of points A and B. You see both A and B use the variable m in the values for their y-coordinates. You also recognize that the radius that starts at center O and ends at point B runs parallel with the y-axis. So, if you can figure out the difference between the y-coordinates of O and B, you know the value of the radius of the circle. Because it is on the same level as A, O has a y-coordinate $m + 3$, just like A. B has a y-coordinate m. The difference between O and B is $3(m + 3 - m)$, so the radius of the circle is 3.

Now, to determine C's x-coordinate, just add the length of two radii (6) to A's x-coordinate (1): $1 + 6 = 7$. C's x-coordinate is 7.

To determine C's y-coordinate, subtract one radius (3) from D's y-coordinate (8), and you get $8 - 3 = 5$. The answer is (7, 5).

5. **B.** The equation $3x - 5y = 15$ intersects the y-axis at the point whose x-coordinate is 0, or $x = 0$. Therefore, simply substitute 0 for x in the given equation and you have:

$$3(0) - 5y = 15$$

This simplifies to $-5y = 15$. Divide both sides by -5 and $y = -3$. The point's coordinates are $(0, -3)$.

You could also solve (or check your answer to) this problem using the equation of a line and plugging in answer choices. The equation of a line is expressed as $y = mx + b$. Convert the equation given in the problem to the format of the equation of a line by subtracting $3x$ from each side of the equation: $-5y = -3x + 15$. Then substitute the x- and y-coordinates in the answer choices into the equation to see which answer satisfies the equation. For B the x-coordinate is 0 and the y-coordinate is -3. Plug the x- and y-coordinates in the equation.

$$-5(-3) = -3(0) + 15$$
$$-15 = 15$$

B works!

D is incorrect because it substitutes 0 for y instead of x. A has the right answer, but the coordinates are switched around.

6. **E.** When you have three terms in each parenthesis, distribute (multiply) one term at a time from one set of parenthesis to the other three terms in the other set of parenthesis.

So, in this problem, you multiply the first term of the first set (x) by the first term of the second set (x) and get x^2. Then multiply the first term of the first set by the second term of the second set and get xy. Then multiply the first term of the first set by the third term of the second set and get $4x$.

You do the same for the second term of the first set to get yx, y^2, and $4y$. And you finish off with the third term of the first set to get $4x$, $4y$, and 16. After you complete the last set of multiplication, the expression will look like this:

$$x^2 + xy - 4x + yx + y^2 - 4y + 4x + 4y - 16.$$

Then combine your like terms. Remember that xy and yx are the same and recognize that the $4x$'s and $4y$'s cancel each other out, because one $4x$ is negative and one is positive (same with the $4y$'s). So, after combining, you get $x^2 + 2xy + y^2 - 16$.

At this point, you should recognize the familiar expression: $x^2 + 2xy + y^2$, which simplifies to $(x + y)^2$. Your answer is $(x + y)^2 - 16$.

7. **C.** This is really just a substitution problem. Look at what they give you in the first part of the problem.

They say "$10x^4 = 5$." If you can recognize that the $10x^4$ in the second part is the same as the $10x^4$ in the first, you can substitute and solve this problem like nobody's business.

$5(10x^4)^2$ becomes $5(5)^2$

Now, follow the order of operations. Raise 5 to the 2nd power before you multiply, and the expression simplifies to $5(25)$, which equals 125.

By the way, another way to solve it from this point is to recognize that first 5 can be thought of as 5^1; and $5^1 \times 5^2 = 5^3$, which also equals 125.

8. **C.** If two parallel lines are cut by a *transversal* (a line cutting across them), then the corresponding angles are *congruent* (equal). Additionally, *supplementary angles* (the ones next to each other along a straight line) add up to 180°. If you need a refresher on angles, check out Chapter 6.

Given your knowledge of angles, draw a picture something like Figure 17-1 with x and y representing the measure of the two "exterior angles on the same side of t."

Now recognize that angle x is congruent with the angle just above y, because they are "corresponding angles."

The last step is to recognize that x and y are supplementary angles and add up to 180°.

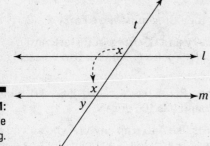

Figure 17-1: Sample drawing.

9. **E.** There are two ways to solve simultaneous equations.

   ✔ Probably the easiest way is to stack them one on top of the other and add them.

   ✔ The other way is to solve the first equation for a "in terms of b," and then substitute what you get for a into the second equation.

To see which method you like best, try them both. Go for the stacking method first.

$a + b = -3$

$a - b = 1$

The $+b$ and $-b$ cancel out, and you are left with $2a = -2$. From there, you can divide both sides by 2 and get $a = -1$. Finally, substitute -1 for a into one of the original equations (just use the first for now), and you get $-1 + b = -3$. To solve for b, add 1 to both sides and you get $b = -2$. For good measure, substitute -1 in for a into the second equation. If you get the same answer for b, you know you are doing something right.

Now try the other way.

In the first equation, to solve for a, subtract b from both sides, and you get $a = -3 - b$. You now have a value for a "in terms of b." Now substitute $-3 - b$ for a in the second equation.

$(-3 - b) - b = 1$

$-3 - b - b = 1$

Combine like terms.

$-3 - 2b = 1.$

To solve the equation, add 3 to both sides to get the term with the variable by itself, which leaves you with $-2b = 4$. Finally, divide both sides by 2 and you get $b = -2$.

TIP

When solving simultaneous equations, plug your solution(s) into both original equations to check your answer.

10. **A.** The problem states that it takes the cube root of the cube root of a number and gets –2. In this case, you are taking some number (call it #1) and taking #1's cube root and getting another number (call it #2), and then taking #2's cube root to get –2.

To solve this problem, then, start with the number –2 and work backward. The question becomes –2 is the cube root of what?

Well that's easy, $-2 \times -2 \times -2 = -8$. Now, repeat the process with –8. –8 is the cube root of what? Multiply $-8 \times -8 \times -8$ and you get –512.

11. **D.** If each opposite side has half of the number of squares shaded, add up the number of squares you see shaded on the three sides and add half that number to it. You should see $4 + 8 + 12 = 24$ squares shaded. Add to that number 12 squares ($\frac{1}{2} \times 24 = 12$) and you get $24 + 12 = 36$.

12. **C.** If Larry has $\frac{2}{3}$ as many dogs as Greg does, then Greg must have Larry $\times 3/2$.

You can check this mathematically. Let L = Larry and G = Greg. You start with L = $\frac{2}{3}$G. To solve for G, multiply both sides by $\frac{3}{2}$ and you get $\frac{3}{2} \times$ L = G. Thus, if Larry has 50 dogs, Greg has $\frac{3}{2} \times 50 = 75$ dogs.

Next, if Greg has $\frac{3}{4}$ as many dogs as Ralph does, Ralph has $\frac{4}{3} \times$ Greg. If G = $\frac{3}{4}$R, multiply both sides by $\frac{4}{3}$ and you get $\frac{4}{3}$G = R. $\frac{4}{3} \times 75 = 100$. Ralph has 100 dogs.

The SAT II test makers throw in answer choices like A to trap anyone who solved the equations in the opposite direction. So, if you solved "Larry has $\frac{2}{3}$ as many dogs as does Greg" as L $\times \frac{2}{3}$ and "Greg has $\frac{3}{4}$ as many dogs as Ralph does" as G $\times \frac{3}{4}$, you would come up with the answer in A. And the option would be there for you, so you wouldn't question your math. Pretty tricky! You won't be tricked, however, if you make sure you translate word problems into the correct math equivalent.

13. **C.** When ray *OA* and ray *OB* form a straight line, they will form a straight angle and measure 180°.

A straight angle measures 180°.

If angle AOC is 29° less than the straight angle (counterclockwise = less than according to the diagram given in the problem), it must measure 151°: $180 - 29 = 151$.

E is incorrect because it adds 29° to 180° instead of subtracting it.

14. **E.** The fastest way to solve this problem is to recognize that when you multiply numbers with the same exponents together, the exponent stays the same.

Therefore, in this multiplication problem, all of the exponents on either side of the equation must have the same value.

$35^c = (5 \cdot 7)^c = 5^c \cdot 7^c$

c must be 2.

The other way to figure this out is to simplify the right side of the equation, and then play with your calculator to find out what power of 35 that is. The right side of the equation simplifies as follows: $5^2 \cdot 7^2 = 25 \cdot 49 = 1,225$. So, $35^c = 1,225$. Turns out 1,225 is 35×35 or 35²; $c = 2$.

You may be tempted to think that using your calculator is always the fastest way to answer a question. Question 14 shows you that this is not always true. It is much quicker to realize that *c* must be 2 because the exponents on the other side of the equation equal 2 than it is to play around with multiples of 35 on your calculator. Always check to see whether there is a simpler way to solve a problem before you resort to your calculator, especially in the first half of the IC test.

15. **B.** Given that you want to solve for the value of $2y - x$, and given that there is a $2y$ and x in the equation, your best bet seems to be to manipulate the given equation so that you have $2y - x$ on one side. Then, whatever is left on the other is your answer.

Begin by moving the x to the other side of the equation, so it is with the $2y$. You do this by subtracting x from both sides.

$2y - x + 4y - y = 0$

Now simplify $4y - y$, but don't include $2y$. You want to see it as part of the equation.

$2y - x + 3y = 0$

Now, subtract $3y$ from both sides of the equation.

$2y - x = 3y$

The answer is $-3y$.

16. **D.** To determine in what percent of the trials there were more than two 3s rolled, you have to divide the number of trials with more than two 3s by the total number of trials, which is 100.

Starting with the middle bar on the graph and working to the right, the number of trials where there were more than two 3s rolled is $10 + 3 + 2$, which equals 15 trials.

Divide by 100.

$^{15}/_{100} = 0.15$ or 15%.

17. **E.** The function says that its value is $1 \div 1.5x$ for any value of x you plug in that is greater than zero.

When you plug in 0.25, you end up with $1 \div [1.5(0.25)]$.

Take a look at your answer choices. All of them are in the form of a fraction. Because the end result should be a fraction, it is easier to solve this function using fractions. Convert the decimals to fractions

$$\dfrac{1}{\left[\left(\dfrac{3}{2}\right)\left(\dfrac{1}{4}\right)\right]}$$

Multiply the fractions, and you get $\dfrac{1}{\frac{3}{8}}$.

When you divide 1 by a fraction, all you do is flip the numerator and denominator of the fraction, so the answer is $\frac{8}{3}$.

18. **C.** The little half-square symbol in angle ECF indicates that it is a right angle measuring 90°. Therefore, triangle ECF is a right triangle.

The first thing you should notice is that segments *CE* and *CF* are both radii of the circle and therefore have the same length *r*. This tells you that $\triangle$ECF is an isosceles right triangle because is has two equal sides. The angles opposite the equal sides must be equal as well, so you know additionally that the triangle is a 45:45:90 triangle.

The ratio of the sides of a 45:45:90 triangle is $s : s : s\sqrt{2}$.

Chord *EF*, with a length of 15 units, is the hypotenuse of the right triangle. To find the length of its legs, divide 5 by $\sqrt{2}$. From glancing at the answer choices, you know that you need your answer in decimal form; so divide and you get 10.61.

If you forget the ratio of a 45:45:90, you can apply the Pythagorean theorem to solve for *r*. But this will use up precious time on the test, so memorize the ratios given to you in Chapter 6 (and on the Cheat Sheet) for special triangles.

If you needed to use the Pythagorean theorem to solve, here is how you would do it.

$r^2 + r^2 = EF^2$

$2r^2 = 15^2$

Square 15 first (225), then divide both sides by 2.

$r^2 = 112.5$

Take the square root of both sides.

$r = 10.61$

19. **B.** Remember, the absolute value of a number is always positive. It simply refers to a number's absolute distance from 0. Therefore, the absolute value bars will make the expression $z - 4$ positive in all cases. Just be careful!

Start with the values that make $|z - 4| = 5$. They are 9 and –1. Those two numbers become your limits. 9 and higher or –1 and lower make the absolute value equal to or greater than 5. The problem asks for the values that make it less than 5. Any value that is less than 9 but greater than –1 will work. Therefore, the answer is $-1 < z < 9$.

20. **D.** In this arrangement with repetition problem, you have 2 choices for the first station (flour or corn tortillas), 4 choices for the second station (steak, chicken, pork, or guacamole), 2 choices for the third station (black or pinto beans), and s number of choices for the last station (s number of salsas). The problem states that there are 48 different burrito possibilities, so you can set up an equation to solve for s.

Use the multiplication principle.

$2 \times 4 \times 2 \times s = 48$

Simplify.

$16s = 48$

Divide both sides by 16.

$s = 3$

Probably hot, medium, and mild.

Answer choice A is there to draw in anyone who added 2, 4, and 2 instead of multiplying them.

21. **A.** The graph shows that at the beginning of the week Chris is 200 miles from home. Sunday, he travels further away, presumably to another client, with whom he spends Monday and Tuesday. Wednesday he travels back to 0 miles from his home, in other words, he drives home.

Answer choice B is meant to trap those who didn't include the time he stayed over at the location of the second client.

22. **D.** To determine the perimeter of triangle ADG, you need to add up segments *AD*, *DG*, and *AG*. In order to do this, you need to solve for the lengths of their sub-segments.

Given that triangle CDE is equilateral and *AD* and *DG* are straight lines, triangles DCE, BDF, and ADG are all similar, and thus all equilateral.

Therefore, segments *BD* and *DF* both equal 4 because *BF* = 4. Consequently, *AD* and *DG* = 6 because BD(4) + AB(2) = 6. The same calculations apply to *DG* (*FG* is also 2). *AG* also must equal 6, because triangle ADG is equilateral. So, the perimeter must equal 6 + 6 + 6 = 18.

23. **B.** This is a substitution problem. Substitute –1 for i^2. The equation simplifies to $[(-1)^3]^2 = (-1)^2 = 1$.

–1 raised to an odd power = –1; –1 raised to an even power = 1

24. **C.** First you need to determine how long it takes the lighthouse to rotate one turn.

The equation for distance problems is distance = rate × time or $D = RT$. You need to find the "how long" or the rate, so modify the equation: $R = D \div T$.

Determine the distance. An entire circle measures 360°, so $D = 360°$. It takes 6 seconds to complete one full rotation; that's your time (T). So, the lighthouse rotates 360° ÷ 6 = 60° per second. Multiply this by 2 seconds and you get 60° × 2 = 120°. It turns 120° in two rotations.

The fact that the boat is 4 miles from the lighthouse is irrelevant, but the information is included to confuse you. Make sure that you focus only on the information you need to solve the problem.

25. **E.** Recognizing that the triangle is right, you should use the following equation to solve:

$\sin \theta$ = opposite ÷ hypotenuse.

Plug in the correct value.

$\sin 35° = 15 \div x$.

Solve.

$0.573576\ldots = 15 \div x$.

Cross-multiply.

$0.573576x = 15$

Finally, divide both sides by 0.57357.

$x = 26.15$

Most of the other incorrect answer choices result from common errors. C incorrectly uses cosine, and D incorrectly uses tangent. A multiplies 15 by sin 35 instead of dividing it.

26. **A.** To solve this function, simply substitute 20 for $h(t)$ on the left side of the equation and then solve for t.

$h(t) = -5t^2 + 20t + 60$

$20 = -5t^2 + 20t + 60$

Subtract 20 from both sides.

$0 = -5t^2 + 20t + 40$

Divide both sides by –5 to make your life easier.

$0 = t^2 - 4t - 8$

Now use the quadratic equation to solve.

$\dfrac{-b \pm \sqrt{b^2 - 4ac}}{2a}$ ($a = 1$; $b = -4$; $c = -8$)

$\dfrac{4 \pm \sqrt{(-4)^2 - 4(1)(-8)}}{2(1)}$

Simplify.

$\dfrac{4 \pm \sqrt{48}}{2} = 5.46$

$\dfrac{4 \pm \sqrt{48}}{2}$ also equals –1.46, but you can cross out because it is negative, and an arrow would not fly for a negative number of seconds.

27. **D.** To solve this problem, use the distance formula derived from the properties of the right triangle in the coordinate plane. To find out more about the distance formula, read Chapter 7.

$$d = \sqrt{(y_2 - y_1)^2 + (x_2 - x_1)^2}$$

Solve for d.

$$d_{OA} = \sqrt{[-1 - (-1)]^2 + (1 - 2)^2}$$

$$d_{OA} = \sqrt{(0)^2 + (-1)^2}$$

$$d_{OA} = \sqrt{0 + 1} = \sqrt{1} = 1$$

If you follow the same procedure for OB and AB, you find that $OB = \sqrt{10} = 3.16$ and $AB = \sqrt{5} = 2.24$.

Therefore, the order from least to greatest is OA, AB, OB.

28. **B.** By definition, an average (or arithmetic mean) equals the sum of all the values divided by the number of values. For a more detailed discussion of averages and means, read Chapter 10.

The equation for determining averages is

$$average = \frac{sum}{number\ of\ terms}$$

On the left of the equation is the arithmetic mean of the values, which is what the problem is asking for. On the right is the sum of all the batting averages (put them in parenthesis) divided by the total number of players, which is 20.

The problem tells you that the average of the nine starters is 0.252. Therefore, in your equation, the sum of the first nine scores can be represented by 9×0.252. Similarly, the sum of the last 11 batting averages (the reserves) can be represented by 11×0.225. The equation looks like this.

$$average_{team} = \frac{(9 \times 0.252 + 11 \times 0.225)}{20} = 0.237$$

You can't just average the two different batting averages $(0.252 + 0.225)/2$. You must account for each individual score.

29. **C.** Drawing a diagram is your best strategy for this problem. See Figure 17-2.

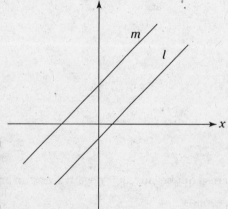

Figure 17-2:
Sample
drawing.

As you can see in Figure 17-2, if line ℓ has a positive slope and a positive x-intercept (to the right of the y-axis), its y-intercept has to be negative (below the x-axis). If line m is parallel to line ℓ, but has a positive y-intercept (above the x-axis), then its x-intercept must be negative (left of the y-axis) and is clearly less than that of line l.

You can think of positive slope as "uphill" from left to right and negative slope as "downhill" from left to right.

30. **E.** Take your time, use your calculator, evaluate each expression, and choose the least value. For this question, a good calculator is essential. Use scientific notation (see Chapter 3) to more easily deal with the large numbers. The number with the smallest exponent is your answer. This may be one you skip and come back to if you have time at the end of the section.

After you have finished your calculations, you will find the following:

$A = 3.7 \times 10^{10}$

$B = 5.6 \times 10^{15}$

$C = 7.6 \times 10^{12}$

$D = 3.3 \times 10^{9}$

$E = 7.1 \times 10^{8} \ (707{,}347{,}971)$

The smallest exponent belongs to E.

31. **D.** When you think of a plane, it helps to think of the two-dimensional space created by a line and one additional point.

For this problem, take the first two vertices given to you in each answer choice and picture a line between them. Add the third vertex to imagine the plane containing the line and the additional point. Then, determine whether the fourth point is in the same plane as the line and additional point. It works only for *T, X, Y,* and *U.*

For questions like this one that require you to consider every answer choice to answer them, the SAT II Math often places the correct answer at D or E to give you more work to do. Beat them at their game by starting from the bottom of the answer choice list. If the correct answer is D or E, you won't have to consider as many answer choices as you would have had you begun with A.

32. **A.** The expression factors into $(x + 5)(x - 3)$, which yields roots of –5 and 3, the sum of which is –2 and the product of which is –15.

33. **C.** Don't let the diagonal line confuse you. Think of a mirror. The near points stay near and the far points stay far, and the distances from the reflection line remain the same.

The only difference here is that, because the reflection line runs on a 45-degree angle through the origin, the x- and y-coordinates will be reflected in the quadrant on the other side of the reflection line, which is the upper left quadrant.

Given that the circle has a radius of 2, to determine the coordinates of *A,* add 2 to the x-coordinate of the center of the circle: (4, –3) becomes (6, –3).

Now, reflect that in the opposite quadrant: (6 becomes –6 and –3 becomes +3: (–6, 3).

34. **E.** To solve this problem, draw a picture. Create a right triangle with segment *EG* as the hypotenuse; Leg 1 will be the segment created by point *E* and the unnamed point directly below it (name it something for clarity, say *P*). Leg 2 will be the segment created by *P* and *G.* Mark the two legs in with dashes so you can see them. Compare your drawing to Figure 17-3.

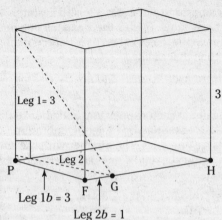

Figure 17-3:
Sample
drawing.

Because it is an edge of the cube, the distance of Leg 1 of the right triangle is 3 units. The distance of Leg 2 is trickier to determine. You actually need to set up a second right triangle to solve for that. Leg 1b of this second right triangle is the segment created by P and point F, and Leg 2b will be segment FG.

You know these two values: Leg 1b is an edge of the cube and measures 3 units; Leg 2b is segment FG, which measures 1 unit because G is one-third of the distance from F to G and $(\frac{1}{3}) \times 3 = 1$.

Use the Pythagorean theorem ($c^2 = a^2 + b^2$) to solve for the length of Leg 2 using what you know about Leg 1b and Leg 2b.

$c^2 = 3^2 + 1^2$

$c^2 = 9 + 1$

$c^2 = 10$

Take the square root of both sides, and you get $c = \sqrt{10}$

Segment EG is the hypotenuse of a right triangle with legs measuring 2 and $\sqrt{10}$. Again, apply the work of Mr. Pythagoras.

$c^2 = 3^2 + \left(\sqrt{10}\right)^2$

$c^2 = 9 + 10$

$c^2 = 19$

Take the square root of both sides and you get $c = \sqrt{19} = 4.36$.

35. **D.** One of these triangles is flipped around like a mirror image of the other. They are different sized triangles, but they are still considered similar triangles. Because of the mirror flip, you just have to be careful about which sides correspond to which.

In this case w corresponds to 4, but y corresponds to x (not to 7). This is actually quite helpful as the problem asks you to evaluate y/x.

To solve, set up a proportion, keeping clear about the corresponding sides.

$\frac{y}{x} = \frac{w}{4}$, and w = 4 so the answer is: $\frac{6}{4} = \frac{3}{2}$.

36. **B.** For the sines of x and y to be equal, the ratios of the sides need to be equal.

Sine = opposite/hypotenuse, so EF/DF has to equal BC/AC (which is $\frac{8}{10} = \frac{4}{5} = 0.8$). As it stands, EF/DF = $\frac{16}{18} = 0.89$, so they don't equal each other.

Where to go from here may seem terribly complicated, but before you despair, take a look at the answer choices. All of the answer choices involve *DF*, so you know you have to do something to *DF*.

Your best bet is to set up a proportion to ensure that the ratio of the sides is equal.

$$\frac{8}{16} = \frac{10}{x}$$

Cross multiply.

$8x = 160$

Divide both sides by 8.

$x = 20$

20 is 2 more than the current 18, so the answer is to increase *DF* by 2.

37. **D.** When you subtract a negative, you are really adding a positive.

By definition, $\sin^2 \Theta + \cos^2 \Theta = 1$, but you have to look closely for that in this problem. Never fear. You have the power.

The expression simplifies as follows:

$$\left[\cos^2 \Theta + 6 - \left(-\sin^2 \Theta + 2\right)\right]^2 =$$
$$\left[\cos^2 \Theta + 6 + \sin^2 \Theta - 2\right]^2$$
$$\left[\cos^2 \Theta + \sin^2 \Theta + 6 - 2\right]^2$$
$$\left[1 + 6 - 2\right]^2$$

$$[5]^2 = 25$$

If you chose E, you forgot to apply the subtraction outside the parenthesis to the 2.

38. **C.** You can rough in the graph of this problem, and it will help you visualize the problem; but if you cannot solve this problem visually and graphing eats up precious time, it's best to rely on solving equations.

This problem involves two equations and substitution. The first equation is that of the line:

$y = 4$.

The second equation is that of the circle.

The general equation for a circle is $(x - h)^2 + (y - k)^2 = r^2$, with (h, k) being the *x*- and *y*-coordinates of the center of the circle and *r* being its radius.

So, the equation for this specific circle is as follows:

$$(x - 2)^2 + (y - 3)^2 = r^2$$

Finally, the problem tells you that, at the intersection point in question, the *x*-coordinate is 5. You now have all the information you need to solve for *r*.

Start with the equation for the circle, and then plug in the *x*- and *y*-coordinates of the intersection point. The *x*-coordinate is 5 and the *y*-coordinate is 4, as given by the equation of the line $y = 4$.

$$(x - 2)^2 + (y - 3)^2 = r^2$$

becomes

$$(5 - 2)^2 + (4 - 3)^2 = r^2$$

Simplify:

$(3)^2 + (1)^2 = r^2$

$9 + 1 = r^2$

$10 = r^2$

Take the square root of both sides.

$r = \sqrt{10} = 3.16$

39. **E.** For this problem, use the typical formula for exponential growth:

$NV = IV \times (1 + P)^Y$, where NV = new value, IV = initial value, P = percent increase, and Y = number of years.

In this problem IV = four pairs = 8; $P = 0.12$ (remember to convert to a decimal); $Y = 2004 - 1987 = 17$.

Now, plug the numbers into the formula.

$NV = 8 \times (1 + 0.12)^{17}$

$NV = 8 \times (1.12)^{17}$

$NV = 8 \times 6.86604 \ldots$ (leave this number in your calculator)

$NV = 54.9$.

As 0.9 of a wolf is not a whole wolf, so the answer is 54. (Perhaps, one of the mothers is pregnant with a wolf cub and she is almost due. That's a good thing for an endangered species!)

40. **B.** The sum of the interior angles of a polygon is = $180° \times (n - 2)$, where n is the number of sides.

If you can't remember that formula, divide the given shape into the minimum number of triangles possible. Each triangle measures 180°. So, for each triangle add 180°, and you get the sum of all the angles in the polygon.

In this problem, with 5 sides, you get a sum of 540°.

Now look at the first half of the given statement.

If $r + s$ is more than 160°, then the lowest it can be is 160° and the highest $t + u + v$ can be is $540 - 160 = 380$.

Thus, $t + u + v < 380$.

The second half of the equation says that $r + s$ has to be less than 240°.

If $r + s$ is less than 240°, the highest it can be is 240° and the lowest $t + u + v$ can be is $540 - 240 = 300$.

Therefore, the second half of the answer is $t + u + v > 300$.

Combine those two deductions.

$t + u + v > 300$ is the same as saying $300 < t + u + v$, so $300 < t + u + v < 380$.

A, C, D, and E are all incorrect as they incorrectly use 360°, 720°, 900°, and 1,080°, respectively, as the total for the interior angles of the polygon instead of 540°.

41. **A.** Thales' theorem tells you that an inscribed angle in a semicircle (an angle with all three points on the circumference of a circle) is a right angle. For more info on inscribed figures, be sure to read Chapter 6.

Now that you know that the triangle has a right angle, you can use "ol' reliable," the Pythagorean theorem, to solve for segment *AC* using the values for the radius of the two legs.

$c^2 = a^2 + b^2$

$c^2 = 3^2 + 4^2$

$c^2 = 9 + 16$

$c^2 = 25$

Take the square root of both sides and you get $c = 5$.

Finally, *AC* is the diameter of the circle because it contains the circle's center. To solve for the radius, divide *AC* by two: 2.5.

42. **E.** –2 from the domain actually produces the highest value of the range, which is 81 [–2 – 1 = –3 and $(-3)^4 = 81$]. 2 from the domain gives you 1 in the range $(2 - 1)^4 = 1^4 = 1$. But do not stop there. Check the other values.

That the lowest number of the domain gave you a higher value for the range than the highest value of the domain should have tipped you off that this problem is a bit bizarro. Your best bet is to try all of the integers. Turns out that 1 in the domain yields 0 in the range $(1 - 1 = 0$ and $0^4 = 0)$. Thus the answer is $0 \le f(x) \le 81$.

43. **C.** The area of a parallelogram equals $b \times h$. Check out Chapter 6 for more information.

The problem gives us the base, which is 9 (5 + 4). To find the height (*FG*), you have to use the right triangle that they suggest.

Angle HEF, measuring 130°, is your starting point.

In a parallelogram, opposite angles are congruent (equal) and all four angles add up to 360°.

Therefore, angle HGF also measures 130°, which means that angle EFG (and angle EHG) measure 50°.

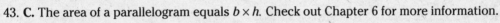

The special properties of this right triangle, where angle EFG = 50°, adjacent side = 5, and the hypotenuse is *FG*, will help you determine the height of the parallelogram.

Use the formula for tangent: tangent = opposite/adjacent

$\tan 50 = \dfrac{o}{5}$

$o = 5\tan 50 = 5 \times 1.191753 = 5.95876\ldots$

Plug into your formula for area.

$A = B \times H$

$A = 9 \times 5.95876$

$A = 53.63$

Answer choice A is what you get if you incorrectly use the formula for cosine. B is the result if you incorrectly use the formula for sine.

44. **B.** Be sure to answer what the question asks. In this case, you need to find the circumference, not the area.

To find the circumference in this problem, you need to know the area. First, use the information on the sunflowers to solve for the area of the garden.

10 flowers per package, times 7 packages, minus 1 flower can be represented by $(10 \times 7) - 1$, which equals 69 sunflowers.

Given that you know that there are 69 sunflowers, you can determine the area of the garden. Keep the square foot units on top as you are solving for area.

In a word problem, paying attention to your units will help you get the right answer.

$$69 \; flowers \times \frac{1 \; ft^2}{5 \; flowers} = 13.8 \; ft^2$$

Wait! You're not done yet! The problem asks you to solve for the circumference of the garden, not the area. So, use the area to solve for the radius, and then use the radius to solve for the circumference.

$A = \pi r^2$

$13.8 = \pi r^2$

Divide by π

$13.8 \div \pi = r^2$

Take the square root of both sides

$r = \sqrt{\frac{13.8}{\pi}} = 2.09587\ldots$ (keep this number in your calculator; don't round off!)

Now solve for circumference:

$C = 2\pi r$

$C = 2\pi(2.09587)$

$C = 13.17$

45. **E.** The area of a triangle is $A = \frac{1}{2}BH$.

You know the length of the base, which is just the diameter XY ($d = 2r$) of the circle, which is $2 \times 4 = 8$. To solve this problem, create a right triangle by drawing in altitude OC, and then solve for the length of this altitude (the height of the triangle). Because the larger triangle (XYO) is equilateral, you know that segment YO measures 8 units and that angle XYO is 60°. Compare your diagram to Figure 17-4.

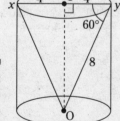

Figure 17-4:
Sample
drawing.

Notice that you now have two 30:60:90 triangles. The formula for the ratio of the sides of 30:60:90 triangles is $s:s\sqrt{3}:2s$, where s is the length of the shortest side and $2s$ is the hypotenuse. The shortest side of one of the triangles is 4, so $s = 4$. That means the measure of the altitude is $4\sqrt{3}$ or 6.928. Now that you know the altitude, you can find the area of the equilateral triangle:

$A = \frac{1}{2}(8 \times 6.928)$

$A = 27.71$

46. **C.** Using a quick number substitution for each possible answer is the fastest way to solve this problem. Try the answer choices that are easiest to plug in first.

    Stop when plugging in 2, 3, and 4 fails to give you 2, 6, and 12.

    $2n$ (Answer A) would give you 4 squares when you plug in 2 for the first set of squares $[2(2) = 4]$, but there are only 2 squares. A doesn't work.

    $2n^2$ (Answer B) gives you 8 squares for the first set $[2(2^2) = 8]$, so B is wrong.

    $n(n - 1)$ (Answer C) works for $2[2(2 - 1) = 2]$; it also works for $3[3(3 - 1) = 6]$, and for $4[4(4 - 1) = 12]$. You can stop now and go on to the next question.

    If you try $n(n + 1)$ and $2(2n - 1)$, you will confirm that D and E don't work.

47. **B.** To solve for an inverse function, you replace $f(x)$ with y, and then solve for x in terms of y, in effect reversing the function.

    Therefore,

    $y = 2x^4 + 2$

    Subtract 2 from both sides.

    $y - 2 = 2x^4$

    Divide both sides by 2.

    $(y - 2) \div 2 = x^4$

    Take the fourth root of both sides (you must use your calculator).

    $\sqrt[4]{\dfrac{(y - 2)}{2}} = x$

    Now plug 5 in for y and solve for x.

    $\sqrt[4]{\dfrac{(5 - 2)}{2}} = x$

    $\sqrt[4]{\dfrac{3}{2}} = x$

    $\sqrt[4]{1.5} = x$

    $x = 1.11$

    E is incorrect because it fails to reverse the function and plugs 5 into the function as is. Be familiar with all the functions on your calculator.

48. **D.** This problem presents two simultaneous equations. $s \cdot t$ means $s = t^3$ and $t \cdot u$ means $t = u^3$. Because you are solving for s in terms of u, you have to make the t's disappear.

    The second equation already has been solved for t in terms of u, so use that and substitute u^3 for t in the first equation. You will have your answer for s in terms of u.

    $s = t^3$

    $s = (u^3)^3$

    $s = u^9$

    If you chose A, B, C, or E, review your rules for exponents in Chapter 3.

49. **C.** The area of the shaded region bound by the graph of the parabola $x = f(y)$ and the y-axis is still 6. Subtracting 3 from $f(y)$ does nothing other than move the parabola 3 units down, which does not change its distance to the y-axis. Therefore, the area remains the same. Don't let the presence of the x-axis fool you; it does not affect the area described in this problem.

50. **A.** You want the concentrate to end up being 27% of the total beverage.

Remember that percentages can be expressed as ratios. The best way to solve the problem is to set up a proportion that will have concentrate on top and the total liquid in the beverage on the bottom, on both sides of the equal sign.

On the left side of the equation, put $\frac{27}{100}$, because you want lemonade to be 27% of the total or 27 parts per 100.

On the right side, on top, you need to put the total amount of concentrate that will end up in the beverage, which is 3 liters.

On the bottom is the total amount of combined liquid. Represent that with $3 + x$, which is the three liters of concentrate plus x amount of water.

Your proportion looks like this:

$$\frac{27}{100} = \frac{3}{3 + x}$$

Now, cross multiply and distribute the 27.

$81 + 27x = 300$

Subtract 81 from both sides.

$27x = 219$

Divide both sides by 27.

$x = 8.11$

Answer Sheet

Begin with Number 1. If the test has fewer than 100 questions, leave the extra spaces blank.

1. Ⓐ Ⓑ Ⓒ Ⓓ Ⓔ	21. Ⓐ Ⓑ Ⓒ Ⓓ Ⓔ	41. Ⓐ Ⓑ Ⓒ Ⓓ Ⓔ	61. Ⓐ Ⓑ Ⓒ Ⓓ Ⓔ	81. Ⓐ Ⓑ Ⓒ Ⓓ Ⓔ
2. Ⓐ Ⓑ Ⓒ Ⓓ Ⓔ	22. Ⓐ Ⓑ Ⓒ Ⓓ Ⓔ	42. Ⓐ Ⓑ Ⓒ Ⓓ Ⓔ	62. Ⓐ Ⓑ Ⓒ Ⓓ Ⓔ	82. Ⓐ Ⓑ Ⓒ Ⓓ Ⓔ
3. Ⓐ Ⓑ Ⓒ Ⓓ Ⓔ	23. Ⓐ Ⓑ Ⓒ Ⓓ Ⓔ	43. Ⓐ Ⓑ Ⓒ Ⓓ Ⓔ	63. Ⓐ Ⓑ Ⓒ Ⓓ Ⓔ	83. Ⓐ Ⓑ Ⓒ Ⓓ Ⓔ
4. Ⓐ Ⓑ Ⓒ Ⓓ Ⓔ	24. Ⓐ Ⓑ Ⓒ Ⓓ Ⓔ	44. Ⓐ Ⓑ Ⓒ Ⓓ Ⓔ	64. Ⓐ Ⓑ Ⓒ Ⓓ Ⓔ	84. Ⓐ Ⓑ Ⓒ Ⓓ Ⓔ
5. Ⓐ Ⓑ Ⓒ Ⓓ Ⓔ	25. Ⓐ Ⓑ Ⓒ Ⓓ Ⓔ	45. Ⓐ Ⓑ Ⓒ Ⓓ Ⓔ	65. Ⓐ Ⓑ Ⓒ Ⓓ Ⓔ	85. Ⓐ Ⓑ Ⓒ Ⓓ Ⓔ
6. Ⓐ Ⓑ Ⓒ Ⓓ Ⓔ	26. Ⓐ Ⓑ Ⓒ Ⓓ Ⓔ	46. Ⓐ Ⓑ Ⓒ Ⓓ Ⓔ	66. Ⓐ Ⓑ Ⓒ Ⓓ Ⓔ	86. Ⓐ Ⓑ Ⓒ Ⓓ Ⓔ
7. Ⓐ Ⓑ Ⓒ Ⓓ Ⓔ	27. Ⓐ Ⓑ Ⓒ Ⓓ Ⓔ	47. Ⓐ Ⓑ Ⓒ Ⓓ Ⓔ	67. Ⓐ Ⓑ Ⓒ Ⓓ Ⓔ	87. Ⓐ Ⓑ Ⓒ Ⓓ Ⓔ
8. Ⓐ Ⓑ Ⓒ Ⓓ Ⓔ	28. Ⓐ Ⓑ Ⓒ Ⓓ Ⓔ	48. Ⓐ Ⓑ Ⓒ Ⓓ Ⓔ	68. Ⓐ Ⓑ Ⓒ Ⓓ Ⓔ	88. Ⓐ Ⓑ Ⓒ Ⓓ Ⓔ
9. Ⓐ Ⓑ Ⓒ Ⓓ Ⓔ	29. Ⓐ Ⓑ Ⓒ Ⓓ Ⓔ	49. Ⓐ Ⓑ Ⓒ Ⓓ Ⓔ	69. Ⓐ Ⓑ Ⓒ Ⓓ Ⓔ	89. Ⓐ Ⓑ Ⓒ Ⓓ Ⓔ
10. Ⓐ Ⓑ Ⓒ Ⓓ Ⓔ	30. Ⓐ Ⓑ Ⓒ Ⓓ Ⓔ	50. Ⓐ Ⓑ Ⓒ Ⓓ Ⓔ	70. Ⓐ Ⓑ Ⓒ Ⓓ Ⓔ	90. Ⓐ Ⓑ Ⓒ Ⓓ Ⓔ
11. Ⓐ Ⓑ Ⓒ Ⓓ Ⓔ	31. Ⓐ Ⓑ Ⓒ Ⓓ Ⓔ	51. Ⓐ Ⓑ Ⓒ Ⓓ Ⓔ	71. Ⓐ Ⓑ Ⓒ Ⓓ Ⓔ	91. Ⓐ Ⓑ Ⓒ Ⓓ Ⓔ
12. Ⓐ Ⓑ Ⓒ Ⓓ Ⓔ	32. Ⓐ Ⓑ Ⓒ Ⓓ Ⓔ	52. Ⓐ Ⓑ Ⓒ Ⓓ Ⓔ	72. Ⓐ Ⓑ Ⓒ Ⓓ Ⓔ	92. Ⓐ Ⓑ Ⓒ Ⓓ Ⓔ
13. Ⓐ Ⓑ Ⓒ Ⓓ Ⓔ	33. Ⓐ Ⓑ Ⓒ Ⓓ Ⓔ	53. Ⓐ Ⓑ Ⓒ Ⓓ Ⓔ	73. Ⓐ Ⓑ Ⓒ Ⓓ Ⓔ	93. Ⓐ Ⓑ Ⓒ Ⓓ Ⓔ
14. Ⓐ Ⓑ Ⓒ Ⓓ Ⓔ	34. Ⓐ Ⓑ Ⓒ Ⓓ Ⓔ	54. Ⓐ Ⓑ Ⓒ Ⓓ Ⓔ	74. Ⓐ Ⓑ Ⓒ Ⓓ Ⓔ	94. Ⓐ Ⓑ Ⓒ Ⓓ Ⓔ
15. Ⓐ Ⓑ Ⓒ Ⓓ Ⓔ	35. Ⓐ Ⓑ Ⓒ Ⓓ Ⓔ	55. Ⓐ Ⓑ Ⓒ Ⓓ Ⓔ	75. Ⓐ Ⓑ Ⓒ Ⓓ Ⓔ	95. Ⓐ Ⓑ Ⓒ Ⓓ Ⓔ
16. Ⓐ Ⓑ Ⓒ Ⓓ Ⓔ	36. Ⓐ Ⓑ Ⓒ Ⓓ Ⓔ	56. Ⓐ Ⓑ Ⓒ Ⓓ Ⓔ	76. Ⓐ Ⓑ Ⓒ Ⓓ Ⓔ	96. Ⓐ Ⓑ Ⓒ Ⓓ Ⓔ
17. Ⓐ Ⓑ Ⓒ Ⓓ Ⓔ	37. Ⓐ Ⓑ Ⓒ Ⓓ Ⓔ	57. Ⓐ Ⓑ Ⓒ Ⓓ Ⓔ	77. Ⓐ Ⓑ Ⓒ Ⓓ Ⓔ	97. Ⓐ Ⓑ Ⓒ Ⓓ Ⓔ
18. Ⓐ Ⓑ Ⓒ Ⓓ Ⓔ	38. Ⓐ Ⓑ Ⓒ Ⓓ Ⓔ	58. Ⓐ Ⓑ Ⓒ Ⓓ Ⓔ	78. Ⓐ Ⓑ Ⓒ Ⓓ Ⓔ	98. Ⓐ Ⓑ Ⓒ Ⓓ Ⓔ
19. Ⓐ Ⓑ Ⓒ Ⓓ Ⓔ	39. Ⓐ Ⓑ Ⓒ Ⓓ Ⓔ	59. Ⓐ Ⓑ Ⓒ Ⓓ Ⓔ	79. Ⓐ Ⓑ Ⓒ Ⓓ Ⓔ	99. Ⓐ Ⓑ Ⓒ Ⓓ Ⓔ
20. Ⓐ Ⓑ Ⓒ Ⓓ Ⓔ	40. Ⓐ Ⓑ Ⓒ Ⓓ Ⓔ	60. Ⓐ Ⓑ Ⓒ Ⓓ Ⓔ	80. Ⓐ Ⓑ Ⓒ Ⓓ Ⓔ	100. Ⓐ Ⓑ Ⓒ Ⓓ Ⓔ

Chapter 18

Practice Test 4, Level IIC

● ●

*O*kay, you've studied your sine, cosine, and tangent. Now is your chance to let 'em know you know your stuff. The following exam is a 50-question, multiple-choice test. You have one hour to complete it.

To make the most of this practice exam, take the test under similar conditions to the actual test.

1. **Find a place where you won't be distracted (preferably as far from your younger sibling as possible).**

2. **If possible, take the practice test at approximately the same time of day as that of your real SAT II.**

3. **Set an alarm for 60 minutes.**

4. **Mark your answers on the provided answer grid.**

5. **If you finish before time runs out, go back and check your answers.**

6. **When your 60 minutes is over, put your pencil down.**

After you finish, check your answers on the answer key at the end of the test. Use the scoring chart to find out your final score.

Read through all of the explanations in Chapter 19. You learn more by examining the answers to the questions than you do by almost any other method.

Table 18-1	For Your Reference

You can use the following information for reference when you are answering the test questions.

The volume of a right circular cone with radius r and height h: $V = \frac{1}{3}\pi r^2 h$

The lateral area of a right circular cone with circumference of the base c and slant height l: $S = \frac{1}{2}cl$

The volume of a sphere with radius r: $V = \frac{4}{3}\pi r^2$

The surface area of a sphere with radius r: $S = 4\pi r^2$

The volume of a pyramid with base area B and height h: $V = \frac{1}{3}Bh$

For each of the following 50 questions, choose the best answer from the choices provided. If no answer choice provides the exact numerical value, choose the answer that is the best approximate value. Fill in the corresponding oval on the answer grid.

Notes:

1. **You must use a calculator to answer some but not all of the questions.**

   You can use programmable calculators and graphing calculators. You should use at least a scientific calculator.

2. **Make sure your calculator is in degree mode because the only angle measure on this test is degree measure.**

3. **Figures accompanying problems are for reference only.**

   Figures are drawn as accurately as possible and lie in a plane unless the text indicates otherwise.

4. **The domain of any function f is assumed to be the set of all real numbers x for which $f(x)$ is a real number unless indicated otherwise.**

5. **You can use the table on the previous page for reference in this test.**

1. If $2 + \frac{3}{a} = 4 + 5a$, for $a > 0$, then $a =$

   (A) ⅗ (B) ½ (C) ⅗ (D) ¾ (E) 1

2. $\dfrac{a}{\left(\dfrac{1}{b} - \dfrac{1}{c}\right)} =$

   (A) 0

   (B) $-abc$

   (C) $ab - ac$

   (D) $\dfrac{abc}{c - b}$

   (E) $\dfrac{ac - ab}{bc}$

3. What are the coordinates for the minimum value of the first full cycle, starting at $x = 0$, of the function $y = 2\cos 2x$?

   (A) $\left(\dfrac{\pi}{4}, -\pi\right)$

   (B) $\left(\dfrac{\pi}{4}, -2\right)$

   (C) $\left(\dfrac{\pi}{4}, -1\right)$

   (D) $\left(\dfrac{\pi}{2}, -2\right)$

   (E) $\left(\dfrac{\pi}{2}, -1\right)$

4. If $\sqrt{7x + 1} = 5.21$, then $x =$

   (A) 4.36 (B) 3.73 (C) 3.13 (D) 2.53 (E) 1.26

5. In the figure, $z \tan \theta =$

   (A) z^2

   (B) $z(x + y)$

   (C) zxy

   (D) $\dfrac{zx}{y}$

   (E) $\dfrac{zy}{x}$

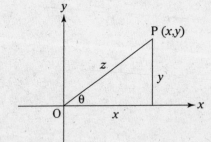

6. If a, b, and c are nonzero real numbers and if $a^5 b^4 c^3 = \dfrac{a^5 b^2}{ab^{-2} c^{-4}}$, then the expression $a^5 b^4 c^3$, in its simplest form, equals:

   (A) $b^4 c^8$ (B) $\dfrac{b^4}{c^2}$ (C) $\dfrac{c^8}{b^{16}}$ (D) $\dfrac{b^4}{c^3}$ (E) c

Go on to next page

7. What are the zeros of the polynomial $4x^2 + 8x - 32$?

   (A) $-4, -2$

   (B) $-4, 2$

   (C) $4, -2$

   (D) $4, 2$

   (E) $8, -1$

8. If the sun's angle of elevation, Θ, measures $29°$, and if the shadow of the tree measures 148 feet, how tall, in feet, is the tree? (Assume that the tree is straight and the ground is level.)

   (A) 62

   (B) 72

   (C) 82

   (D) 92

   (E) 129

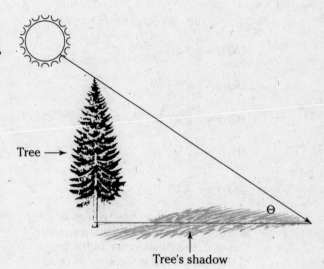

Tree →

Tree's shadow

9. The graph of $\begin{cases} x = 4r - 2 \\ y = -3r + 1 \end{cases}$ is a straight line whose x-intercept is

   (A) $-\frac{2}{3}$

   (B) $-\frac{1}{2}$

   (C) 1

   (D) $\frac{1}{2}$

   (E) $\frac{2}{3}$

10. If $f(x) = \sqrt{0.5x^2 + 2x}$ and $g(x) = \dfrac{x-2}{x+2}$, then $g(f(5)) =$

    (A) 0.34 (B) 0.41 (C) 0.69 (D) 0.83 (E) 0.97

11. Where defined, $\cot^2(4\theta)\tan^2(4\theta) =$

    (A) 0

    (B) $\cot^4(4\theta)$

    (C) $\tan^4(4\theta)$

    (D) 1

    (E) $2\tan^2(4\theta)$

Go on to next page

12. If $f(x) = 4x - 3$ and $g(f(2)) = -3$, which of the following could be $g(x)$?

   (A) $-2x + 1$

   (B) $-2x - 1$

   (C) $-x + 2$

   (D) $-x - 2$

   (E) $x + 2$

13. In the figure, an equation of line ℓ is

   (A) $y = \frac{3}{5}x + 3$

   (B) $y = -\frac{5}{3}x + 5$

   (C) $y = -\frac{3}{5}x + 5$

   (D) $y = -\frac{5}{3}x + 3$

   (E) $y = -\frac{3}{5}x + 3$

14. The figure shows a cube with an edge of length 6 centimeters and a circle, with center C, inscribed in one of the bases of the cube. What is the area, in square centimeters, of triangle ABC?

   (A) 4.24

   (B) 12.73

   (C) 15.91

   (D) 19.10

   (E) 25.46

15. The mean age of the 11 members of a marketing focus group was 28. A group of new members, which had a mean age of 22, was added to the existing group. As a result, the mean age of the marketing focus group decreased to 26.40. After the addition of new members, how many total members were in the marketing focus group?

   (A) 6

   (B) 14

   (C) 15

   (D) 16

   (E) 17

Go on to next page

16. If $-\frac{\pi}{2} < x < \frac{\pi}{2}$ and $\tan x = 12$, what is the value of $\sin\left(\frac{x}{3}\right)$?

    (A) 0.028

    (B) 0.332

    (C) 0.476

    (D) 0.541

    (E) 0.880

17. Yuan-Jen, a pharmacist, wants to strengthen a mixture of 45% saline solution to 75% saline solution. How many liters of 85% saline solution should be added to 20 liters of the 45% mixture to achieve this goal?

    (A) 60

    (B) 70

    (C) 80

    (D) 90

    (E) 100

18. If $f(x) = \dfrac{x^2 - 6x - 27}{x - 9}$, what value does $f(x)$ approach as x approaches 9?

    (A) 0

    (B) 3

    (C) 6

    (D) 9

    (E) 12

19. What number should be subtracted from each of the three numbers 11, 35, and 131 so that the resulting three numbers form a geometric progression?

    (A) 0 (B) 1 (C) 2 (D) 3 (E) 4

20. If $3^x = 4^y$ and $5^y = 6^z$, then $\frac{x}{y} =$

    (A) 1.70

    (B) 1.40

    (C) 1.01

    (D) 0.71

    (E) 0.36

21. What is the degree measure of the smallest angle of the triangle in the figure that has sides of lengths 3, 4, and 6?

    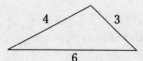

    (A) 63.51

    (B) 53.77

    (C) 36.23

    (D) 26.49

    (E) 19.75

Go on to next page

22. The figure shows a portion of the graph of $y = (1.5)^x$ and the circle with center (3, 1). The two intersect as illustrated. What is the area of the circle?

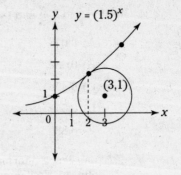

(A) 8.05

(B) 9.06

(C) 10.06

(D) 12.98

(E) 15.90

23. Which of the following lines are asymptotes of the graph of $y = \frac{x}{1-x}$?

   I. $x = 1$

   II. $y = 1$

   III. $y = -1$

(A) I only

(B) II only

(C) I and II only

(D) I and III only

(E) I, II, and III

24. If, in the figure, points A and B are on the line with the equation $3x - y = -6$, what is the slope of segment BC?

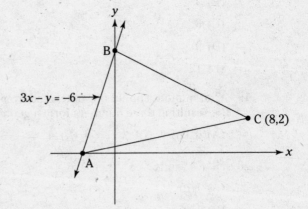

(A) -2

(B) -1

(C) $-\frac{1}{2}$

(D) $\frac{1}{2}$

(E) It cannot be determined from the information given.

25. What is the domain of $f(x) = \sqrt[2]{14 - 3x^2}$?

(A) $x \geq 0$

(B) $x \leq 2.16$

(C) $-2.16 \leq x \leq 2.16$

(D) $-3.74 \leq x \leq 3.74$

(E) all real numbers

Go on to next page

26. The graph $y = |f(x)|$ is shown in the figure. Which of the following could be a graph of $y = f(x)$?

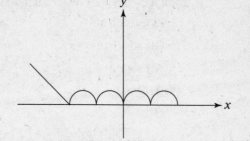

(A)

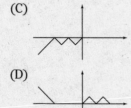

(B)

(C)

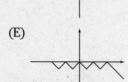

(D)

(E)

27. If sin x = tan x, which of the following is a possible radian value of x?

(A) −1.00

(B) −0.50

(C) 0

(D) 0.50

(E) 1.00

28. The number of cases of the Mumps disease is declining. The number of cases that exist at time t years after the year 1990 can be represented by the function: $P(t) = 5300e^{-0.315t}$. Assuming this is accurate, in what year would there be only 100 cases left?

(A) 1995

(B) 1997

(C) 1999

(D) 2002

(E) 2005

Go on to next page

29. If $f(2 - 3x) = 2 + 3x$ for all real numbers x, then $f(x) =$

   (A) $-x + 4$

   (B) $-x + 3$

   (C) $-x + 2$

   (D) $x + 2$

   (E) $1 - 2x$

30. Which of the following could be the coordinates of the center of a circle tangent to the positive y-axis and the negative x-axis?

   (A) $(-4, 4)$

   (B) $(0, -4)$

   (C) $(4, -3)$

   (D) $(4, -4)$

   (E) $(4, 0)$

31. If $4x - 2y = 1$ and $x^2 - 5y = 0$ for $x \geq 1$, then $x =$

   (A) 0.26

   (B) 3.25

   (C) 6.24

   (D) 9.74

   (E) 10.24

32. If $f(x) = \ln x$ for $x > 0$, then $f^{-1}(x) =$

   (A) x^e

   (B) e^x

   (C) 10^x

   (D) x^{10}

   (E) $\ln_x e$

33. What is the range of the function defined by $f(x) = \begin{cases} x + 1, x > 1 \\ \sqrt{4x}, x \leq 1 \end{cases}$?

   (A) $y < 0$

   (B) $y \geq 1$

   (C) $y \geq 0$

   (D) $0 \leq y < 1$

   (E) all real numbers

Go on to next page

34. If $x_0 = 1$ and $x_{n+1} = (x_n + 1)^2$, then $x_4 =$

    (A) 25

    (B) 676

    (C) 458,329

    (D) 2.10×10^{11}

    (E) 4.41×10^{22}

35. The figure shows a triangle inscribed in the top half of a circle with center C. What is the area of the whole circle in terms of x?

    (A) πx

    (B) $2\pi x$

    (C) πx^2

    (D) $\frac{1}{2}\pi x^2$

    (E) $\frac{1}{4}\pi x^2$

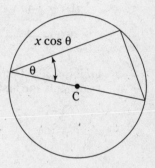

36. In a deck of cards there are 52 total cards, which include 4 of each number (1–10) and face card (Jack, Queen, King, Ace) and 13 of each of the four suits (Hearts, Diamonds, Clubs, and Spades). After shuffling the cards thoroughly and turning over the top card, what is the probability of turning over a King or a Club?

    (A) 0.019

    (B) 0.077

    (C) 0.250

    (D) 0.308

    (E) 0.327

37. If the magnitudes of vectors x and y are 11 and 6, respectively, then the magnitude of vector $(x + y)$ could NOT be

    (A) 4

    (B) 5

    (C) 6

    (D) 11

    (E) 17

38. If $(2.41)^a = (1.67)^b$ and $a + b = 3$, then what is the value of a?

    (A) 0.57

    (B) 0.58

    (C) 1.10

    (D) 1.90

    (E) 4.19

Go on to next page

39. $\dfrac{\left[(n-2)!\right]^2}{\left[(n-1)!\right]^2} =$

   (A) n^2

   (B) $\dfrac{1}{n^2}$

   (C) $n^2 - 4n + 4$

   (D) $n^2 - 4n + 4$

   (E) $\dfrac{1}{n^2 - 2n + 1}$

40. If the 30th term of an arithmetic sequence is 110 and the 50th term of the sequence is 230, what is the first term of the sequence?

   (A) –70

   (B) –64

   (C) 6

   (D) 64

   (E) 70

41. If arctan $(\tan x) = 1$, and $-\dfrac{\pi}{2} < x < \dfrac{\pi}{2}$, then x could equal

   (A) 0

   (B) 1

   (C) $\dfrac{\pi}{4}$

   (D) $\dfrac{\pi}{2}$

   (E) π

42. For all θ, $[\sin \theta \cdot \sin (-\theta)] + [\cos \theta \cdot \cos (-\theta)] =$

   (A) $2\cos \theta$

   (B) $2\sin \theta$

   (C) $\cos^2 \theta - \sin^2 \theta$

   (D) $\sin^2 \theta - \cos^2$

   (E) 1

43. The figure shows a cube with an edge of length 7. What is the volume of the pyramid ABCD?

   (A) 16.33

   (B) 24.50

   (C) 57.17

   (D) 85.75

   (E) 114.33

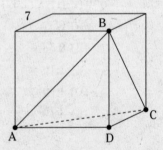

Go on to next page

44. The radius of the base of a right circular cone is 8 and the radius of a parallel cross-section is 3. If the distance between the base and cross-section is 10, what is the lateral area of the cone?

    (A) 300

    (B) 350

    (C) 400

    (D) 450

    (E) 500

45. A direct proof of the statement "If $a < b$ and $b < c$, then $a < c$" could begin with the assumption that

    (A) $a < c$ is true

    (B) $a < c$ is not true

    (C) $a \geq c$

    (D) $b < c$ is true

    (E) $a < b$ is true

46. Suppose the graph $f(x) = -x^2 + 1$ is translated 2 units left and 3 units down. If the resulting graph represents $g(x)$, what is the value of $g(-2.5)$?

    (A) −22.25

    (B) −16.25

    (C) −9.25

    (D) −2.25

    (E) 3.75

47. A little league coach is sitting down to determine the starting batting order for tomorrow's game. If there are 12 players on the team and 9 spots in the batting order, how many different batting orders are possible?

    (A) 79,833,600

    (B) 39,917,460

    (C) 1,320

    (D) 770

    (E) 220

48. What is the length of the minor axis of the ellipse whose equation is $15(x-4)^2 + 40(y+2)^2 = 120$?

    (A) 5.66

    (B) 3.46

    (C) 2.83

    (D) 1.73

    (E) It cannot be determined from the information given.

Go on to next page

49. Under which of the following conditions is $\dfrac{x+y}{\left(\frac{x}{y}\right)}$ definitely positive?

    (A) $x < 0 < y$

    (B) $y < 0 < x$

    (C) $0 < x < y$

    (D) $x < y < 0$

    (E) none of the above

50. If you have a box of 10 identical-looking chocolates, half filled with caramel and half filled with maraschino cherries, what is the probability of picking and eating two caramels in a row?

    (A) 0.200

    (B) 0.222

    (C) 0.250

    (D) 0.278

    (E) none of the above

Answer Key for Practice Exam 4

1.	C	26.	B
2.	D	27.	C
3.	D	28.	D
4.	B	29.	A
5.	E	30.	A
6.	A	31.	D
7.	B	32.	B
8.	C	33.	C
9.	A	34.	C
10.	B	35.	E
11.	D	36.	D
12.	C	37.	A
13.	E	38.	C
14.	B	39.	E
15.	C	40.	A
16.	C	41.	B
17.	A	42.	C
18.	E	43.	C
19.	D	44.	D
20.	B	45.	E
21.	D	46.	D
22.	A	47.	A
23.	D	48.	B
24.	C	49.	C
25.	C	50.	B

Scoring Your Exam

First, calculate your "raw score." On this exam, you get one point for each correct answer, and you lose ¼ (0.25) of a point for each wrong answer. Your raw score, then, is determined by the following formula: Raw score = number of correct answers – (0.25 × number of wrong answers). Feel free to use this Raw Score Wizard to help you determine your raw score:

Raw Score Wizard

1. Use the answer key to count the number of right answers   1) _____

2. Use the answer key to count the number of wrong answers   2) _____

3. Multiply the number of wrong answers by 0.25 (Line 2 × 0.25)   3) _____

4. Subtract this number from the number of right answers
   (Line 1 – Line 3)   4) _____

5. Round this number to the nearest whole number
   (Round Line 4 up or down)   5) _____

   This final number is your raw score.

Use Table 18-2 on the next page to convert your raw score to one of the College Board Scaled Scores, which range from 200–800. Ultimately, when you take the test for real, this scaled score will be the one the College Board reports to you and the colleges/universities you request.

Note: The College Board explains that it converts students' raw scores to scaled scores in order to make sure that a score earned on one particular subject test is comparable to the same scaled score on all other versions of the same subject test. Scaled scores are adjusted so that they indicate the same level of performance regardless of the individual test taken and the ability of the specific group of students that takes it. In other words, a score of 500 on one version of SAT II Math Level IIC administered at one time and place indicates the same level of achievement as a score of 500 on a different version of the test administered at a different time and place.

Table 18-2		Scaled Score Conversion Table, Mathematics Level IIC Test			
Raw	Scaled	Raw	Scaled	Raw	Scaled
50	800	29	600	8	410
49	790	28	590	7	400
48	780	27	580	6	390
47	780	26	570	5	380
46	770	25	560	4	380
45	750	24	550	3	370
44	740	23	540	2	360
43	740	22	530	1	350
42	730	21	520	0	340
41	720	20	510	−1	340
40	710	19	500	−2	330
39	710	18	490	−3	320
38	700	17	480	−4	310
37	690	16	470	−5	300
36	680	15	460	−6	300
35	670	14	460	−7	280
34	660	13	450	−8	270
33	650	12	440	−9	260
32	640	11	430	−10	260
31	630	10	420	−11	250
30	620	9	420	−12	240

Chapter 19

Explaining the Answers to Practice Test 4, Level IIC

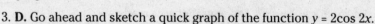

The following explanations are for Practice Test 4, Level IIC, which is in Chapter 18. Make sure you read through all of the explanations. You may pick up some valuable information.

1. **C.** Solve the equation for a! Subtract $\frac{3}{a}$ from both sides, and then subtract 4 from both sides. This leaves you with:

$$-2 = 5a - \frac{3}{a}$$

Now find the lowest common denominator for the expression on the right side, which is a. When you apply the LCD, the expression changes to

$$-2 = \frac{5a^2}{a} - \frac{3}{a}$$

or

$$-2 = \frac{5a^2 - 3}{a}$$

Now, cross-multiply.

$$5a^2 - 3 = -2a \text{ or } 5a^2 + 2a - 3 = 0$$

Factor the quadratic equation.

$$(5a - 3)(a + 1) = 0$$

So the solutions are $a = \frac{3}{5}$ and $a = -1$, but you can eliminate the second one because the problem states that $a > 0$.

2. **D.** To evaluate the expression, first find the lowest common denominator for the expression in the denominator. The LCD is bc. The expression now looks like this:

$$\frac{a}{\left(\frac{b-c}{bc}\right)} =$$

Now, instead of dividing by the fraction, multiply by the reciprocal:

$$\frac{abc}{b-c}$$

3. **D.** Go ahead and sketch a quick graph of the function $y = 2\cos 2x$.

When graphing trigonometric functions, changing the coefficient changes the amplitude and changing the angle changes the period.

In this case, the coefficient 2 doubles the amplitude; on the y-axis the function will now range from -2 to 2 (instead of -1 to 1). Meanwhile, doubling the angle (from x to $2x$) cuts the period in half, so instead of the period going from 0 to 2π, it goes only from 0 to π.

Finally, when sketching the graph, keep in mind that when cosine is at 0 on the x-axis, it is in the middle of its cycle and at its maximum (as opposed to sine which is at the beginning of its cycle and a 0). Examine Figure 19-1 and you get your answer: $y = 2\cos 2x$ is at its minimum value at $\left(\frac{\pi}{2}, -2\right)$.

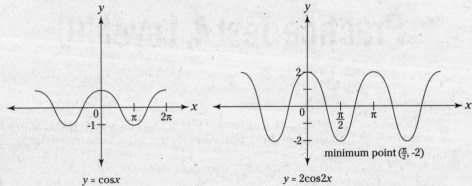

$y = \cos x$ minimum point $\left(\frac{\pi}{2}, -2\right)$

$y = 2\cos 2x$

4. **B.** You have to "break out" the variable x from the radical "jail" in which it is trapped. Square both sides of the equation. You end up with:

$$\sqrt{7x+1} \times \sqrt{7x+1} = 5.21 \times 5.21$$

which simplifies to $7x + 1 = 27.1441$.

Subtract 1 from both sides of the equation, divide both sides of the equation by 7, and you get $x = 3.73$

If there is more than one term under the radical symbol, you may *not* separate them. If you guessed D, you separated the 1 prematurely.

5. **E.** $\tan\theta = \dfrac{opposite}{adjacent} = \dfrac{y}{x}$. Therefore, $z\tan\theta = z \times \dfrac{y}{x} = \dfrac{zy}{x}$. If you answered anything else, look over your trigonometric identities (see Chapter 9).

6. **A.** For a reminder of how to multiply and divide like numbers (bases) with different exponents (powers), as well as how to use negative exponents, see Chapter 15.

To get the expression to its simplest form, solve for one of the variables in terms of the other two. Examining the possible answers, you see that the SAT II gurus want the answers in terms of b and c, which means you have to solve for a.

When in doubt, examine the possible answers for clues as to how to answer the question.

Ask yourself this question. "Can I solve for one of these variables easily?" In this case, the answer is yes, as you will be able to isolate a (or c, but we will use a). One way to get a by itself is to cancel a5 on both sides of the equation, which will leave you with just a in the denominator of the right side of the equation. Another way is to recognize that, on the right side of the equation, $\dfrac{a^5}{a} = a^4$. The equation now reads:

$$a^5 b^4 c^3 = \frac{a^4 b^2}{b^{-2}c^{-4}}$$

Divide both sides of the equation by a^4 and you are left with just a on the left side.

$$ab^4 c^3 = \frac{b^2}{b^{-2}c^{-4}}$$

Now, recognize that b^{-2} in the denominator translates to b^2 in the numerator and that $b^2 \times b^2 = b^4$ and that c^{-4} in the denominator translates to c^4 in the numerator:

$$ab^4c^3 = b^4c^8$$

Now divide both sides by b^4 and c^3.

$$a = c$$

Finally, simplify the expression as the problem requests.

Substitute c for a in the expression.

$$c^5b^4c^3 = b^4c^8$$

7. **B.** Set the polynomial equal to zero, then factor, and you have your zeros.

When factoring a polynomial, see whether you can simplify the coefficients.

In this case, if you divide both sides of the equation $4x^2 + 8x - 32 = 0$ by 4, it then simplifies to $x^2 + 2x - 8 = 0$. That factors to $(x + 4)(x - 2) = 0$. The zeros are -4 and 2. If you answered, A, C, or D, review Chapter 4.

8. **C.** If you were flying over the figure for this question, out of your window on the right side of the airplane, you'd notice you're flying over a right triangle. Its vertices are the top and bottom of the real tree and the right end of the shadow. As the problem explains, the sun's angle of elevation is $29°$ and the tree's shadow is 148 feet long. Therefore, to determine the height of the tree, use the formula for tangent:

$$\tan 29 = \frac{opposite}{adjacent} = \frac{x}{148}$$

Cross-multiply and you get $\tan 29° \times 148 = x$. So, $x = 82$.

9. **A.** When you first looked at this problem, you were probably thinking "What?! How do I graph these? I don't see x and y, and I see two equations, not one; so how could it result in a straight line?" The answer is simple: simultaneous equations! The fun never stops!

To get rid of that meddlesome r, you can either solve one equation for r and substitute what you get into the other equation, or just solve both equations for r and set the other sides equal to each other. We'll try the latter.

Solve both equations for r.

$$r = \frac{x+2}{4} \text{ and } r = \frac{y-1}{-3}.$$

So, $\frac{x+2}{4} = \frac{y-1}{-3}$

Cross-multiply and distribute and you get $-3x - 6 = 4y - 4$.

To solve for the x-intercept, substitute 0 for y.

$$-3x - 6 = 4(0) - 4$$

Simplify.

$$-3x - 6 = -4$$
$$-3x = 2$$
$$x = -\frac{2}{3}$$

10. **B.** The best way to solve the types of problems that combine functions is to work inside out (that's good because you're used to working this way, given the order of operations).

The problem asks you to find *g* of *f* of 5. In other words, solve *f(x)* for *f(5)*, and then plug what you get for *x* in the other equation. Plug in 5 for *x* in *f(x)*.

$$f(5) = \sqrt{0.5(5)^2 + 2(5)} = 4.74$$

Be careful to follow the proper order of operations!

Now substitute 4.74 for *x* in the equation *g(x)*.

$$g(4.74) = \frac{4.74 - 2}{4.74 + 2} = 0.41$$

If you chose E, you went the wrong way on a one-way street. You solved *f(g(20))* instead of *g(f(20))*.

11. **D.** In addition to sine, cosine, and tangent, there are also cosecant (csc), secant (sec), and cotangent (cot). These three are essentially defined as the reciprocals ($1/x$) of the sine, cosine, and tangent functions, respectively — and that is the best way to remember them. You can review their formulas in Chapter 9 or Chapter 15.

Substitute 1/tan for cotan and recognize that the two $\tan^2 4\theta$'s will cancel.

$$\left(\frac{1}{\tan^2 4\theta}\right) \times \tan^2 4\theta = 1$$

12. **C.** When combining functions, work from the inside out. Find *f(2)* first.

$$f(2) = 4(2) - 3 = 5$$

Then plug 5 into the potential answers for *g(x)* and see if you get –3. When you plug 5 into –*x* + 2, you get –(5) + 2 = –3. Bingo!

13. **E.** One of the standard forms for the equation of a line is the slope/intercept form or $y = mx + b$, where *m* is the slope and *b* is the *y*-intercept ($x = 0$). To review this concept, read Chapter 7.

As the figure indicates, you have two points on the line, one of which is the *y*-intercept; you're in good shape.

$slope = \frac{y_2 - y_1}{x_2 - x_1}$ or the change in *y*-coordinates divided by the change in *x*-coordinates.

It does not matter which point you start with, provided that you start with the same point for both the *x*- and *y*-coordinates.

$$slope = \frac{y_2 - y_1}{x_2 - x_1} = \frac{0 - 3}{5 - 0} = \frac{-3}{5} = -\frac{3}{5}$$

Combine that with the *y*-intercept 3, and you get the equation for the line.

$$y = -\frac{3}{5}x + 3$$

14. **B.** At this point, it must seem like the right triangle is that friend of yours that has outstayed his welcome. Oh, well, you must grin and bear it, because your friend the triangle has made the SAT II his home.

To find the area of triangle ABC, first determine the length of its base and height. The height is *AB* which equals 6, the length of the edge of the cube. To figure out base *BC*, draw a right triangle, as shown in Figure 19-2. Both legs measure 3 because one is half of the edge length and the other is the radius of the circle (which are both 3).

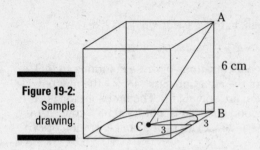

Figure 19-2:
Sample
drawing.

6 cm

A

B

C 3 3

Now, all hail Señor Pythagoras and his theorem, muy muy famoso:

$$c^2 = a^2 + b^2 = (3)^2 + (3)^2 = 9 + 9 = 18$$

Now, take the square root of both sides, and you get $c = \sqrt{18} = 4.24$. Now solve for the area.

A = ½bh = ½(4.24)(6) = 12.73

It does not matter which leg you use for the base and which leg you use for the height. Just make sure the two form a right angle.

15. **C.** By definition, an arithmetic mean equals the sum of all the values divided by the actual number of values. For a more detailed discussion of averages and means, read Chapter 10. Set up an equation and plug in what you know.

$$average = \frac{sum}{number\ of\ terms}$$

On the left of the equation is the arithmetic mean of the ages of the group, which the problem tells you is 26.4. On the right is the sum of all the ages (put them in parenthesis), divided by the total number of people.

The problem tells you that the average of these first 11 ages was 28; therefore, in your equation, the sum of the first 11 ages can be represented by 11×28. Add to that the sum of the ages of the new group, which will be x times 22 or $22x$ (let x = the number of newbies [new people] and multiply it by the average age, which is 22). Divide the left side by the total number of people, which is $11 + x$.

$$26.4 = \frac{(11 \times 28 + x \times 22)}{11 + x}$$

Now, multiply both sides of the equation by $(11 + x)$ and simplify within the parenthesis on the right side.

$$26.4(11 + x) = 308 + 22x$$

Distribute and solve.

$$290.4 + 26.4x = 308 + 22x$$

$$4.4x = 17.6$$

$$x = 4$$

Wait! You aren't done yet. The problem asks you how big the new group is, so add 4 to 11 and you get 15.

16. **C.** Take the inverse tangent of both sides and you get that $x = 85.24$. Now divide that by 3 and you get 28.41. Take the sine of that and you get 0.476.

Sine, cosine, and tangent functions are cyclical. They repeat themselves in a pattern on a graph. That is why the problem limits the period to a small range.

17. **A.** Set up an equation that represents the problem. In this case, it will look like this:

    The first saline solution + the second saline solution = the mixture of the two

    Keep in mind that you are dealing with the salt concentrations of these solutions so that fact should be reflected in the individual terms of the equation. First, let x = the number of liters in the second solution, as that is what you are solving for. The first solution is 20 liters at a 45% concentration or 0.45(20). The second solution is x liters at 85% concentration or 0.85x. Finally, the mixture of the two will be (20 + x) liters at a concentration of 75% or 0.75(20 + x). You now have:

    $0.45(20) + 0.85x = 0.75(20 + x)$

    Have at it. Multiply out and solve.

    $9 + 0.85x = 15 + 0.75x$

    $0.10x = 6$

    $x = 60$

18. **E.** In this problem $f(x)$ is undefined for x = 9 (can't divide by 0), so what $f(x)$, meaning y, is doing as x approaches 9 is a good question. But, you can't just plug in 9 for x, because dividing by zero is undefined. Therefore, you have to look for a way that $(x - 9)$ will cancel out. Turns out it is a factor of the quadratic expression on top. The numerator factors to $(x - 9)(x + 3)$. Then cancel out the two $(x - 9)$'s on the top and bottom, and you are left with $(x + 3)$. Now plug in 9 for x and you get 9 + 3 = 12.

19. **D.** A geometric progression is a sequence of numbers where each term after the first is found by *multiplying* the previous number by a fixed term called a common ratio. When you subtract 3 from 11, 35, and 131, you get 8, 32, and 128, whose common ratio is 4.

20. **B.** Ooooohh: simultaneous equations that require logs. Cool. Okay, maybe not, but definitely doable. Solve both equations for y and then set them equal to each other. Then, manipulate the equation so that you are solving for $\frac{x}{z}$. To solve $3^x = 4^y$ and $5^y = 6^z$, take the log of both sides and solve for y.

    $3^x = 4^y$

    $x \log 3 = y \log 4$

    $\dfrac{x \log 3}{\log 4} = y$

    Similarly

    $5^y = 6^z$

    $y \log 5 = z \log 6$

    $y = \dfrac{z \log 6}{\log 5}$

    Set the equations equal to each other.

    $\dfrac{x \log 3}{\log 4} = \dfrac{z \log 6}{\log 5}$

    To solve for $\frac{x}{z}$, divide both sides by z, and then by log3 and multiply both sides by log4 and log5.

    $4^2 = x^2 + z^2$

21. **D.** The key to mountains is altitude. The key to finding angle measure is also altitude. Drop an altitude dividing the regular triangle into two right triangles. (For a review of this concept, read Chapter 6.) You now have two triangles and several unknowns, but don't panic. First, draw a diagram and label everything you know and assign a letter variable to everything you don't. Compare your diagram to Figure 19-3.

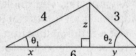

Figure 19-3:
Sample
drawing.

Eventually, you will have two equations and two unknowns. The first equation is easy: $x + y = 6$. To derive the second equation, you have another two sets of equations and two (well three, actually) unknowns. Use the Pythagorean theorem for the two right triangles.

$4^2 = x^2 + z^2$ and $3^2 = y^2 + z^2$

Solve for z^2 in both and set the equations equal to each other. You get $4 - x^2 = 3^2 - y^2$. With that and $x + y = 6$, you are in more familiar territory. Solve for x in the first equation: $x = 6 - y$. Now substitute $6 - y$ in for x in the other equation and solve for y.

$4^2 - (6 - y)^2 = 3^2 - y^2$

$4^2 - (36 - 12y + y^2) = 3^2 - y^2$

$16 - 36 + 12y - y^2 = 3^2 - y^2$ (Be careful as you distribute the negative sign!)

$-20 + 12y = 9$ ($-y^2$ cancels out on both sides)

$12y = 29$

$y = 2.41$

$x = 6 - y = 6 - 2.42 = 3.58$

Now, use the equation for cosine to find the measure of the angles.

$\cos\theta_1 = \dfrac{adjacent}{hypotenuse} = \dfrac{3.58}{4} = 0.895$, so $\theta_1 = 26.49$

$\cos\theta_2 = \dfrac{adjacent}{hypotenuse} = \dfrac{2.41}{3} = 0.803$, so $\theta_2 = 36.58$

Given these two small angles, it is clear that the third angle of the triangle will be large, so the answer is 26.49.

TIP

22. **A.** The area of the circle is $A = \pi r^2$. So all you have to do is find the radius of the circle.

According to the figure, the radius is the distance from $(3, 1)$ to the point on the circle that intersects with the graph. Find the coordinates of that point, and you can use your handy dandy distance formula.

The x-coordinate is obvious. The dotted line indicates that it is 2. The y-coordinate, then, is what you get when you plug 2 in for x in the equation of the curve $y = (1.5)^x$.

$y = (1.5)^2 = 2.25$

So the coordinates of the second point are $(2, 2.25)$

Now use the distance formula to find the radius.

$d = \sqrt{(1 - 2.25)^2 + (3 - 2)^2} = \sqrt{1.5625 + 1} = 1.60$

$y = \dfrac{x}{1 - x}$

Now, go back to the formula for area.

$A = \pi r^2 = \pi(1.60)^2 = 8.05$

23. **D.** This one is a little tricky. Normally, when you think of asymptotes, you think of vertical asymptotes, which are essentially the "zeros" of the denominator of a rational function. The idea is that you cannot have zero in the denominator of a rational function (because when you divide by zero the result is undefined). So, if you set the denominator of the fraction equal to zero and solve, this will inform you as to the values that x cannot be. There are also horizontal asymptotes, which have different properties than do the vertical ones, but for SAT II purposes you only need to know that they exist. If the problem does not specify, like this one does not, you are responsible for both vertical and horizontal asymptotes.

The vertical asymptotes are easy. Set the denominator equal to zero and solve for x.

$1 - x = 0$ or $x = 1$.

To get the horizontal asymptote, solve for x in terms of y (finding the inverse function), and then repeat the vertical asymptote process with the new denominator. Start with $\frac{y}{1+y} = x$.

Cross-multiply and solve for x.

$y(1 - x) = x$

$y - yx = x$

$y = x + yx$

$y = x(1 + y)$ (factored out the x)

$\frac{y}{1+y} = x$

Finally, set the denominator equal to zero and you get the horizontal asymptote $y = -1$ in addition to $x = 1$.

24. **C.** Back to the slippery "slope." Determine the coordinates of point B and you can determine the slope. B is the y-intercept of the line with the equation $3x - y = -6$. So, plug in 0 for x and solve for y.

$3(0) - y = -6$

$-y = -6$ and $y = 6$.

So, the two points you are using are: $(0, 6)$ and $(8, 2)$.

$slope = \frac{y_2 - y_1}{x_2 - x_1} = \frac{2 - 6}{8 - 0} = \frac{-4}{8} = -\frac{1}{2}$

25. **C.** To determine the domain of a function, start with all real numbers, and then exclude any members of this group that would make the range undefined. Because you can't take the square root of a negative number, the expression under the radical sign must be greater than or equal to zero.

Set up an equation that expresses just that. $14 - 3x^2 \geq 0$. Now solve.

$14 \geq 3x^2$ or $x^2 \leq 4.67$

Take the square root of both sides, and be sure to switch the less than to a greater than for the negative solution:

$x \leq 2.16$ and $x \geq -2.16$

When you have x^2 on one side of an inequality, be aware that after taking the square root of both sides, you are, in effect, dealing with the absolute value of x and have to flip the inequality accordingly. See the two following examples:

$x^2 \leq 4$

$|x| \leq 2$

$x \leq 2$ and $x \geq -2$ ($-2 \leq x \leq 2$)

$x^2 \geq 4$

$|x| \geq 2$

$x \geq 2$ and $x \leq -2$ $(-2 \geq x \geq 2)$

26. **B.** The absolute value of a number is simply its absolute distance from zero (like on a number line) and will always be positive. You could think of it as the positive version of a number.

    This figure, then, is the positive version of something. The question is what that something is. There are several possibilities, but any correct answer will have the current positive version reflected below the *x*-axis (the *y*-axis is irrelevant here). B is correct as it does just this. A and E are wrong because the straight line section has mysteriously jumped to the other side. C would have been right if it had two more humps below the *x*-axis on the right side of the *y*-axis. D is way off.

27. **C.** You can switch your calculator to Radian mode for this one (the word "radian" in the problem is your clue!).

    To solve the problem, solve for *x*.

    Remember that $\tan x = \frac{\sin x}{\cos x}$. Next, substitute that into the original equation.

    $\sin x = \frac{\sin x}{\cos x}$

    The two sin *x*'s cancel out and you are left with

    $P(t) = 5300e^{-0.315t}$

    Cross-multiply and you get that cos *x* = 1. Take the inverse cosine of both sides, with your calculator in Radian mode, and you get that *x* = 0.

    Remember to switch your calculator back to Degree mode.

28. **D.** Plug in and solve.

    And remember that the use of the real number constant *e* requires that you use the natural log function ("ln" button) on your calculator instead of the usual base 10 log function ("log" button).

    Starting with $P(t) = 5300e^{-0.315t}$ and plugging in you get

    $100 = 5300e^{-0.315t}$

    Divide both sides by 5,300.

    $0.01886\ldots = e^{-0.315t}$

    Take the natural log of both sides.

    $\ln(0.01886) = -0.315t$ or $-3.9702 = -0.315t$

    Divide both sides by –0.315.

    $t = 12.60$ or about twelve and a half years.

    Add that to 1990 and you get the middle of 2002.

    If you chose A, you used the log base 10 instead of the natural log.

29. **A.** Simply substitute 2 – 3*x* for *x* in each of possible answers for *f*(*x*) and see which one spits out the answer 2 + 3*x*. Your clue is that the 3*x* changes signs but keeps the coefficient of 3. Turns out, when you substitute 2 – 3*x* for A, –*x* + 4, you get –(2 – 3*x*) + 4 = –2 + 3*x* + 4 = 2 + 3*x*.

30. **A.** No doubt, you quickly sketched a picture of this one and placed the circle snuggly in the crook of the two axes in the upper left quadrant. Two of its points (its lowermost and right-most points) are touching the axes. You can further deduce that because the center is

equidistant (the same distance) from all points on the circle, the center must be equidistant from the axes themselves. The only answer that satisfies all these requirements is (–4, 4).

31. **D.** To solve these simultaneous equations, solve for one variable in terms of the other, substitute the answer into the other equation and then use the quadratic formula. Also, pay attention to the specific requirement "for $x \geq 0$" explicitly stated by the problem. Solve the second equation for y.

$$x^2 = 5y$$

$$y = \frac{x^2}{5}$$

Substitute y into the first equation.

$$4x - 2\left(\frac{x^2}{5}\right) = 1$$

Simplify.

$$4x - \frac{2}{5}x^2 - 1 = 0$$

At this point, you can multiply the equation by –5, and then reorder the terms to avoid dealing with fractions and negative squared terms.

$$-20x + 2x^2 + 5 = 0$$

$$2x^2 - 20x + 5 = 0$$

Plug in the coefficients to the quadratic equation.

$$x = \frac{20 \pm \sqrt{(20)^2 - 4(2)(5)}}{2(2)}$$

Simplify.

$$x = \frac{20 \pm \sqrt{360}}{4}$$

$x = 9.74$ and $x = 0.26$

The problem says that $x \geq 1$, so you discount the second one.

The other answers involve a myriad of errors involving the application of the quadratic formula. As always, be careful.

32. **B.** As a review, $f^{-1}(x)$ represents the inverse function, or what happens when you take your function and invert it, solving for x in terms of y. $f(x) = \ln x$ becomes $y = \ln x$.

Now, take the inverse ln of both sides. As the natural log utilizes base e, you get

$$e^y = e^{\ln x}$$

Simplify the right side.

$$e^y = x$$

Finally, it is customary to switch back x and y, so that the equation takes the form of the usual function, and the answer is $y = e^x$.

33. **C.** The easiest way to solve this problem is to use your graphing calculator. Have it graph both functions at the same time (don't worry about the domain restrictions yet). After you have the graph in front of you, see what each is doing for its designated x's (now observe the domain restrictions). You will quickly see that while the straight line $y = x + 1$ includes all possible real numbers for its portion of the domain $x > 1$, the equation with the radical $y = \sqrt{4x}$ ends abruptly at the x-axis ($y = 0$) during its portion $x \leq 0$ because you can't take the square root of a negative number. When the two are considered as one, the answer is $y \geq 0$.

34. **C.** The problem tells you the value of the first term (x_0) in this sequence and an equation that represents the value of one term (x_{n+1}) in terms of the previous term (x_n) To solve this sequence problem, then, start with the second term (x_1) and go from there.

$$x_1 = (x_0 + 1)^2$$

$$x_1 = (1 + 1)^2 = 4$$

Now that you know the value of x_1, you can repeat the process for x_2.

$$x_2 = (x_1 + 1)^2$$

$$x_2 = (4 + 1)^2 = 25$$

Repeat the process for x_3.

$$x_3 = (x_2 + 1)^2$$

$$x_3 = (25 + 1)^2 = 676$$

Finally, repeat the process for x_4.

$$x_4 = (x_3 + 1)^2$$

$$x_4 = (676 + 1)^2 = 458,329$$

35. **E.** An *inscribed angle* (an angle with all three points on the circumference of a circle) in a semicircle is a right angle. Given that the triangle is right and that its hypotenuse is the diameter of the circle, you can solve for the area of the circle by solving for the diameter, dividing by 2 to get the radius and plugging that into the area formula.

As the adjacent side equals $x \cos \theta$, you can use the cosine formula to solve for the hypotenuse which you will call h:

$$\cos \theta = \frac{adjacent}{hypotenuse} = \frac{x \cos \theta}{h}$$

Notice that the two $\cos \theta$'s will cancel and you are left with

$$1 = \frac{x}{h}$$

When you cross-multiply, you get $h = x$.

To derive the radius from the diameter, remember to divide by 2.

If the diameter is x, then the radius is $\frac{x}{2}$.

Plugging $\frac{x}{2}$ into the area formula you get

$$A = \pi r^2 = \pi \left(\frac{x}{2}\right)^2 = \pi \frac{x^2}{4} = \frac{1}{4}\pi x^2$$

If you chose C, you used the diameter instead of the radius. If you chose D, you forgot to square the 2 in the denominator of the radius.

36. **D.** This is an overlapping EITHER/OR probability. You need to calculate the probability of picking EITHER a King OR a Club, but you have account for the overlap between the two possibilities, namely the King of Clubs.

When you combine probabilities using either/or you add the probabilities together — but in the case of overlapping sets, you have to subtract out the probability of the overlap. To review these concepts, see Chapter 10.

In this case, here is how you set up the equation (P stands for probability — original, huh?)

$$P_{\text{A King or a Club}} = P_{\text{King}} + P_{\text{Club}} - P_{\text{The King of Clubs}}$$

Plug in and solve.

$$P = \frac{4}{52} + \frac{13}{52} - \frac{1}{52} = \frac{16}{52} = 0.308$$

37. **A.** When you combine two vectors, their maximum sum is one plus the other (if they are headed in the same directions), and the least they can add up to is one minus the other (if they are headed in opposite directions).

    In this case the most these vectors could add up to is $11 + 6 = 17$ and the least is $11 - 6 = 5$.

38. **C.** More simultaneous equations using logs! Can this day get any better? Use the easier equation to solve for b in terms of a (because you want to eventually solve for a), and then plug what you get into the harder equation. If $a + b = 3$, then $b = 3 - a$. Plugging that into the harder equation, you get:

    $$\log(2.41)^a + \log(1.67)^{3-a}$$

    which becomes

    $$a\log(2.41) = (3 - a)\log(1.67)$$

    Because you want to solve for a, divide both sides by $(3 - a)$ and $\log(2.41)$.

    $$\frac{a}{(3-a)} = \frac{\log(1.67)}{\log(2.41)}$$

    Simplify.

    $$\frac{a}{(3-a)} = 0.583$$

    Cross-multiply and solve.

    $$0.583(3 - a) = a$$

    $$1.749 - 0.583a = a$$

    $$1.749 = 1.583a$$

    $$1.10 = a$$

39. **E.** The factorial of the natural number n (indicated by "$n!$"), is defined as the product of n and all the positive integers less than n ($3! = 3 \times 2 \times 1$; $4! = 4 \times 3 \times 2 \times 1$; and so on).

    The key to this problem is to expand the expression only as much as you need to and then let equal items on the top and bottom cancel each other out. Remember that the factorial symbol "!" indicates that you keep multiplying the expression by one less than the previous number (in other words, $n - 1$).

    $$\frac{\left[(n-2)!\right]^2}{\left[(n-1)!\right]^2}$$

    can be expanded to

    $$\frac{(n-2)!(n-2)!}{(n-1)(n-1)(n-2)!(n-2)!}$$

    At this point, the four $(n-2)!$'s cancel each other out and you are left with

    $$\frac{1}{(n-1)(n-1)} \text{ or } \frac{1}{(n-1)^2}$$

    The last step is to recognize that $(n-1)^2 = n^2 - 2n + 1$.

40. **A.** An *arithmetic sequence* (or *arithmetic progression*) is a sequence of real numbers for which each term is equal to the previous term plus a constant (called the *common difference*). For example, starting with 1 and using a common difference of 4, you get the arithmetic sequence 1, 5, 9, 13, and so on. In general, then, you can create a formula for an arithmetic sequence $a_n = a_0 + nd$, where n is the number of the current term, a_0 is the value of the first term, and d is the common difference.

With this equation and the information in the problem, you can set up simultaneous equations and solve for the unknowns. Using the first piece of information, that the 30th term of an arithmetic sequence is 110, you can generate the equations $110 = a_0 + 30d$. Using the second piece of information from the problem, that the 50th term of the sequence is 230, you can generate the equation $230 = a_0 + 50d$. Solve both equations for a_0 and then set them equal to each other:

$110 = a_0 + 30d \qquad 230 = a_0 + 50d$

$a_0 = 110 - 30d \qquad a_0 = 230 - 50d$

Therefore,

$110 - 30d = 230 - 50d$

Combine your like terms:

$20d = 120$

$d = 6$

So, the common difference, or the numerical difference from one term and the next, is 6. The last step of the problem is to plug this answer for d back into one of the two initial equations to solve for a_0.

$110 = a_0 + 30d$

$110 = a_0 + 30(6)$

$110 = a_0 + 180$

$a_0 = -70$

41. **B.** The arctangent is another way of saying the inverse tangent or $\tan^{-1}$ (the same goes for arcsin and arccos).

The arctan (tan x) is going to equal plain-old x because you are taking the inverse tangent of the tangent of x. Here it is mathematically:

arctan (tan x) = 1

$x = 1$

Note that the limited domain in the problem assures that x has no other values. (The trigonometric functions are cyclical and repeat themselves, repeat themselves, repeat themselves.)

42. **C.** This problem harkens back to your trigonometric identities, specifically, the reduction formulas. This problem seeks to determine how the value of sine, cosine, and tangent are changed when you apply the function to the opposite of θ (or negative θ). The two formulas that apply in this case are:

$\sin(-\theta) = -\sin\theta$

$\cos(-\theta) = \cos\theta$

Now, using the two formulas just given, substitute these values in for the original equation as follows.

$[\sin\theta \cdot \sin(-\theta)] + [\cos\theta \cdot \cos(-\theta)]$

becomes

$[\sin\theta \cdot (-\sin\theta)] + [\cos\theta \cdot (\cos\theta)]$

Simplify.

$-\sin^2\theta + \cos^2\theta$ or $\cos^2\theta - \sin^2\theta$

Be careful; the temptation here may have been to answer E, which is 1, because you know that $\sin^2\theta + \cos^2\theta = 1$. But the final expression does not add the two but subtracts them.

43. **C.** You are given the formula for the volume of a pyramid at the beginning of the exam. It is $V = \frac{1}{3}Bh$, where B is the area of the base of the pyramid and h is the height of the pyramid. The height of this pyramid you already know is 7 or the length of the edge of the cube. The area of the base is the area of a triangle, which is $A = \frac{1}{2}bh$. It just so happens that both base and height are also 7 as they are also both edges of the cube. See, the two edges of the cube form a right angle, and when you add the diagonal from C to A, you form a right triangle with the edges as the base and height.

$A = \frac{1}{2}bh = \frac{1}{2} \times 7 \times 7 = 24.5$

Plug this in for B in the volume equation.

$V = \frac{1}{3}bh = \frac{1}{3} \times 24.5 \times 7 = 57.17$

44. **D.** As per the instructions at the beginning of the exam, the formula for lateral area is $S = \frac{1}{2}cl$, where c is the circumference of the base and l is the slant height (the diagonal edge of the cone from top to bottom). Solving for the circumference is easy — like taking candy from a baby: $c = 2\pi r = 2 \times \pi \times 8 = 50.265$.

TIP

To solve for the slant height, draw a picture. You should recognize that this problem involves using similar triangles, and you will be able to solve for the height h of the cone using a proportion. In turn, you can solve for the slant height l using the Pythagorean theorem. Your picture, according to the problem, should look like Figure 19-4. (The letters are added for reference.)

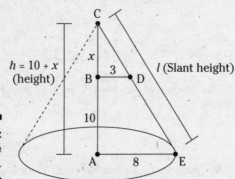

Figure 19-4: Sample drawing.

Because the two cross-sections are parallel, triangles ACE and BCD must be similar. Therefore, you can set up a proportion as follows:

$$\frac{BD}{AE} = \frac{BC}{AC}$$

In other words, the respective sides of the two triangles are in the same direct proportion to one another.

Fill in the values, and then add x for unknown segment BC.

$$\frac{3}{8} = \frac{x}{10 + x}$$

Cross-multiply.

$30(10 + x) = 8x$

Simplify.

$30 + 3x = 8x$

$30 = 5x$

$6 = x$

If $x = 6$, the height of the cone is $6 + 10 = 16$.

Next, use the Pythagorean theorem to solve for the slant height.

$c^2 = a^2 + b^2 = (8)^2 + (16)^2 = 64 + 256 = 320$

Take the square root of both sides, and you get $c = \sqrt{320} = 17.889$.

Finally, go back to the original equation for lateral area and plug in the numbers.

$S = \frac{1}{2} \times 50.265 \times 17.889 = 450$

The other possible answers involve a variety of errors including labeling the segments in your picture incorrectly, setting up your proportion wrong, and making mathematical errors in your solving and calculations. Make sure you don't rush through the problem.

45. **E.** By definition, to perform a direct proof, assume that the premises are true and, if the conclusion can be logically deduced from them, you have proved your case. Thus, the first step to this direct proof is to assume that the first premise, "$a < b$," is true. D is not quite right as it is the second premise and not the first, so it would not be where you "begin."

46. **D.** A parabola is the graph of a quadratic function. For a review on how to shift or translate parabolas based on changes in the standard form of their equations, see Chapter 7.

In this problem, you start with $f(x) = -x^2 + 1$. When you translate (or shift) it 2 units to the left, the equation changes to $-(x + 2)^2 + 1$. And when you translate it 3 units down, the equation changes to $-(x)^2 - 2$ (as $1 - 3 = -2$). Finally, the problem tells you that this new graph represents $g(x)$, and then asks you to find the value of $g(-2.5)$. Substitute -2.5 for x in this new equation.

$-(-2.5 + 2)^2 - 2 = -(-0.5)^2 - 2 = -0.25 - 2 = -2.25$

The other possible answers involve changing the equation of the parabola in the wrong directions, adding when you should be subtracting, and vice-versa.

47. **A.** This is a straightforward permutation problem. For a review on permutations, see Chapter 11.

The formula for a permutation is $_nP_r = \frac{n!}{(n-r)!}$ where n is the number of total items from which you are choosing and r is the number of items you choose. Another way of reading it is "the number of permutations of n items taken r at a time."

Now just plug in and solve.

$_nP_r = \frac{n!}{(n-r)!}$

$_nP_r = \frac{12!}{(12-9)!}$

Simplify the parenthesis.

$_nP_r = \frac{12!}{(3)!}$

Expand your greatest factorial (in this case 12!) so you can cancel and simplify.

$_nP_r = \frac{12 \times 11 \times 10 \times 9 \times 8 \times 7 \times 6 \times 5 \times 4 \times 3!}{3!}$

Cancel your 3! and solve.

$_nP_r = \frac{12 \times 11 \times 10 \times 9 \times 8 \times 7 \times 6 \times 5 \times 4}{1} = 79,833,600$

That's a lot of batting orders. We hope the coach does not consider each one individually!

48. **B.** The standard equation for an ellipse is $\dfrac{(x-h)^2}{a^2} + \dfrac{(y-k)^2}{b^2} = 1$; where (h, k) is the center (or intersection point of the major and minor axis), a is the length of the semi-major axis (half of the major axis); and b is the length of the semi-minor axis (half of the minor axis). As the two words indicate, the major axis is the longer of the two and the minor is the smaller. It follows, then, that if the ellipse is horizontal, a will end up under the term with the x-coordinate (the first one), and if the ellipse is vertical, a will end up under the term with the y-coordinate (the second one).

To solve this problem, manipulate the equation until it is in standard form, and then glean (meaning "to collect") the appropriate information. The equation of this ellipse is

$$15(x-4)^2 + 40(y+2)^2 = 120$$

To get the ellipse in standard form, the right side has to equal 1, so divide both sides of the equation by 120.

$$\dfrac{(x-4)^2}{8} + \dfrac{(y+2)^2}{3} = 1$$

Now, you should be able to tell that this is a horizontal ellipse and that $8 = a^2$, the square of the length of the semi-major axis, and $3 = b^2$, the square of the length of the semi-minor axis. Because the problem asks for the length of the minor axis, you have to solve for b and then multiply the answer by 2. If $b^2 = 3$, $b = 1.73$. Multiply that by 2.

$$1.73 \times 2 = 3.46$$

49. **C.** This problem refers to the various rules of combining signs for the various mathematical operations (see Chapter 3). The bottom line is that for the quotient $\dfrac{x+y}{\left(\dfrac{x}{y}\right)}$ to be positive, either both top and bottom have to be positive or both top and bottom have to be negative.

Go through each of the possible answers and see what effect each has on the quotient. According to answer choice A, x is negative and y is positive. That would make the denominator negative. Now what would it do to the numerator? The truth is you don't know. It depends on which absolute value (x or y) is greater. That is probably why the problem says "definitely" positive.

In answer choice B, y is negative and x is positive. B yields the exact same uncertain situation as A. In C, x and y are both positive. That sounds promising. In fact, if both x and y are positive, the sum in the numerator and the quotient in the denominator will be positive; therefore, the quotient from the problem will be positive. This choice is correct. To be sure, evaluate D. According to D, x and y are both negative. This will make the sum in the numerator negative but the quotient in the denominator positive; therefore, the quotient from the problem will be negative. Yep, C is right!

50. **B.** This question provides an example of a dependent AND probability. Both events have to occur (the first AND the second) and the two events are dependent.

If two events are dependent, that means the outcome of the first event affects the probability of the second. For a review of probability, see Chapter 11.

In this case, both the first and second events have to be caramel. When you pick the first and it is a caramel, that affects the probability that the next one will be a caramel. The formula for dependent AND probabilities looks like this (with A and B as the two events):

$$P_{A \text{ and } B} = P_A \times P_{B \text{ given } A}$$

"$P_{B \text{ given } A}$" means the probability of B occurring given that A has already occurred. Fill in what you know.

$$P = \frac{5}{10} \times \frac{4}{9} = \frac{20}{90} = 0.222$$

In other words, for the first event, you have a 5 out of 10 chance of choosing a caramel because there are 5 caramels and 10 total chocolates. After the first event, however, you are down to 4 caramels and 9 total chocolates, so the probability of choosing the second caramel is 4 out of 9. Simply multiply the two together, and you are set.

Part VI
The Part of Tens

"For the next month, instead of practicing on the baseball diamond, we'll be practicing on a baseball trapezoid. At least until everyone passes the geometry questions on the SAT II Math test."

In this part . . .

We give you an awful lot of information in each of the review chapters. This part synthesizes that information into lists of the ten most important formulas and definitions to remember for the test. You may want to memorize these lists for test day.

Chapter 20

Ten Formulas You Need to Know on Test Day

The SAT II Math tests provide you with several formulas at the beginning of each test, but those aren't the only procedures you're going to need to know to do well on the test. This chapter provides you with a synopsis of the equations you should know for the test that aren't listed on the Cheat Sheet at the front of this book.

A Formula for Doing Algebraic Work Problems

$$\text{Production} = \text{rate of work} \times \text{time}$$

Production stands for the amount of work that gets done. This formula works well for word problems. For example, two writers, Sarah and Martha, are working on a book. Sarah writes 16 pages per day, while Martha writes 20 pages per day. If they each work 8-hour days, how many pages can two of them write in one hour, assuming they maintain a steady rate? You can use the formula to find out how much each of them writes per day (16 + 20), and then divide that by 8 hours.

Another Handy Formula to Apply to Some Distance Problems

The formula for figuring out distance or speed problems in algebra is

$$\text{distance} = \text{rate} \times \text{time}$$

You can solve any problem involving distance, speed, or time spent traveling with this equation.

The Equation for Average or Mean

To find the average or mean of a set of values, use this formula.

$$A = \frac{Sum\ of\ all\ numbers}{Amount\ of\ numbers\ that\ make\ up\ the\ sum}$$

The Quadratic Formula

When you can't simply solve a quadratic equation by factoring, you may have to use the quadratic formula, which is a rearrangement of the classic $ax^2 + bx + c = 0$ to become

$$x = \frac{-b \pm \sqrt{b^2 - 4ac}}{2a}$$

The Slope-Intercept Formula

The characteristics of a line can be demonstrated with a formula. The equation of a line generally shows y as a function of x.

$$y = mx + b$$

In the slope-intercept formula, m is a constant that indicates the slope of the line, and the constant b is the y-intercept. So a line with a formula $y = 2x + 3$ has a slope of 2 and a y-intercept of 3.

The Point-Slope Form

If you know the slope of a line and a point located on the line, you can recognize the graph of that line or be able to give the equation for the line using the slope-intercept formula. The point-slope form comes in handy for this type of question. The point-slope form of the equation for a line is stated as

$$y - b = m(x - a)$$

where m is the slope and (a, b) is the known point.

The Equation for a Circle

The equation for a circle is

$$(x - h)^2 + (y - k)^2 = r^2$$

with (h, k) being the x- and y-coordinates of the center of the circle and r being its radius.

Two Formulas for the Equation of the Parabola

The most you'll really need to remember about the parabola for SAT II is summed up in the two formulas for the equation of the parabola.

- ✔ **The standard form:** $y = a(x - h)^2 + k$ (where $a \neq 0$)
- ✔ **The general form:** $y = ax^2 + bx + c$ (where $a \neq 0$)

A Formula for the Equation of an Ellipse

Here is the formula for the ellipse with its center at the origin.

$$\frac{x^2}{a^2} + \frac{y^2}{b^2} = 1$$

where a and b are the lengths represented by the semi-major axis and the semi-minor axis.

The Formula for a Hyperbola

The formula for a hyperbola is nearly the same as the one for the ellipse. But instead of adding the distances to the fixed points together, you subtract the distances from two fixed points.

Here is the formula for the equation of a hyperbola having the origin as its center.

$$\frac{x^2}{a^2} - \frac{y^2}{b^2} = 1$$

Chapter 21

Ten Definitions You Shouldn't Leave Home without Knowing

▶ Synthesizing a bunch of information into several important terms

▶ Compiling valuable conversation topics for your next party

You just have to know some math terminology to make it on the SAT II Math. This chapter lists those terms that you'll see in questions that you need to know for the question to make sense.

Absolute Value

The absolute value of any real number is that same number without a negative sign. Absolute value is marked by a number inside those two little bar like things: $|4|$. So $|-8| = 8$.

Real Numbers, Integers, and Rational Numbers

SAT II Math questions contain terminology that refers to specific types of numbers.

Real numbers

Real numbers include all whole numbers, fractions, rational, and irrational numbers. They are pretty much limitless.

Integers

Integers are all positive and negative whole numbers and 0. Integers are not fractions or decimals or portions of a number.

Rational numbers

A rational number is any number that can be expressed as a fraction. Rational numbers include all positive and negative numbers, plus all fractions and decimals that either come to an end or repeat infinitely.

Types of Triangles Defined

The SAT II questions mention different types of triangles, and you need to know what they are.

Equilateral triangle

An equilateral triangle is simply equal; its three sides are equal and its three angles are equal. It's a very diplomatic shape.

Acute triangle

The three angles of an acute triangle all measure less than 90°.

Obtuse triangle

An obtuse triangle has one angle that measures more than 90°.

Isosceles triangle

The isosceles triangle has two equal sides with two corresponding equal angles.

Coordinates

The coordinates of a coordinate plane are the two numbers that show the distance of a point from the origin. The horizontal (x) coordinate is always listed first and the vertical coordinate (y) is listed second.

Parabola

The parabola is a curve that opens either upward or downward or to the right or left on the coordinate plane.

Interior Angles

Interior angles are the angles that are inside a many-sided figure.

Chord

When you think of a chord as it regards the SAT II, you may think of a heavy rope that you need to tie you to your seat to maintain focus during the test, but that's not what the SAT II has in mind when it asks you a question about chords. In geometry, a chord is a line segment that cuts across a circle and connects two points on the edge of a circle. Those two points at the end of the chord are also the endpoints of an arc of the circle. The rope may be more useful, however.

Domain

When dealing with functions, the domain is all of the possible values for the independent variable, or x in $f(x)$.

Radian

A radian is just another way of measuring fractions of a circle instead of degrees. When you see radian in an SAT II Math question, you know your calculator should be in radian mode.

Asymptotes

An asymptote is a straight line that the function comes very close to but never quite touches. Kind of like getting a score of 800 on the SAT II Math. Some will get close, but few will actually touch it.

Index

• •

• *Numbers* •

• *A* •

• W •

• X •

• Y •

• Z •

BUSINESS, CAREERS & PERSONAL FINANCE

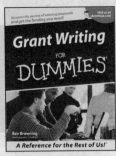

Grant Writing FOR DUMMIES
A Reference for the Rest of Us!
0-7645-5307-0

Home Buying FOR DUMMIES
A Reference for the Rest of Us!
0-7645-5331-3 *†

Also available:
- Accounting For Dummies †
  0-7645-5314-3
- Business Plans Kit For Dummies †
  0-7645-5365-8
- Cover Letters For Dummies
  0-7645-5224-4
- Frugal Living For Dummies
  0-7645-5403-4
- Leadership For Dummies
  0-7645-5176-0
- Managing For Dummies
  0-7645-1771-6

- Marketing For Dummies
  0-7645-5600-2
- Personal Finance For Dummies *
  0-7645-2590-5
- Project Management For Dummies
  0-7645-5283-X
- Resumes For Dummies †
  0-7645-5471-9
- Selling For Dummies
  0-7645-5363-1
- Small Business Kit For Dummies *†
  0-7645-5093-4

HOME & BUSINESS COMPUTER BASICS

Windows XP FOR DUMMIES
A Reference for the Rest of Us!
0-7645-4074-2

Microsoft Office Excel 2003 ALL-IN-ONE DESK REFERENCE FOR DUMMIES
0-7645-3758-X

Also available:
- ACT! 6 For Dummies
  0-7645-2645-6
- iLife '04 All-in-One Desk Reference For Dummies
  0-7645-7347-0
- iPAQ For Dummies
  0-7645-6769-1
- Mac OS X Panther Timesaving Techniques For Dummies
  0-7645-5812-9
- Macs For Dummies
  0-7645-5656-8
- Microsoft Money 2004 For Dummies
  0-7645-4195-1

- Office 2003 All-in-One Desk Reference For Dummies
  0-7645-3883-7
- Outlook 2003 For Dummies
  0-7645-3759-8
- PCs For Dummies
  0-7645-4074-2
- TiVo For Dummies
  0-7645-6923-6
- Upgrading and Fixing PCs For Dummies
  0-7645-1665-5
- Windows XP Timesaving Techniques For Dummies
  0-7645-3748-2

FOOD, HOME, GARDEN, HOBBIES, MUSIC & PETS

Feng Shui FOR DUMMIES
A Reference for the Rest of Us!
0-7645-5295-3

Poker FOR DUMMIES
A Reference for the Rest of Us!
0-7645-5232-5

Also available:
- Bass Guitar For Dummies
  0-7645-2487-9
- Diabetes Cookbook For Dummies
  0-7645-5230-9
- Gardening For Dummies *
  0-7645-5130-2
- Guitar For Dummies
  0-7645-5106-X
- Holiday Decorating For Dummies
  0-7645-2570-0
- Home Improvement All-in-One For Dummies
  0-7645-5680-0

- Knitting For Dummies
  0-7645-5395-X
- Piano For Dummies
  0-7645-5105-1
- Puppies For Dummies
  0-7645-5255-4
- Scrapbooking For Dummies
  0-7645-7208-3
- Senior Dogs For Dummies
  0-7645-5818-8
- Singing For Dummies
  0-7645-2475-5
- 30-Minute Meals For Dummies
  0-7645-2589-1

INTERNET & DIGITAL MEDIA

Digital Photography FOR DUMMIES
A Reference for the Rest of Us!
0-7645-1664-7

Starting an eBay Business FOR DUMMIES
A Reference for the Rest of Us!
0-7645-6924-4

Also available:
- 2005 Online Shopping Directory For Dummies
  0-7645-7495-7
- CD & DVD Recording For Dummies
  0-7645-5956-7
- eBay For Dummies
  0-7645-5654-1
- Fighting Spam For Dummies
  0-7645-5965-6
- Genealogy Online For Dummies
  0-7645-5964-8
- Google For Dummies
  0-7645-4420-9

- Home Recording For Musicians For Dummies
  0-7645-1634-5
- The Internet For Dummies
  0-7645-4173-0
- iPod & iTunes For Dummies
  0-7645-7772-7
- Preventing Identity Theft For Dummies
  0-7645-7336-5
- Pro Tools All-in-One Desk Reference For Dummies
  0-7645-5714-9
- Roxio Easy Media Creator For Dummies
  0-7645-7131-1

* Separate Canadian edition also available
† Separate U.K. edition also available

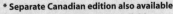

Available wherever books are sold. For more information or to order direct: U.S. customers visit www.dummies.com or call 1-877-762-2974.
U.K. customers visit www.wileyeurope.com or call 0800 243407. Canadian customers visit www.wiley.ca or call 1-800-567-4797.

SPORTS, FITNESS, PARENTING, RELIGION & SPIRITUALITY

0-7645-5146-9

0-7645-5418-2

Also available:
- Adoption For Dummies
  0-7645-5488-3
- Basketball For Dummies
  0-7645-5248-1
- The Bible For Dummies
  0-7645-5296-1
- Buddhism For Dummies
  0-7645-5359-3
- Catholicism For Dummies
  0-7645-5391-7
- Hockey For Dummies
  0-7645-5228-7

- Judaism For Dummies
  0-7645-5299-6
- Martial Arts For Dummies
  0-7645-5358-5
- Pilates For Dummies
  0-7645-5397-6
- Religion For Dummies
  0-7645-5264-3
- Teaching Kids to Read For Dummies
  0-7645-4043-2
- Weight Training For Dummies
  0-7645-5168-X
- Yoga For Dummies
  0-7645-5117-5

TRAVEL

0-7645-5438-7

0-7645-5453-0

Also available:
- Alaska For Dummies
  0-7645-1761-9
- Arizona For Dummies
  0-7645-6938-4
- Cancún and the Yucatán For Dummies
  0-7645-2437-2
- Cruise Vacations For Dummies
  0-7645-6941-4
- Europe For Dummies
  0-7645-5456-5
- Ireland For Dummies
  0-7645-5455-7

- Las Vegas For Dummies
  0-7645-5448-4
- London For Dummies
  0-7645-4277-X
- New York City For Dummies
  0-7645-6945-7
- Paris For Dummies
  0-7645-5494-8
- RV Vacations For Dummies
  0-7645-5443-3
- Walt Disney World & Orlando For Dummies
  0-7645-6943-0

GRAPHICS, DESIGN & WEB DEVELOPMENT

0-7645-4345-8

0-7645-5589-8

Also available:
- Adobe Acrobat 6 PDF For Dummies
  0-7645-3760-1
- Building a Web Site For Dummies
  0-7645-7144-3
- Dreamweaver MX 2004 For Dummies
  0-7645-4342-3
- FrontPage 2003 For Dummies
  0-7645-3882-9
- HTML 4 For Dummies
  0-7645-1995-6
- Illustrator CS For Dummies
  0-7645-4084-X

- Macromedia Flash MX 2004 For Dummies
  0-7645-4358-X
- Photoshop 7 All-in-One Desk Reference
  For Dummies
  0-7645-1667-1
- Photoshop CS Timesaving Techniques
  For Dummies
  0-7645-6782-9
- PHP 5 For Dummies
  0-7645-4166-8
- PowerPoint 2003 For Dummies
  0-7645-3908-6
- QuarkXPress 6 For Dummies
  0-7645-2593-X

NETWORKING, SECURITY, PROGRAMMING & DATABASES

0-7645-6852-3

0-7645-5784-X

Also available:
- A+ Certification For Dummies
  0-7645-4187-0
- Access 2003 All-in-One Desk Reference
  For Dummies
  0-7645-3988-4
- Beginning Programming For Dummies
  0-7645-4997-9
- C For Dummies
  0-7645-7068-4
- Firewalls For Dummies
  0-7645-4048-3
- Home Networking For Dummies
  0-7645-42796

- Network Security For Dummies
  0-7645-1679-5
- Networking For Dummies
  0-7645-1677-9
- TCP/IP For Dummies
  0-7645-1760-0
- VBA For Dummies
  0-7645-3989-2
- Wireless All In-One Desk Reference
  For Dummies
  0-7645-7496-5
- Wireless Home Networking For Dummies
  0-7645-3910-8